P9-CKY-376

Springer Series in Information Sciences 1

Springer Series in Information Sciences

Editors: Thomas S. Huang   Manfred R. Schroeder

Teuvo Kohonen

Content-Addressable Memories

Second Edition

With 123 Figures

Springer-Verlag Berlin Heidelberg New York
London Paris Tokyo

6eflort

Professor Teuvo Kohonen

Department of Technical Physics, Helsinki University of Technology
SF-02150 Espoo 15, Finland

Series Editors:

Professor Thomas S. Huang

School of Electrical Engineering, Purdue University
West Lafayette, IN 47907, USA

Professor Dr. Manfred R. Schroeder

Drittes Physikalisches Institut, Universität Göttingen, Bürgerstrasse 42–44,
D-3400 Göttingen, Fed. Rep. of Germany

ISBN 3-540-17625-X 2. Auflage   Springer-Verlag Berlin Heidelberg New York
ISBN 0-387-17625-X 2nd edition   Springer-Verlag New York Berlin Heidelberg

ISBN 3-540-09823-2 1. Auflage   Springer-Verlag Berlin Heidelberg New York
ISBN 0-387-09823-2 1nd edition   Springer-Verlag New York Berlin Heidelberg

Library of Congress Cataloging in Publication Data. Kohonen, Teuvo, Content-addressable memories. (Springer series in information sciences; 1) Bibliography: p.   Includes index. 1. Associative storage. 2. Information storage and retrieval systems. I. Title. II. Series. TK7895.M4K63   1987   004.5   87-4765

This work is subject to copyright. All rights are reserved, whether the whole or part of the material is concerned, specifically the rights of translation, reprinting, re-use of illustrations, recitation, broadcasting, reproduction on microfilms or in other ways, and storage in data banks. Duplication of this publication or parts thereof is only permitted under the previsions of the German Copyright Law of September 9, 1965, in its version of June 24, 1985, and a copyright fee must always be paid. Violations fall under the prosecution act of the German Copyright Law.

© Springer-Verlag Berlin Heidelberg 1980 and 1987
Printed in Germany

The use of registered names, trademarks, etc. in this publication does not imply, even in the absence of a specific statement, that such names are exempt from the relevant protective laws and regulations and therefore free for general use.

Offset printing and bookbinding: Brühlsche Universitätsdruckerei, Giessen
2153/3150-543210

LIBRARY
The University of Texas
at San Antonio

Preface to the Second Edition

Due to continual progress in the large-scale integration of semiconductor circuits, parallel computing principles can already be met in low-cost systems: numerous examples exist in image processing, for which special hardware is implementable with quite modest resources even by nonprofessional designers. Principles of content addressing, if thoroughly understood, can thereby be applied effectively using standard components. On the other hand, mass storage based on associative principles still exists only in the long-term plans of computer technologists. This situation is somewhat confused by the fact that certain expectations are held for the development of new storage media such as optical memories and "spin glasses" (metal alloys with low-density magnetic impurities). Their technologies, however, may not ripen until after "fifth generation" computers have been built.

It seems that software methods for content addressing, especially those based on hash coding principles, are still holding their position firmly, and a few innovations have been developed recently. As they need no special hardware, one might expect that they will spread to a wide circle of users.

This monograph is based on an extensive literature survey, most of which was published in the First Edition. I have added Chap.7, which contains a review of more recent work. This updated book now has references to over 1200 original publications.

In the editing of the new material, I received valuable help from Anneli Heimbürger, M.Sc., and Mrs. Leila Koivisto.

Otaniemi, Finland   *Teuvo Kohonen*
February, 1987

Preface to the First Edition

Designers and users of computer systems have long been aware of the fact
that inclusion of some kind of content-addressable or "associative" functions
in the storage and retrieval mechanisms would allow a more effective and
straightforward organization of data than with the usual addressed memories,
with the result that the computing power would be significantly increased.
However, although the basic principles of content-addressing have been known
for over twenty years, the hardware content-addressable memories (CAMs) have
found their way only to special roles such as small buffer memories and control units. This situation now seems to be changing: Because of the development of new technologies such as very-large-scale integration of semiconductor circuits, charge-coupled devices, magnetic-bubble memories, and certain
devices based on quantum-mechanical effects, an increasing amount of active
searching functions can be transferred to memory units. The prices of the more
complex memory components which earlier were too high to allow the application
of these principles to mass memories will be reduced to a fraction of the total system costs, and this will certainly have a significant impact on the
new computer architectures.

In order to advance the new memory principles and technologies, more information ought to be made accessible to a common user. To date it has been
extremely difficult to gain an overview of content-addressable memories; during the course of their development many different principles have been tried,
and many electronic technologies on which these solutions have been based have
become obsolete. More than one thousand papers have been published on content
addressing, but this material has never been available in book form. Numerous
difficulties have also been caused by the fact that many developments have
been classified for long periods of time, and unfortunately there still exists
material which is unreachable for a common user. The main purpose of this book
has been to overcome these difficulties by presenting most of the relevant
results in a systematic form, including comments concerning their practical
applicability and future development.

The organization of material in this book has been based on a particular
logic: the text commences with basic concepts, and the software methods are
presented first because they are expected to be of interest to all who have
an access to general-purpose computers. Computer designers who are concerned
with the intimate hardware, too, will be interested in the logic principles
and electronic implementations of CAM circuits given in the next chapters; the
highly parallel content-addressable computers which may concern only a few
specialists are reviewed last. Nonetheless it may be envisioned that future
computer systems will increasingly acquire these principles so that the circle of users becoming acquainted with these devices will widen. One may also
notice that content addressing has been understood in this book in a more general way than usual. Although the term "content-addressable memory" in computer engineering has usually been restricted to referring to certain hardware constructs only, nonetheless there have existed pure software solutions,
e.g., the extremely fast searching method named hash coding which for equally
good reasons ought also to be called content-addressing. On the other hand,
while the implementation of semantic and relational data structures has usually been discussed in the context of software searching methods, essentially
the same principles are directly amenable to implementation by special hardware. Such hardware is already being developed for hash coding, and thus the
boundary between the hardware and software approaches is disappearing. It was
therefore felt that both of these approaches should be discussed in the same
volume, in order to facilitate their comparison and to make the issues in
favor of one or the other more clear. Chapter 2 which deals with hash coding
has been made as complete as possible.

This book is an outgrowth of courses given at the Helsinki University of
Technology, as well as research work pursued on associative memory under the
auspices of the Academy of Finland. I am grateful for all material help obtained from the above institutions. The assistance of the following persons
should be appreciated: Mrs. Pirjo Teittinen helped in the collection of literature and typed out my manuscripts. Mr. Erkki Reuhkala contributed to the
compilation of material on hash coding. Mrs. Rauha Tapanainen, as well as Mr.
Heikki Riittinen, prepared the drawings in their final form. Mrs. Maija-Liisa
Hylkilä typed out the Subject Index. Proofreading of the complete book was
done by Mrs. Hylkilä as well as Dr. Erkki Oja; parts of it were checked by
other members of our Laboratory, too.

Otaniemi, Finland   *Teuvo Kohonen*
November, 1979

Contents

Chapter 3 Logic Principles of Content-Addressable Memories

Chapter 7 Review of Research Since 1979

Chapter 1   Associative Memory, Content Addressing, and Associative Recall

1.1 Introduction

The subject area of this book consists of various principles, methods, and devices of computer technology which are often grouped under the heading *associative memory*. A word of caution at this point will be necessary: the field of all phenomena related to associative memory is probably much wider than that ever covered in computer and information sciences. There is no other original model for associative memory than in human memory and thinking; unfortunately and surprisingly the experimental results do not yet allow us to deduce what the detailed memory mechanism thereby applied is. Only indirect evidence is available [1.1]. Similarly, while research in the artificial intelligence techniques claims to deal with correct reasoning and problem-solving, this does not necessarily imply that the real biological information processes should resemble even the best artificial ones. In order that a physical or abstract system could be named "associative memory", one should stipulate at least that its function comply with the phenomenological features of human memory expressed in the *Classical Laws of Association*; we shall revert to them a bit later on in Sect. 1.4.1.

It seems that there exist two common views of associative memory. One of them, popular in computer engineering, refers to a principle of organization and/or management of memories which is also named *content-addressing*, or searching of data on the basis of their contents rather than by their location. The coverage of this book in broad outline coincides with it. The other view is more abstract: memory is thereby understood as a semantic representation of knowledge, usually in terms of *relational structures*. We shall explain this latter concept in Sect. 1.3.

The purpose of this book is in the first place practical. Although it would be very interesting to expound the meaning of associative memory in its most general sense, such discussions can be found elsewhere (cf [1.1,2]). The primary scope of this representation is to serve as a text and reference

to a voluminous research on content-addressable computer memories pursued
during more than 20 years (for general review articles, see [1.3-5]).

1.1.1 Various Motives for the Development of Content-Addressable Memories

The different forms of content-addressable memories used in computer technology may at least partly be explained by the following expectations which
have guided their development:

1) Mankind is faced with the problem of "information explosion". For its
handling it has been suggested that the *filing of documents and data by "associations"*, rather than by indexing, should be used [1.6]. This expectation
has materialized only in certain special applications such as the data structures used in computer-aided design, whereas usual document retrieval is
mostly performed on the basis of mutually nonassociated descriptors.

2) With the invention of certain adaptive or learning elements around
1955-60, such as the Perceptron [1.7], some optimistic views about producing
intelligent functions and systems by the principle of self-organizing were
held. The Perceptron networks were originally named "associative", and with
slight modification they might have been developed to resemble certain content-addressable memory constructs nowadays used for parallel searching. It
turned out later, however, that the information-processing power of these
devices alone is too weak for the implementation of artificial intelligence
of practical value, and for this purpose, implementation of information processes by high-level computer languages was felt necessary. While only a few
artificial intelligence functions have emanated from the Perceptrons, nonetheless the latter have given an impulse for the birth of another important
branch of information science, namely, the *pattern recognition* research.

3) The conventional or von Neumann-type computers, and especially the
principles of addressing computational variables by the indices of their locations, have frequently been regarded as unsatisfactory in large problems.
Many attempts have been made to replace the conventional random-access memories by structures from which the operands could be called out by their
symbolic names or *data contents*, and possibly many operands at a time. The
high-level algorithmic and problem-oriented languages might take significant
advantage of such a feature. There seem to exist at least two reasons for
which development in this direction has not been particularly rapid. One of
them is the much higher price of content-addressable memories compared to the
random-access memories. Another reason is that the problem of encoding the
variables symbolically has already effective software solutions which do not

require special memories (cf Chap. 2). This, however, does not yet bring about
the other feature which would be very desirable in large problems, namely,
retrieving of a great number of variables from the memory simultaneously.

4) While associations were originally considered for the description of
interrelations or cross-references between pieces of information only, it
has later turned out that searching of data by its partial content can effectively be utilized in the manipulation of *arithmetic algorithms*. Such content
addressing can be made in a highly parallel fashion, i.e., simultaneously over
a great number of data elements, usually at a rather low level, referring
to transformations that occur in the binary representations. The problem of
parallel computation has also another direction where content-addressability
is used to control higher-level algorithmic functions in parallel (cf Chap. 6).

5) Content-addressable memory functions have recently been found extremely
useful in the implementation of *buffer memory organizations* (cf Chap. 5)
which in large memory systems have provided very high average performance at
reasonable cost.

6) Small and fast content-addressable memories have been found useful for
the implementation of *programmed sequential control operations* in central
processing units as well as other devices.

7) One task of future computer technology is to make machines interpret
verbal statements given in the natural language. While some solutions for
this exist, nonetheless it may turn out that some newer forms of associative
or content-addressable memory are applicable for its handling in a simpler
and yet more efficient way.

1.1.2 Definitions and Explanations of Some Basic Concepts

In the first extensive review article on content-addressable memories published by HANLON in 1966 [1.3], the following definition can be found: "Associative memories have been generally described as a collection or assem-
blage of elements having data storage capabilities, and which are accessed
simultaneously and in parallel on the basis of data content rather than by
specific address or location." HANLON points out, however, that the correct
meaning of "associative" refers to interrelationships between data, and not
to the storage mechanism; consequently, he recommends that the name *content
addressable memory* (CAM) be used for the latter. Incidentally, it has been
emphasized in several system-theoretical studies (cf [1.1]) that the recall
of information by association is also implementable by continuous or distributed physical systems, and so the content-addressable memories discussed

in this book ought to be understood only as one possible mechanism for the storage and recollection of associations.

PARHAMI [1.5] remarks that even parallelism in searching operations is not essential, as long as the functional operation is concerned. Instead he proposes three definitions of which we shall combine the latter two, and substitute the word "content-addressable" for "associative":

Content-addressable memory: a storage device that stores data in a number of cells. The cells can be accessed or loaded on the basis of their contents.

Content-addressable processor: a content-addressable memory in which more sophisticated data transformations can be performed on the contents of a number of cells selected according to the contents, or a computer or computer system that uses such memory as an essential component for storage or processing, respectively.

It may be necessary to clarify at this point that accessing data on the basis of their content always means some *comparison* of an external *search argument* with part or all of the information stored in all cells. Whether this is done by software, mechanical scanning or parallel electronic circuits, is immaterial in principle; however, a "genuine" content-addressable memory performs all comparisons in parallel. Another fact to emphasize is that comparison by *equality match* between the search argument and the stored item is not the only mode used. If the stored data have numerical values, the purpose of content-addressable search may be to locate all cells the contents of which satisfy certain *magnitude relations* with respect to the search arguments, for instance, being greater than or less than the given limit, or between two specified limits. Content-addressable searching is sometimes performed without reference to an external search argument, for instance, when locating the maximum or minimum in a set of stored numbers. Finally, searching on the basis of *best match* of the search argument with the various stored data, in the sense of some *metric*, may be necessary. This is already very near to the basic task of *pattern recognition*, and we shall revert to it with the various similarity measures in Sect. 1.4.2.

Various alternative names for the CAM have been suggested, e.g., associative store, content-addressed memory [1.8], data-addressed memory [1.9], catalog memory [1.10], multiple instantaneous response file [1.11], parallel search file [1.12], and (parallel) search memory [1.13]. The diversity of content-addressable computers is yet greater, and many names used for them refer only to a specific construct (cf Chap. 6).

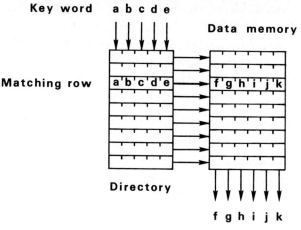

Fig. 1.1. Catalog memory

For further illustration, let a trivial organization which in particular might be named *Catalog Memory* be delineated here (Fig. 1.1). It consists of two parts of which the one on the left is the *directory* and that on the right the *data memory*. This system shall only compare the search argument, here named *key word*, with all words stored in the dictionary. For simplicity it is assumed that all stored words are different so that at most one match can be found. Every word location in the directory contains a register for the stored word, as well as a special combinational logic circuit which compares the key word, broadcast in parallel to all locations, with the stored contents. Each location has an output at which a response is obtained if the key word and the stored word agree. As assumed, not more than one response (exemplified in Fig. 1.1 by a bold arrow) shall be obtained. The data memory is a usual linear-select random-access memory for which the output lines of the directory serve as address lines; no address decoder is thereby needed. To put it in another way, the directory assumes the role of the address decoder, the operation of the latter being dependent on words stored in it. The whole system can be regarded as a special logic circuit with input-output signal delay (access time of the CAM) which is very short, say, 50 ns.

Hardware content-addressable memories will be discussed in detail in Chaps. 3, 4. Two auxiliary features which are commonplace in CAMs are a provision to *mask* parts in the search argument, and an organization which allows the sequential readout of *several responding items*.

1.2 The Two Basic Implementations of Content Addressing

The two main principles of content addressing are the one based on a data-dependent memory mapping, implemented by programming techniques (software), and the other which uses special hardware constructs for the storage and retrieval of data items. It is remarkable that both of these principles were invented almost simultaneously, around 1955; this shows that with the advent of the first commercial computer systems there already existed a strong need for content addressing. While over twenty years have now passed since the introduction of these methods, no essentially new issues have been presented in favor of one or the other. Therefore both principles are reviewed here as they are *currently* standing, and any conclusions about their future status are left to the reader.

1.2.1 Software Implementation: Hash Coding

Before the introduction of the first automatic programming languages such as FORTRAN, a need was already felt for assemblers which could refer to the computational variables by their symbolic names. Translation of a name into the address of a location at which the actual variable is stored means maintenance of a table. Frequent look-ups to this table are then necessary. Instead of scanning all the stored entries every time, or performing a linear search, much quicker methods were invented.

Since every name has a binary representation in the memory, it thereby also has a *numerical value*. In principle, if the address space of the memory system were unlimited, the name could be stored at an address which equals its numerical value, and any data associated with the variable would be retrievable in a single access to the memory. In practice the available memory is almost always much smaller than the range of numbers spanned by all permissible names. Obviously some sort of *compression* of the space of numerical values is necessary.

Assume now that the names are represented by a 26-letter alphabet, and the first two letters only are regarded in determining the numerical value;

<u>Table 1.1.</u> Randomly chosen names, and their numerical values computed from the first two letters (see text)

Sample No.	Name	Value (address in the memory)	Associated data
1	ISABEL	226	D(1)
2	SOPHIE	482	D(2)
3	HOWARD	196	D(3)
4	JULES	254	D(4)
5	EDGAR	107	D(5)
6	JOHN	248	D(6)
7	WILMAR	580	D(7)
8	GEORGE	160	D(8)
9	JESSE	238	D(9)
10	NIGEL	346	D(10)
...	...	...	...
14	HARLOW	182	D(14)
15	WALDEMAR	572	D(15)
16	HANS	182	D(16)

there are $26^2 = 676$ different pairs of letters, and a memory area with a corresponding number of locations is reserved for the table. (With the computers of the 1950s, this would have been a tremendous investment in memory space.) Now assume that names with random beginnings are used; the probability for different names to have different addresses is then not very small. Table 1.1 is a set of randomly chosen first names of persons used as indentifiers; their numerical values are given in the third column. These values are calculated in the following way: denote A = 0, B = 1, ..., Z = 25. A pair of letters is regarded as an integer in a basis of 26, so, e.g., IS = $8 \cdot 26 + 18 = 226$. The address in the table which is defined by the numerical value of the name, by the application of some simple rule, is named *calculated address*. At sample 16 a name which has the same beginning as No. 14 was picked up; it is said that a *conflict* or *collision* occurred. Since both names cannot be stored in the same location, a *reserve location*, easily reachable from the calculated address, shall be found. Chapter 2 deals extensively with this problem. Let it suffice here to mention one possibility:

the *next empty location* (in this case 183) following the calculated address
is used. If a search for an empty location is made cyclically over the table, such a location can be found sooner or later as long as the table is
not completely full. It is striking that if the table is not more than half
full and the names are chosen randomly, there is a high probability for
finding an empty location within a reach of a few locations from the calculated address.

In order to *resolve* whether an entry is stored at the calculated address
or one of its reserve locations, the name or its unique identifier must be
stored together with the data. Assume that the name itself is used. The
correct location is found when the stored name agrees with that used as a
search argument. Table 1.2 exemplifies the contents of part of the memory,
corresponding to the example shown in Table 1.1.

Table 1.2. Partial contents of the memory area used to store the data corresponding to Table 1.1

Address	Contents of the location	
	Name	Data
...	...	...
160	GEORGE	D(8)
...	...	...
182	HARLOW	D(14)
183	HANS	D(16)
...	...	...
196	HOWARD	D(3)
...	...	...

A search for, say, the data associated with HANS is performed easily.
The calculated address of HANS is 182. Since such an identifier is not found
there, a search from the next locations reveals that HANS was stored at
address 183. The associated data, D(16), are then found.

Obviously the average number of trials to find a stored item from the
neighborhood of its calculated address depends on the loading of the table
and also on its size; moreover, the number of trials is smallest if the
items are scattered uniformly in the table. Instead of using sampled letters

as in the above introductory example, a better method is to compute some
function of the whole name which spreads the calculated addresses at random
(but, of course, deterministically for every name) over the available memory
area. Such a method is named *hash coding*, and the address mapping is defined
by the *hashing function*. Chapter 2 contains a thorough analysis of various
aspects of hash coding.

1.2.2 Hardware Implementation: The CAM

As another introductory example we shall consider the parallel searching
from a special content-addressable memory (CAM), corresponding to the directory depicted in Fig. 1.1. Assume that at the word locations, binary patterns
with bit elements 0 or 1 are stored. One word location is depicted in Fig.
1.2. The search argument is broadcast via a set of parallel lines, each one
carrying a bit value, to the respective bit positions of all word locations.
A built-in logic circuitry, consisting of a *logical equivalence* gate at every
bit position, and a *many-input AND* gate that is common to the whole location,
is shown in Fig. 1.2. It will be easy to see that the logic circuit gives a
response if and only if the search argument agrees with the stored binary
pattern. This kind of circuit shall be imagined at every word location of
the directory in Fig. 1.1.

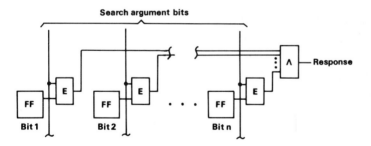

Fig. 1.2. Comparison circuit for one CAM word location. FF = bit-storage
flip-flop, E = logical equivalence gate, ∧ = logical AND gate

Chapter 3 will deal in more detail with logical principles, and Chap. 4
with different hardware realizations developed for the CAMs.

1.3 Associations

1.3.1 Representation and Retrieval of Associated Items

Direct Associations. The most primitive kind of association is formed when the representations of two or more items or events are brought together and stored in memory in direct physical or logical contact. Biological memory very probably operates in this way, and sensory experiences are *conditioned* in memory if they occur coincidentally, i.e., simultaneously or within a short span of time. In computers, items or events have discrete representations which also can be associated by their logical connection, e.g., being various properties of a common object, or having been brought together by definition. Examples of the first kind are documents with a number of related attributes. Associations of the latter kind are entries in various tables, directories, and dictionaries.

Some representations are themselves compound items, e.g., *patterns* consisting of identifiable subpatterns. The subpatterns may then be thought to be mutually associated by their constellation in the main pattern. Conversely, separate patterns can often be associated to form a superpattern which then has an identity and meaning of its own, logically distinguishable from its components.

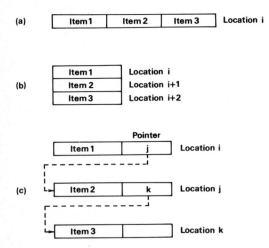

Fig. 1.3a-c. Representation of associated items in computer memory: a) in the same location, b) in consecutive locations, c) through pointers (illustrated by dashed arrows)

It may be proper to call associations described above *direct associations*, to distinguish them from more complex logical constructs discussed later on. Direct associations which consist of discrete items or their unique literal or numerical values are representable as *ordered sets* which, again, are stored as *lists* in computer memories (cf Sect. 2.6.1). For the representation of direct associations there are many possibilities in computers; they are all based on the automatic mechanisms provided for memory access. To show that items are associated, they can be stored in the same memory location (Fig. 1.3a), in consecutive locations which are automatically retrievable in a sequence (Fig. 1.3b), or in a set of arbitrary locations which are connected by cross-references named *pointers* (Fig. 1.3c). A pointer is an address code which is found in the location being currently read out; a small program loop is run to read out the next item from that address.

Association by Inference or Indirect Association. Without going into the details of semantic and logic relations, it may be useful to point out the existence of another kind of association. While the items above were brought together and represented by their direct physical or functional connection, two or more items can also be found mutually relevant if their representations are *indirectly* connected via a chain of cross references which is not traced through until in the recall phase. In this case there must exist several direct associations in memory which share *common* or *identical items*; for example, consider the ordered sets (A,B,C) and (C,D,E) which have been formed and stored independently. Although originally there was no direct connection between, say, B and E, nonetheless these items are no longer independent, as will be seen. If there exists a mechanism such that, e.g., item A used as a search argument will directly retrieve the associated items B and C, and C thus found will evoke D and E, then a simple process has reached B and E which thus have been brought together. Similar chains occur in propositional calculus in which the items consist of elementary propositions. We shall name the connection between items which is formed in the recall phase *indirect association,* or in analogy with logic, *association by inference.*

Relations. Two elements $x \in E$ and $y \in F$ are said to be related by R if (x,y) belongs to R where R is a (defined) subset of the set of all possible pairs of elements, i.e., of $E \times F$ (the Cartesian product of E and F).

The following notations for the *relation* between x and y are used:

$$x \, R \, y \, , \quad x \overset{R}{\rightarrow} y \, .$$

This somewhat abstract definition may become more clear if simple *semantic relations* are considered. Assume that x and y are nouns and R is a verbal construction. Assume two lists of specific cases of x and y :

x :   article A, article B, book C
y :   artificial intelligence, pattern recognition, digital electronics.

Not all pairs (x,y) may have relevance; for instance, paper A may not deal with digital electronics. Meaningful pairs are found upon documentation, whereby the following types of observation must be made:

paper A *deals with* pattern recognition
book C *deals with* artificial intelligence
etc.

The construction *'deals with'* here now defines the *type of relation* and it can be put to correspond to R. It is, of course, possible to construct many kinds of qualifier R.

The representation of a relation in memory can be an ordered triple of the type (A,R,B), e.g., ('paper A', 'deals with', 'pattern recognition') where the quotation marks are used to denote literal contents. Incidentally, this has the format of an association between A, R, and B. We shall a bit later revert to relations when discussing structures of knowledge representable by them. Before that, let us consider how relations can be retrieved from memory.

Representation and Searching of Relations by Content Addressing. There exist many methods for fast searching of listed items. Content addressing is the fastest of them since an item is locatable directly, and sorting of the stored items is not required as in most other methods. This principle is particularly advantageous with large data bases. We shall first exemplify searching of a relation by content addressing using hash coding.

The preliminary example of association by inference showed us that it will be necessary to retrieve a direct association on the basis of any of its component items. In the retrieval of information from relational structures discussed below in Sect. 1.3.2 it will further be found necessary to

search for a relation on the basis of any of the following combinations used
as search argument:

A,R,B, (A,R), (R,B), (A,B), (A,R,B) .

Here an ordered pair or triple is simply understood as the concatenation of
its literal component items. Searching on the basis of (A,R,B) is needed
only to check whether a relation exists in memory or not.

The same relation may thus have seven different types of search argument.
If a relation is stored in memory by hash coding, in fact seven copies of it
could be stored, each one in a separate memory area or table reserved for
this type of search argument. Other methods for the handling of this task
will be discussed in Sect. 2.6.

Retrieval of relations is much more straightforward if the triples can be
stored in the word locations of a content-addressable memory. Using *masking*
of fields as discussed in Chap. 3, any of the combinations shown above can
be defined as a search argument which has to match with the corresponding part
of the stored relation, whereby retrieving of the whole relation is made in a
single parallel readout operation.

A special hardware solution for the representation and retrieval of relations, not only by external search arguments but also using associations by
inference will be presented in Sect. 6.3.

Attributes and Descriptors. The position of an item in an ordered set defines a specific *role* for this item. The position may, for instance, correspond to a particular *attribute,* and the item stored at this position is then
the particular *value* given to this attribute. For instance, in a dictionary
example the first word in an ordered pair may be a type of English form,
e.g., 'horse', and the second that of a French form, namely, 'chevaux'. The
expressions 'English form' and 'French form' are here attributes, whereas
'horse' and 'chevaux' are their *values*, respectively. Perhaps even more
illustrative examples of attributes can be found in various personal records,
an example of which is in Table 1.3.

Table 1.3. An example of attributes and their values

Attributes:	Surname	Given name	Sex	Age	Height	Citizenship
Values:	DOE	John	Male	45	5ft. 7in.	Australian

This example shows that some values may be numerical, while the others are nonnumerical. As an ordered set this record would be written

(DOE, John, Male, 45, 5ft. 7in., Australian) .

The set of values corresponding to attributes should carefully be distinguished from a set of *descriptors* used to specify, e.g., a scientific work. The descriptors have independent values and accordingly they form only an *unordered set*. An "association" can be formed only between this set and the specification of the document, e.g.,

({associative, content-addressable, memory, processor}, THIS BOOK).

1.3.2 Structures of Associations

For the storage and searching of semantic information, "associative" computer languages can be used. Languages for the handling of structures of information based on *relations* were independently invented by SAVITT et al. [1.14, 15] as well as FELDMAN and ROVNER [1.16,17], following earlier work on list processing languages (e.g., McCARTHY et al. [1.18]).

Notice that if the only task were document retrieval on partial specification of the set of attributes or descriptors, this would be a straightforward application of content-addressing methods. Plenty of solutions for it can be found in this book. The basic difficulty will manifest itself when one considers certain problems of computer-aided design and artificial intelligence in which it is often not clear in the beginning how the identity of the data to be searched should be specified. Only association by inference may be a possible mode of retrieval since a closer specification of items may depend on information actually stored in the memory. There will be a more detailed description of such a concrete process in Sect. 6.3; let the discussion here be restricted to fundamentals.

The basic "association" in this kind of language is a relation (A,O,V) where A stands for "attribute", O for "object", and V for "value", respectively. This triple construct has been known earlier [1.19], and at least three components have been considered necessary for the definition of *information structures*. If the "association" is represented as $O \overset{A}{\to} V$, a connection between the general form of relation and an attribute is clearly discernible. Actually A alone does not completely define the attribute; it is the pair of arguments (A,O) which does it. For example, if A = 'color of' and O = 'an apple', then the complete attribute is in fact 'color of an apple', whereby V is its value, e.g., 'red'.

When comparing a relation with a document consisting of an ordered set of attribute values, the former is found *self-documenting*: it explicitly announces the attribute, whereas the attributes in the latter are only implicitly and by convention defined by the position in the set. On the other hand, (A,O,V) might also be regarded as a two-argument "document" if O is regarded as the value of another implicitly given attribute which is the name of the object. Further, if several attributes for an object O shall be considered, it is possible to express a set of relations such as $[(A_1,O,V_1), (A_2,O,V_2), ...]$. The self-documenting property of relations makes it possible to store many different kinds of document in the same memory and to define connections between them as shown in the following subsection.

Structures of information can be formed in two ways. In the more usual of them, as already mentioned earlier, chains and networks of relations are formed if several relations share common items. The second way uses the concept of *compound relation*.

The identity of an item in two or more relations is sufficient to define a structure, and no other references are needed. If information is represented in graph form, using the notation $A \overset{R}{\to} B$ for a relation, then even though items occur in many relations, they are drawn only in one place, and the structure follows from the composite representation. It should be carefully noticed that the graph itself is not shown explicitly in memory: its structure is completely abstract, although it can always be reconstructed by following the cross-references in the relations. Let the following example clarify what is meant. Table 1.4 represents the contents of a memory area

Table 1.4. Examples of relations stored in memory

(year, article A, 1965)
(year, article B, 1978)
(year, book C, 1979)
(topics, article A, AI)
(topics, article A, programming languages)
(topics, article B, AI)
(topics, article B, robotics)
(topics, book C, AI)
(part of, book C, article A)
(part of, book C, article B)

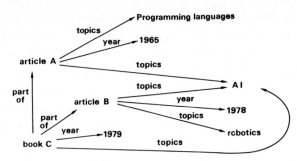

Fig. 1.4. Relational structure which corresponds to Table 1.4

used to store one particular collection of information in the form of relations. The abstract graph that results is shown in Fig. 1.4.

Another construct which also generates structures is the *compound item* which itself is a relation. The compound item may enter any position in a relation, whereby a *compound relation* results. This is a normal way of constructing *list structures* (cf Sect. 2.6.1). Compound relations may have significance in certain implicit statements, e.g., "The output of printer is the result of query", the graph of which is drawn in Fig. 1.5.

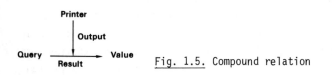

Fig. 1.5. Compound relation

Implicit Definition of a Searching Task. Structures of associations have to be applied if a searching task is defined implicitly. A very simple example of this is: "Find all articles on artificial intelligence published in 1978." (This example, of course, could be handled in many ways.) Here it is noteworthy that the memory contains only the following types of relations

(year, article A_i, B_i)
(topics, article A_i, C_i)

and a *combination search* is needed. The search criterion may be expressed in terms of *control structures* (cf Sect. 6.3) using an unknown item x:

(year, x, 1978),
(topics, x, AI).

In this case, searching is performed in two passes. One of them locates all associations which respond to a search argument (year, 1978), and thus yields a set of tentative values for x. The second pass locates all relations which respond to the search argument (topics, AI), and a second set of values for x is obtained. Finally it is necessary only to form the intersection of the two sets. This is not exactly the way in which processors of associative languages handle the searching: e.g., for more information about LEAP, see [1.16,17,20,21]. Another solution, explained in detail in Sect. 6.3, is based on special parallel hardware operations, but it should be noticed that the same operations are readily implementable by programming, although not in parallel.

1.4 Associative Recall: Extensions of Concepts

In order to set the constructs and methods discussed in this book in their right perspective, it will be necessary to consider some further aspects of associative memory, especially *associative recall*, in order to understand the limitations of present-day content-addressable memories.

First of all it shall be pointed out that the computer-technological "associations" have a completely rigid structure: either there exists a link between associated items or not. This way of definition of association seems to follow from two facts. One of them is a widespread tradition in information sciences and logic to operate only on discrete-valued variables and propositions, whereby it may also be believed that these are the elements of higher-level information processes. (Some newer approaches, like the fuzzy-set theory [1.22] in fact tend away from this tradition). Although it is generally correctly stated that thinking processes operate on symbolic representations, the meaning of "symbolic" is often confused with "provided with a unique discrete code". The second reason which has determined the format of associations in computer science is more practical. There has existed no other powerful tool for the implementation of artificial information processes than the digital computer, and the latter has strongly guided the selection of abstract concepts which are used for the description of information processes. Although some attempts have been made to interpret associations as *collective* or *integral effects*, e.g., [1.23-36], their demonstration by computer simulation has turned out very cumbersome and not rewarding. The demonstrations have mostly dealt with rather trivial examples. In principle, however, this extension of the concept of associative memory is a very important one, and if it were possible to fabricate hardware for collective computations, or to utilize more effectively the potential of the existing computers (an example of which is found in Sect. 2.7), new dimensions in information processing might be developed.

In this section some attempts are made to uncover some of those phenomena which are believed to belong under the heading associative recall, but which have not yet been effectively realized in computer technology. Especially implementations of *graded associations* shall be considered.

1.4.1 The Classical Laws of Association

In a small book entitled *On Memory and Reminiscence* [1.37], the famous Greek philosopher Aristotle (384-322 B.C.) stated a set of observations on human memory which were later compiled as the Classical Laws of Association. The conventional way for their expression is:

> *The Laws of Association*
> Mental items (ideas, perceptions, sensations or feelings) are connected in memory under the following conditions:
> 1) If they occur simultaneously ("spatial contact").
> 2) If they occur in close succession ("temporal contact").
> 3) If they are similar.
> 4) If they are contrary.

Some objections to these laws might be presented by a contemporary scientist. In view of our present acquaintance with various computing devices, it seems necessary to distinguish the phases of operation signified as *writing* (or storage) and *reading* (or recall). Obviously simultaneity or close succession of signals is necessary for their becoming mutually conditioned or encoded in a physical system, whereas in a process of recall, the evoked item (or part of it) might have, e.g., a high positive *correlation* (similarity) or a negative one (contrast) to that item which is used as the input *key* or *search argument*. Consequently, it seems that Laws 1 and 2 relate to writing, and 3 and 4 to reading, respectively. Moreover, these laws seem to neglect one further factor which is so natural that its presence has seldom been considered. This is the *background* or *context* in which primary perceptions occur, and this factor is to a great extent responsible for the high capacity and selectivity of human memory.

The following significant features in the operation of human associative memory shall be noticed: 1) Information is in the first place searched from the memory on the basis of some measure of *similarity* relating to the key pattern. 2) The memory is able to store representations of *structured sequences*. 3) The recollections of information from memory are *dynamic processes*, similar to the behavior of many time-continuous physical systems. In fact, Aristotle made several remarks on the recollections being a *synthesis* of memorized information, not necessarily identical to the original occurrence.

The Laws of Association now give rise to some important issues and they will be discussed below in Sects. 1.4.2-5.

1.4.2 Similarity Measures

The useful capacity of memory for patterned information depends on its ability to recall the wanted items with sufficient *selectivity*. If there existed a unique representation for every occurrence, then content-addressable searching could simply be based on *exact match* between the search argument and the stored representations as in content-addressable computer memories. However, natural patterns and especially their physiological representations in neural realms are always contaminated by several kinds of error and noise, and the separation of such patterns naturally leads to the problematics thoroughly discussed in the subject area of *pattern recognition* research; most of its special problems, however, fall outside the scope of this book.

The genuine process of associative recall thus ought to take into account some extended form of the concept of *similarity*. Several approaches, in an approximately ascending order of generality, to the definition of this concept are made below.

Hamming Distance. Perhaps the best known measure of similarity, or in fact dissimilarity between digital representations is the *Hamming distance*. Originally this measure was defined for binary codes [1.38], but it is readily applicable to comparison of any *ordered sets* which consist of discrete-valued elements.

Consider two ordered sets x and y which consist of distinct, nonnumerical symbols such as the logical 0 and 1, or letters from the English alphabet. Their comparison for dissimilarity may be based on the *number of different symbols in them*. This number is known as the *Hamming distance* ρ_H which can be defined for sequences of equal length only: e.g.,

$$x = (1,0,1,1,1,0)$$
$$y = (1,1,0,1,0,1)$$ $\Big\}$ $\rho_H(x,y) = 4$

and

$$u = (p,a,t,t,e,r,n)$$
$$v = (w,e,s,t,e,r,n)$$ $\Big\}$ $\rho_H(u,v) = 3$.

For binary patterns $x = (\xi_1, \ldots, \xi_n)$ and $y = (\eta_1, \ldots, \eta_n)$, assuming ξ_i and η_i as Boolean variables, the Hamming distance may be expressed formally as a circuit operation

$$\rho_H(x,y) = bitcount \ \{(\overline{\xi_i} \wedge \eta_i) \vee (\xi_i \wedge \overline{\eta_i}) \mid i = 1, \ldots, n\} \qquad (1.1)$$

where the function *bitcount* S determines the number of elements in the set S which attain the value logical 1; the Boolean expression occurring as an element in the above set is the *EXCLUSIVE OR (EXOR)* function of ξ_i and η_i.

The restriction imposed on the lengths of representations, or numbers of elements in sets, can be avoided in many ways (cf, e.g., the definitions of Levenshtein distances a bit later on). As an introductory example, consider two *unordered sets* A and B which consist of distinct, identifiable elements; if they had to represent binary codes, then for these elements one could select, e.g., the indices of all bit positions with value 1. Denote the number of elements in set S by n(S). The following distance measure [cf Sect. 2.7 and (2.36)] has been found to yield a simple and effective resolution between unordered sets:

$$\rho(A,B) = \max \ \{n(A), n(B)\} - n(A \cap B) \ . \qquad (1.2)$$

This measure is related to the Tanimoto similarity measure discussed below.

Correlation. The comparison or detection of *continuous-valued* signals or patterns is often based on their *correlation* which is another trivial measure of similarity. Assume two ordered sets, or sequences of real-valued samples $x = (\xi_1, \xi_2, \ldots, \xi_n)$ and $y = (\eta_1, \eta_2, \ldots, \eta_n)$. Their correlation is

$$C = \sum_{i=1}^{n} \xi_i \eta_i \ . \qquad (1.3)$$

If x and y are understood as Euclidean (real) vectors, then C is their *scalar product*.

In case one of the sequences may be *shifted* with respect to the other by
an arbitrary amount, the comparison can better be based on a translationally
invariant measure, the maximum correlation over a specified interval:

$$C_m = \max_k \; \sum_{i=1}^{n} \xi_i n_{i-k} \; , \quad k = -n, \; -n+1, \; \ldots, \; +n \; . \tag{1.4}$$

In this case the sequences $\{\xi_i\}$ and $\{n_i\}$ are usually defined outside the
range $i = 1, \ldots, n$, too. Of course, shifted comparison can be applied with
any of the methods discussed below.

When similarity is measured in terms of C or C_m, two assumptions are usually involved: 1) The amount of information gained in an elementary comparison
is assumed directly proportional to respective signal intensities. 2) The
amounts of information gained from elementary comparisons are assumed additive.

It will be necessary to emphasize that correlation methods are most suitable for the detection of periodic signals which are contaminated by *Gaussian
noise*; since the distributions of natural patterns may often not be Gaussian,
other criteria of comparison, some of which are discussed below, must be considered, too.

Direction Cosines. If the relevant information in patterns or signals is
contained only in the *relative magnitudes* of their components, then similarity
can often be better measured in terms of *direction cosines* defined in the following way. If $x \in R^n$ and $y \in R^n$ are regarded as Euclidean vectors, then

$$\cos \theta = \frac{< x,y >}{\| x \| \; \| y \|} \tag{1.5}$$

is by definition the cosine of their mutual angle, with $< x,y >$ the scalar
product of x and y, and $\| x \|$ the Euclidean norm of x. Notice that if the
norms of vectors are standardized to unity, then (1.5) complies with (1.3),
or $\cos \theta = C$.

Notice that the value $\cos \theta = 1$ is defined to represent *exact match*;
vector y is then equal to x multiplied by a scalar, $y = \alpha x$ ($\alpha \in R$). On the
other hand, if $\cos \theta = 0$ or $< x,y > = 0$, vectors x and y are said to be
orthogonal.

The quality of results obtained in a comparison by direction cosines,
too, is dependent on the noise being Gaussian. This measure is frequently
used in the identification of acoustic spectra (e.g., in speech recognition).

Euclidean Distance. Another measure of similarity, actually that of *dissimilarity*, closely related to the previous ones, is based on the *Euclidean distance* of x and y defined as

$$\rho_E(x,y) = \| x - y \| = \sqrt{\sum_{i=1}^{n} (\xi_i - \eta_i)^2} \quad . \tag{1.6}$$

Although seemingly natural for the detection of differences, $\rho_E(x,y)$ in reality often yields worse results in comparison than the previous method, on account of its greater sensitivity to the lengths of the vectors to be compared; notice that $\| x - y \|^2 = \| x \|^2 + \| y \|^2 - 2 < x,y >$. On the other hand, if the lengths of the vectors are normalized, the results obtained are identical with those obtained by the previous methods. Often ρ_E is applicable to comparisons made in *parameter spaces.*

Measures of Similarity in the Minkowski Metric. Obviously (1.6) is a special case of distance which defines the *Minkowski metric:*

$$\rho_M(x,y) = \left(\sum_{i=1}^{n} |\xi_i - \eta_i|^\lambda \right)^{1/\lambda} , \quad \lambda \in R \quad . \tag{1.7}$$

The so-called *city-block distance* is obtained with $\lambda = 1$.

Tanimoto Similarity Measure. Some experiments have shown [1.39-43] that determination of similarity between x and y in terms of a measure introduced by TANIMOTO [1.44] yields good results; it may be defined as

$$S_T(x,y) = \frac{< x,y >}{\| x \|^2 + \| y \|^2 - < x,y >} \tag{1.8}$$

The origin of this measure is in the comparison of sets. Assume that A and B are two unordered sets of distinct (nonnumerical) elements, e.g., identifiers or descriptors in documents, or distinct features in patterns. The similarity of A and B may be defined as the ratio of the number of their common elements to the number of all different elements; if $n(X)$ is the number of elements in set X, then the similarity is

$$S_T(A,B) = \frac{n(A \cap B)}{n(A \cup B)} = \frac{n(A \cap B)}{n(A) + n(B) - n(A \cap B)} \quad . \tag{1.9}$$

Notice that if x and y above were binary vectors, with components $\in\{0,1\}$ the value of which corresponds to the exclusion or inclusion of a particular element, respectively, then $< x,y >$, $\| x \|$, and $\| y \|$ would be directly comparable to $n(A \cap B)$, $n(A)$, and $n(B)$, correspondingly. Obviously (1.8) is a generalization of (1.9) for real-valued vectors.

The Tanimoto measure has been used with success in the evaluation of relevance between documents [1.41,42]; the descriptors can thereby be provided with individual weights. If a_{ik} is the weight assigned to the kth descriptor of the ith document, then the similarity of two documents denoted by x_i and x_j is obtained by defining

$$< x_i,x_j > = \sum_k a_{ik}\, a_{jk} = \alpha_{ij} \, ,$$

and

$$S_T (x_i,x_j) = \frac{\alpha_{ij}}{\alpha_{ii} + \alpha_{jj} - \alpha_{ij}} \, . \tag{1.10}$$

Weighted Measures for Similarity. The components of x and y above were assumed independent. In practical applications they may be generated in a stochastic process which defines a statistical dependence between them; it can be shown that for vectors with normally distributed noise the optimal separation is obtained if instead of the scalar product, the *inner product* is defined as

$$< x,y >_\psi = < x,\psi y > \, , \tag{1.11}$$

or the *distance* is defined as

$$\rho_\psi (x,y) = \| x - y \|_\psi = \sqrt{(x - y)^T \psi (x - y)}_+ \, , \tag{1.12}$$

in which the weighting matrix ψ is the *inverse of the covariance matrix* of x and y, and T denotes the *transpose*.

Since ψ is assumed symmetric and positive semidefinite, it can be expressed as $\psi = (\psi^{\frac{1}{2}})^T \psi^{\frac{1}{2}}$, whereby x and y can always be preprocessed before their use as a search argument, or storage in memory, using the transformations $x' = \psi^{\frac{1}{2}}x$, $y' = \psi^{\frac{1}{2}}y$. Now comparison can be based on Euclidean measures (scalar product or distance) on x' and y'.

Unfortunately there are some drawbacks with this method: 1) In order to evaluate the covariance matrix for patterns of high dimensionality (n), an immense number of samples ($\gg n^2$) have to be collected. 2) Computation of matrix-vector products is much heavier than formation of scalar products; notice, however, that during the recall phase only the search argument is to be multiplied by a matrix.

Comparison by Operations of Continuous-Valued Logic. A transition from Boolean logic to the multiple-valued and even continuous-valued one is rather straightforward. The basic operations of multiple-valued logic were first introduced by ŁUKASIEWICZ [1.45] and POST [1.46], and later extensively utilized in the theory of fuzzy sets by ZADEH [1.22,47,48], as well as others. Here we shall adopt only a few concepts, believed to be amenable to fast and simple computation in comparison operations.

The application of continuous-valued logic to comparison operations is here based on the following reasoning. The "amount of information" carried by a scalar signal is assumed proportional to the signal from a certain neutral or zero level. For instance, consider a continuous scale in which ξ_i, $\eta_i \in [0, +1]$. The signal value 1/2 is assumed indeterminate, and the representation of information is regarded the more reliable or determinate the nearer it is to either 0 or +1. The *degree of matching* of scalars ξ_i and η_i is expressed as a generalization of the *logical equivalence*. Let us recall that the equivalence of Boolean variables a and b is

$$(a \equiv b) = (\overline{a} \wedge \overline{b}) \vee (a \wedge b) \ . \tag{1.13}$$

In continuous-valued logic, the logic product ($\wedge$) is replaced by minimum selection (*min*), the logic sum ($\vee$) is replaced by maximum selection (*max*), and the logical inversion is replaced by complementation with respect to the scale ($\overline{a} = 1 - a$). In this way, (1.13) is replaced by "equivalence" $e(\xi,\eta)$,

$$e(\xi,\eta) = max \ \{min \ (\xi,\eta), \ min \ [(1-\xi), \ (1-\eta)]\} \ . \tag{1.14}$$

If (1.14) is applied to the comparison of ξ_i and η_i, the next problem is how to combine the results $e(\xi_i, \eta_i)$. One possibility is to generalize the AND operation shown in Fig. 1.2, using the operation $\underset{i}{min}$. However, the effect of mismatching at a single element would be too fatal. Another possibility is the linear sum of $e(\xi_i, \eta_i)$ as in (1.3). A compromise would be to define the similarity of x and y with the aid of some function which is symmetrical with respect to its arguments, e.g.,

$$S_M(x,y) = \varphi^{-1} \left\{ \sum_{i=1}^{n} \varphi[e(\xi_i, \eta_i)] \right\} \tag{1.15}$$

where φ is a monotonic function, and φ^{-1} its inverse. For instance,

$$S_M(x,y) = \sqrt[p]{\sum_{i=1}^{n} [e(\xi_i, \eta_i)]^p}_{+}, \tag{1.16}$$

with p some real value is one possibility. Notice that with p = 1 the linear sum is obtained, and with $p \to -\infty$, $S_M(x,y)$ will approach $min_i [e(\xi_i, \eta_i)]$.

This method has two particular advantages when compared with, say, the correlation method: 1) Matching or mismatching of low signal values is taken into account. 2) The operations max and min are computationally, by digital or analog means, much simpler than the formation of products needed in correlation methods. For this reason, too, p = 1 in (1.16) might be preferred.

It has turned out in many applications that there is no big difference in the comparison results based on different similarity measures, and so it is the computational simplicity which ought to be taken into account in the first place.

Variational Similarity. The following principle can be combined with many comparison criteria. Its leading idea is that patterned representations are allowed to be marred by deformations or local scale transformations, and the comparison for similarity is then performed by considering only small pieces of patterns at a time. The pieces are shifted relative to each other to find their maximum degree of matching. As the partial patterns must be connected anyway, the matching must be done sequentially, whereby this becomes a kind of *variational problem.*

The matching procedure, here termed *variational similarity*, is illustrated by a symbol string matching example. The three possible types of error that can occur in strings are: 1) Replacement or substitution error (change of a symbol into another one). 2) Insertion error (occurrence of an extra symbol). 3) Deletion error (dropping of a symbol). Errors of the latter two types stretch or constrict the string, respectively, and their effect is analogous with scale transformations.

Assume that one of the strings is a reference, and for simplicity, in the other one to be compared, two or more errors of the same type are not allowed in adjacent symbols. (Several errors may, however, occur in the string distant from each other.) Consider two strings written at the sides of a lattice as shown in Fig. 1.6. A line shall connect lattice points which are selected by the following rule:

A Dynamic Matching Procedure: Assume that the line has already been defined to go through the lattice point (i.j); the next point shall be selected from (i+1,j), (i+1,j+1), and (i+1,j+2). Compare the symbol pairs corresponding to these three points. If there is one match only, take the corresponding point for the next point on the line. If there is no match, select (i+1,j+1) for the next point on the line. If there are matches at (i+1,j+1) and some other point, select (i+1,j+1). If there are matches only at (i+1,j) and (i+1,j+2), select for the next point (i+1,j+2).

A *matching score* is now determined by counting the number of matching pairs of symbols along the above line.

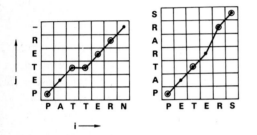

Fig. 1.6. Two examples of variational matching

The two examples shown in Fig. 1.6 may be self-explanatory. Naturally, many other and more effective local matching criteria can be used; this example will only convey the basic idea. For instance, instead of exact match, some measure of similarity between nonidentical symbols may be defined; this principle is usual in the matching of phonemic strings [1.49,50]. Sequences of numerical values can also be matched in an analogous way, by dividing the sequences into smaller segments, and comparing the respective segments.

Levenshtein or Edit Distance. Closely related to the variational similarity is a dissimilarity measure for symbol strings introduced in the theory of coding errors by LEVENSHTEIN [1.51]. It is assumed that string y can be derived from string x by application of a sequence of *editing operations*; assume that this is done by p_i replacements, q_i insertions, and r_i deletions.

The *Levenshtein distance* (LD) or *edit distance* between x and y is defined as

$$LD(x,y) = \min_i \ (p_i + q_i + r_i) \ \ .$$

(1.17)

Since the different types of error may occur with different frequencies, an improved measure is the *weighted Levenshtein distance* (WLD) defined as

$$WLD(x,y) = \min_i \ (a \ p_i + b \ q_i + c \ r_i)$$

(1.18)

with a, b, and c suitable constants, depending on application.

Determination of WLD can be made by a dynamic programming method (cf, e.g., [1.52]).

Similarity by Invariant Features. One great problem, especially in pattern recognition and experimental psychology, is to define similarity in a way which takes into account only *invariant features*. For instance, different images may be recognizable as similar, or to represent the same object or theme, although they were only derivable from each other by some transformation like displacement, rotation, or affine transformation. There are two principally different approaches to this problem: 1) Similarity is defined in terms of certain *transformation functions* instead of the original patterns; e.g., the modulus of the complex two-dimensional Fourier transform is invariant with respect to an arbitrary displacement (translation) of the original image in its plane, and the modulus of the two-dimensional Mellin transform is invariant with respect to an arbitrary affine transformation. 2) The original pattern is represented by a set of local patterns (*local features*) extracted from it, each one of the local features being encoded independently of its position, orientation, and size. The topological relations of these local patterns may be expressed syntactically, e.g., using the relational structures discussed in Sect. 1.3.2 (cf [1.53]).

There are almost endless possibilities to define *features* for patterns; notice that the transformation functions are only samples of them. Extraction of features is one central theme in pattern recognition research.

We shall present in Sect. 2.7 a very fast string matching method which is based on feature comparison; this method may have applications with other types of pattern, too.

1.4.3 The Problem of Infinite Memory

When dealing with samples of natural images or signals, one big problem is the filling up of any available memory capacity sooner or later. By advanced instrumentation it is easy to gather a great deal more information than can ever be represented in the memory systems, at least in their random-access parts. This is the *problem of infinite memory*, as it is often referred to. This problem has been known in systems theory, especially in the *theory of adaptive (optimal) filters*. The central principle in the latter is to represent a dynamical system or filtering operation by a finite set of parameters. These parameters are *recursively updated* by all received signals whereby they can be regarded as a kind of *memory* for all received information.

A characteristic of all constructs discussed in this book is that the memory shall be able to *recall* the stored representations as faithfully as possible, with a selectivity depending on the measure of similarity thereby applied. However, mere memory-dependent responses or classifications may not be regarded sufficient to represent the operation of associative memory. It seems that the available memory capacity could be utilized more efficiently if it were possible to represent only significant details accurately, and for those parts which are more common, to use average or stereotypic representations. It seems that the human memory to a great extent operates in this way; although a single occurrence can cause clear memory traces, nonetheless recollections are often only stereotypic, affected by earlier experiences. This must be due to the fact that biological systems almost without exception tend to optimize their resources, and this principle must also be reflected in the utilization of memory.

One elementary way for the description of "infinite" memory is to assume that there are representations in the memory for only a finite set of items $\{x^{(i)}\}$, but corresponding to every item there exists a state variable $y^{(i)}$ which is supposed to average over a great number of realizations of $x^{(i)}$. Consider, for simplicity, the occurrences of $x^{(i)}$ as discrete items $x_k^{(i)}$. The state variable is assumed to change as

$$y_{k+1}^{(i)} = f(y_k^{(i)}, x_k^{(i)}) , \qquad (1.19)$$

where f can attain many possible forms; the simplest of them is a weighted sum of $y_k^{(i)}$ and $x_k^{(i)}$ whereby a moving arithmetic average is formed over the sequence of $\{x_k^{(i)}\}$. In a more complex case the recursion may also involve scale transformations. When something is *recalled* from memory, then it may be thought that an external search argument x is compared with the set of representations of different items $\{y_t^{(i)}\}$ existing in memory at time t; the one for which some similarity measure $S(x,y_t^{(i)})$ is maximum, will be picked up and recalled. Notice that this process of memorization and recall is not only able to smooth over an arbitrary number of representations of an item but also to *synthesize* new information. For instance, the "averaged" representation $y_t^{(i)}$ may have resulted from realizations $x_k^{(i)}$ with inherently different contents.

Although a process described by (1.19) is easily amenable to computing, it is not certain that this is exactly the way in which information is stored in biological memory. In particular one should take into account the *spatial distribution* of memory traces in the brain for which there exists plenty of experimental evidence. We shall briefly touch this question in the next subsection.

1.4.4 Distributed Memory and Optimal Associative Mappings

There exists an intriguing feature in the physiological properties of biological memory: although various kinds of lesion in the brain tissue are known to cause characteristic deficiencies in the performance of memory, nonetheless it has turned out extremely difficult to demonstrate memory effects in the individual neural components. This discrepancy may be partly explained by the fact that the operation of the brain requires cooperation of its various parts; it may be as easy to cause drastic malfunctions in a complex computer system by breaking some of its lines or components. But another possible explanation is that memory traces are *distributed* over the neural realm in the form of some transformation function, and traces from many different occurrences are *superimposed* on the same tissue. There is some evidence for such distributed superposition in the classical works of LASHLEY [1.54]; many newer findings confirm that the degree of performance in various kinds of sensory functions or motor skills is directly proportional to the volume of neural tissue dedicated to it.

Although the scope of this book is far from biological issues, nonetheless the simplest possible physical embodiments of associative memory are regarded so relevant to the present discussion that at least the basic system models developed for distributed memory ought to be mentioned. More detailed discourses can be found, e.g., in [1.1]. As a byproduct, this presentation yields some partial solutions to the problem of infinite memory mentioned above, especially to the automatic (adaptive) formation of clustered representations in memory.

The fundamental feature in a system-theoretical approach to associative recall is that *no memory locations, and accordingly, no addressing system need be assumed.* The *selectivity* of output recollections in relation to the input patterns used as the search argument must come from the properties of transformation functions used to represent the contents of the memory; although the following analogy does not deal with the same thing, it may be used for comparison. A simultaneous (superpositive) transmission over a single line is possible using *orthogonal functions* as carriers of signals. In a similar way, the individual memory traces on a distributed memory medium can in a way be orthogonalized, and in spite of their superimposition on the same elements, the loss of information may be negligible.

Optimal Linear Associative Mapping. Consider the system block denoted by M in Fig. 1.7; in its basic physical embodiment it may be regarded as some kind of signal network. A set of simultaneous signals may be considered as

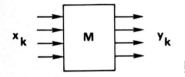

Fig. 1.7. Accociative mapping

a real (Euclidean) vector with n components, $x_k \in R^n$ where the subscript k identifies this set. Another set of signals which is obtained with insignificant delay with respect to those of x_k is denoted $y_k \in R^m$. The network is assumed to perform a static transformation or mapping of x_k into y_k, and if in the simplest case a linear mapping is considered, then M must be regarded as a transfer operator, a matrix $M \in R^{m \times n}$.

The Central Problem of Selectivity reads: Assume a finite set of pairs (x_k, y_k), $k \in S = \{1, 2, \ldots, N\}$. Does there exist an M such that

for all $k \in S$, $y_k = Mx_k$? (1.20)

This is one of the basic problems of linear algebra, and it has a simple answer:

Lemma [1.1]: If all the x_k, $k \in S$ are linearly independent (no one can be expressed as a linear combination of the others), then a unique solution of (1.20) exists and it is

$$M = Y(X^TX)^{-1}X^T \qquad (1.21)$$

where $X = [x_1,...,x_N]$ and $Y = [y_1,...,y_N]$ are matrices with the x_k and y_k as their columns, respectively, and the superscript T denotes the transpose of a matrix. If the x_k are linearly dependent, then there exists a unique approximative solution in the sense of least squares

$$\hat{M} = YX^+ \qquad (1.22)$$

where X^+ is the *pseudoinverse* of X [1.1].

Incidentally, $\hat{M}$ has the form of the *best linear unbiased estimator* (BLUE) which is a kind of *Gauss-Markov estimator*. Theoretically, even if there was an infinite number of pairs (x_k,y_k), but they were *clustered*, nonetheless there would exist an M which defines an "infinite" associative memory in the sense of (1.20). Such a matrix can be computed recursively, whereby its formation is equivalent to an *adaptive process*.

Associative Recall by M. Assume that an input pattern $x \in R^n$, corresponding to a *search argument*, is operated upon by M. An output $y = Mx \in R^m$, corresponding to a *recollection* from "memory" M, is thereby obtained. In the case that the x_k were linearly independent, the recollection has an important form [1.1]:

$$y = Mx = \sum_{k=1}^{N} \gamma_k \, y_k \, , \qquad (1.23)$$

where the γ_k are scalar coefficients representing the *linear regression* of x on the set $\{x_k\}$. In other words, y is a *linear mixture* of the "memorized" patterns y_k, each one of the latter occurring in the mixture with a certain optimal weight. Notice that if $x = x_{k_0} \in S$, then $\gamma_{k_0} = 1$ and $\gamma_k = 0$, $k \neq k_0$; an exact match will thus produce an error-free recollection y_{k_0}.

In case the x_k are linearly dependent, especially in the case of "infinite memory", (1.23) would still have the same form but the γ_k have a different and more complex interpretation.

Nonlinear Associative Mappings. The selectivity in associative recall, especially with respect to a set of linearly dependent input patterns, can be improved if instead of the linear transform, a nonlinear one (e.g., a polynomial mapping) is taken [1.55]. It should be noticed, however, that there are infinitely many forms of nonlinear transformations. The patterns can also be stored in spatially separate locations and recalled by a matching process; this is also a nonlinear mapping, and the performance in recall is finally dependent only on the similarity criterion applied (cf Sect. 1.4.3).

It should also be pointed out that there exist *nonlinear estimators* corresponding to (1.22) which are specified by pairs of data (x_k, y_k) and which thus implement a kind of associative recall, too. These estimators can be constructed, e.g., by *stochastic approximation* techniques (cf, e.g., [1.56]).

Autoassociative Recall by Projection Operators. Very interesting associative recollections are obtained if the optimal linear associative mappings are considered in the case that $y_k = x_k \in R^n$. In this case (1.22) becomes

$$M = XX^+ . \qquad (1.24)$$

Incidentally, this is the so-called *orthogonal projection operator* or *projector* with some interesting properties.

It is said that the vectors x_k *span a linear subspace* $\mathcal{L} \subset R^n$, or they constitute the basis vectors of $\mathcal{L}$. An arbitrary vector $x \in R^n$ can always uniquely be decomposed as a sum of two component vectors,

$$x = \hat{x} + \tilde{x} \qquad (1.25)$$

such that $\hat{x}$ is the *linear regression* (best linear combination) of the x_k on x in the sense of least squares, and $\tilde{x}$ is the residual. In fact, $\hat{x}$ and $\tilde{x}$ are mutually orthogonal. It is said that $\hat{x}$ is the *optimal autoassociative recollection* relative to the stored information $\{x_k\}$ and the search argument x.

Demonstrations of the associative properties of the projector XX^+ have been made by computer simulations; Fig. 1.8 represents results of an experiment in which the patterns consisted of digitized pictures with 3024 pixels (picture elements), and there were 100 pictures which defined M. A key pattern x, with a part missing produced a recollection in which this part was present [1.1].

It ought to be emphasized that the representation of projectors in computer memories, and the computation of the orthogonal projections need not be performed in matrix form; representation of an orthogonal basis of $\mathcal{L}$, and direct computation of $\hat{x}$ can be performed by the *Gram-Schmidt orthogonalizing process* [1.1].

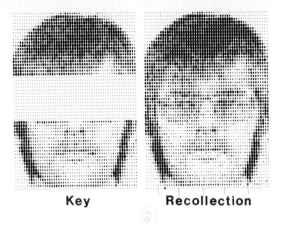

Key   **Recollection**

Fig. 1.8. Demonstration of autoassociative recall by XX^+

1.4.5 Sequential Recollections

We shall now consider another important aspect in the theory of associative memory, namely, the mechanism by which timed sequences of representations can be memorized and reconstructed. This is often called the problem of *temporal associations* or *temporal recall*. We shall further emphasize that the sequences of recollections are usually *structured*, i.e., they may branch into alternative successive sequences depending on some kind of background or context information; the sequences may also become cyclic.

The theory of automata (cf, e.g., [1.57]) deals with machines which are intended for production of sequences of machine states. The fundamental machine, the *finite-state machine* is an abstract system wich has a finite set of internal states S, it accepts one element of a finite set of "input states" I (e.g., patterns of input signals), and it produces another element of a finite set of "output states" O (e.g., patterns of output signals) thereby undergoing a state transition. This transition is a function of the present input state as well as the internal state; the output is another function of the input state and internal state, respectively. The new internal state of the system and the output state are formally defined by a pair of mappings (f,g) such that

$$f: I \times S \rightarrow S , \quad g: I \times S \rightarrow O , \tag{1.26}$$

where × denotes the Cartesian product. These expressions may also be written more explicitly. If $s_i \in S$ is one of the internal states, if subscript i is used to denote a *discrete time index*, if the present input is $i_i \in I$, and the

34

present output is $o_i \ \varepsilon \ O$, respectively, then one can write

$$f: s_{i+1} = f(s_i, i_i), \quad g: o_i = g(s_i, i_i) . \tag{1.27}$$

If the system variables are signals in digital computing circuits, a sequentially operating system is obtained by feedback of output signals of a combinational circuit to its inputs as shown in Fig. 1.9. The effective output signals are obtained by another combinational circuit. Time delays

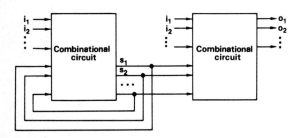

Fig. 1.9. Sequential circuit, basic form

usually exist in the internal signal paths or in the feedback; it is the amount of feedback delay which sets the time scale in the dynamic (sequential) process.

One of the central objectives in the theory of automata is to devise mappings f and g, such that maximum-length output sequences can be produced by a minimum number of state variables. This problem has practical importance in the design of computing circuits.

We shall approach the problem of temporal associations using a simple system model which also has the appearance of a finite-state machine (Fig. 1.10). The central block in it is some embodiment of *content-addressable memory* (CAM), with two types of input: external inputs A and B, and feedback input D. The system has an output part for recollection C. The feedback D is derived from C via a block with unit time delay; in the simplest case we may have $D(t) = C(t-1)$, in which t is a discrete-time variable. The CAM has two modes of operation, *writing* and *reading*, explained below.

There is a characteristic feature in the system of Fig. 1.10 which distinguishes it from other sequential circuits. The internal states are not arbitrary; on the contrary, as the central block represents a memory, its internal state shall contain replicas of the pairs of patterns (A,B) received at the input. Furthermore, the feedback pattern D shall be associated

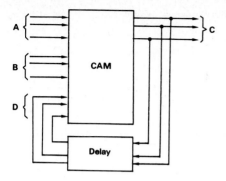

Fig. 1.10. Associative memory for structured sequences

with the input so that a set of *ordered triples*

$$\{(A(t), B(t), D(t))\}$$

is in fact stored in the CAM in the *writing mode*.

The two external inputs A and B are assumed to have specialized roles in the information process, although the CAM makes no difference between them. A sequence of inputs $\{A(t)\}$ is assumed to comprise the *central temporal pattern*, whereas B assumes the role of *background* and is stationary. For different A-sequences, the prevailing background B shall be different; this makes it possible to distinguish between different sequences in spite of the same $A(t)$ patterns occurring in them.

One further specification shall be made: for the encoding of a temporal process in the CAM, it will be necessary to generate an output $C(t)$ and a feedback pattern $D(t)$ during the writing of each triple. In some CAM circuits it is possible to obtain an output during the writing process (cf, e.g., Fig. 4.1), but every writing operation may also be accompanied by a subsequent reading. In the simplest case we thus have $C(t) = A(t)$.

Assume now that the sequence of external inputs $\{A(1),B), (A(2),B), \ldots, (A(N),B\}$ has been received and stored. When the first input $(A(1),B)$ arrives, the D input has no signals at it; an input with no signal shall be denoted by the symbol $\emptyset$ ("don't care"). Upon content-addressable search, the value $\emptyset$ shall be neglected. According to the above consideration, the *effective input sequence* shall now be

$$S = \{(A(1),B,\emptyset), (A(2),B,A(1)), \ldots, (A(N),B,A(N-1))\} \tag{1.28}$$

which is then a set of triples stored as such.

The *associative recall* of a memorized sequence is now made in the *reading mode* of the CAM, e.g., by the application of a *key input* (A(1),B,∅). For the recollection of the rest of the sequence, no further keys A(k), k = 2,3,... shall be necessary. (The reading could also be started in the middle of a sequence.) The CAM is assumed to produce a recollection A(1) at its output port; this pattern will be mediated to the D input by a unit time delay as D(2) = A(1). Reading will now be continued automatically, but the A(k) input, k = 2,3,... thereafter attains an "empty" or ∅ value. The next input is (∅,B,A(1)) which will match in its specified part with the second term in the sequence (1.28). Consequently, it will produce the output A(2), and a continued process will retrieve the rest of the sequence A(3), A(4), ..., A(N).

It may be clear that the CAM is able to store a great number of independent sequences, each one being retrievable by its own key input, consisting of an (A,B) pattern.

We have intentionally avoided the mention of one further problem. It may occur that some input pattern, say, (∅,B,A(k)) may match several triples stored in the memory. This is the *multiple-match* situation, and in pattern recognition tasks it would be named *reject*. In usual information retrieval this is a normal situation since all matching items will have to be found. If content-addressable memories are used to generate structured sequences, then the output C, however, must always be *unique*. The following possibilities for the handling of this problem exist:

1) It may be forbidden to store triples which would cause rejects.
2) One may apply similarity measures with a higher resolution in order that the probability for rejects become negligible.
3) One may arbitrarily choose one of the responding items, e.g., in the order of some priority applied in readout (cf Sect. 3.3.2), or at random.
4) The output may be some synthesis of the multiple responding items.

Each of the four procedures mentioned above has a different philosophy. The first of them may be applied, e.g., when designing automatic, programmable control circuits based on the CAM. The second method is typical for sequential pattern recognition in which special associative memories are used. The last two methods have more theoretical interest and they may relate to simulation of heuristic behavior in automatic problem solving tasks.

It should be noticed that the system model of Fig. 1.10 can have many different embodiments: the CAM may be a hardware circuit, a hash-coding scheme, or a distributed memory, and the matching criterion applied in recall may be based on any similarity measure described in Sect. 1.4.2. Finally this scheme may also combine the operations discussed in Sect. 1.4.3 whereby the representations stored in the memory are gradually and adaptively updated to produce recollections which are some "averages" of repetitive input sequences.

Recently, BOHN [1.58] has presented an interesting related associative scheme for timed associations.

Chapter 2   Content Addressing by Software

This chapter contains a review of certain programmed searching methods which
might properly be called associative or content-addressable, because an entry
(record, document, file, associated word or data, etc.) can be accessed directly on the basis of a keyword or a specified part of the entry. It is to
be noted that all of these methods are connected with the use of general-
purpose computers, and the entries are thereby stored in standard random-
access (addressable) memories. The methods described in Sects. 2.2-7 are generally known under the name *hash coding*. Alternative names used for the
same techniques are *hashing, hash addressing, hash-table method, scatter-
storage techniques, key transformation, key-to-address transformation, com-
puted-entry table method, randomization, random addressing, scrambling,
direct-access method,* and possibly others. Another content-addressable organization of data named *TRIE* is described in Sect. 2.8.

2.1 Hash Coding and Formatted Data Structures

Hash coding is the fastest programmed searching method known, especially
relating to large files. When one uses it, the keywords need not be ordered
or sorted in any way. A minor drawback of hashing is a slightly higher demand of memory space when compared to the other methods; in view of the fact
that the prices of memory components have been radically decreasing over many
years, and this tendency continues, it seems reasonable to waste a little
memory when plenty of benefits are thereby gained. Moreover, the need for
additional memory space is indeed not very big: with a 50 percent reserva-
tion, entries can be located by hash coding in a number of accesses to memory
which on the average only slightly exceeds unity.

The extremely high speed of hash coding in searching operations is because
entries are stored in and retrieved from *memory locations, the addresses of
which are computed as some simple arithmetic functions of the contents of
their key words.* The contemporary digital computers can perform such arith-

40

metic computations very quickly compared with memory accesses, especially those made to the mass memories.

The set of all possible keywords is generally called the *name space*, and the set of addresses in memory into which the keywords are mapped is named the *address space*. Accordingly, hash coding implements the transformation of keywords into *hash addresses*. In general, this transformation is not an injection since the keywords have to be chosen freely, and there is no possibility of keeping a record of those hash addresses which have already been used. Hence there remains a nonzero probability for two different keywords to be mapped onto the same address, and this instance is named *collision* (or *conflict*). The problem of collisions can be handled quite easily and without significantly wasting time for it: the remedy is to reserve some place for the overflowing entry which is directly derivable from the place of collision. For instance, locations which have consecutive addresses with respect to the hash address may be tried until an empty one is found; for other methods, see Sect. 2.3. However, it should be noticed that when one or more reserve locations are made to depend on a common hash address, there becomes a need to have extra means for the identification of the overflowing entries with their keywords, since the hash address is the same, and the exact place of the reserve location cannot be predicted. The usual solution is to store the keyword (or some unique function of it) in conjunction with the corresponding entry, and during retrieval, the keyword used as the *search argument* must then be compared with keywords stored in possible reserve locations in order to find the correct entry (Fig. 2.1).

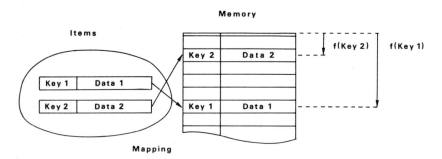

Fig. 2.1. Mapping of entries in hash-coding (*Hash table*)

If the entry should be locatable by several alternative keywords, a separate hash address must be computed for each of them, and an identical copy

of the entry might be stored at all of these addresses. In order to avoid
wasting memory for multiple copies of the same entry which possibly may be
very long, the latter may be represented in the hash table by some shorter
identifier, for instance, address reference to a memory area where the proper
entry is stored (cf discussion of the *hash index table* in Sect. 2.4). Such
references are named *pointers*; they facilitate a logical manipulation of information without the need of retrieving real entries.

By far the most important applications of hash coding are found in systems
programming. Hash tables were being used already around 1954 in IBM assemblers (cf Sect. 2.9), and practically all *symbol tables of assemblers and
compilers* are nowadays implemented by hashing techniques. This practice has
been dictated by the cost of computing which is centrally dependent on compiling; since symbol tables during compilations as well as computations are
accessed frequently and in an unpredictable order, hash coding has proven most
efficient in their maintenance.

Another fruitful area of application for hash coding is in the *management
of data bases*. Especially in multiple-keyword searching operations, some
effective implementations based on hash coding can be found (cf Sect. 2.6).
Since data bases or data banks are normally constructed using multilevel
storages, also named hierarchical memories, some extra points of view in the
construction of hash tables must thereby be taken into account. It should be
emphasized, however, that even with multilevel storages the entries can be
located by hash coding in a time which, in principle at least, is almost
independent of the size of memory, and this should be an especially important
factor with very large data bases.

There is one particular aspect in the applicability of hash coding which
has not been discussed much in literature. A preassumption of the usual hashing methods is that the keywords must be *unique* and *error-free*. This restriction has actually not been regarded as a serious one by most programmers
since they have learned that their programs must always be formally flawless.
A similar completeness of names, words, number codes, and other descriptors,
however, may be difficult to achieve with masses of data which are collected
in large data bases. It is a known fact that to err is very human; the rate of
errors made in simple routine operations such as keying may amount to as high
as one percent and this kind of error is not always detectable. Moreover,
ambiguities in data may be caused by other reasons, too. There are names
and transliterations which even theoretically can occur in several versions,
and sometimes only the phonemic appearance of a keyword is known. There are
also applications in which it would be desirable to let different versions of

the same item span a class which then should be recognizable in an invariant way. This kind of demand may at first seem to restrict seriously the applicability of hash coding. Fortunately, this need not be the case; although it will be necessary, for the achievement of error-tolerance, to add some redundancy to the memorized information as well as to the computing operations, the growth of costs remains modest, and significant benefits can still be gained by the use of hash coding. The demonstration given in Sect. 2.7 is intended to point out that hash coding may constitute a straightforward solution to searching, even in the case of incomplete key information.

There exist some searching tasks which, however, do not easily lend themselves to hash coding. The most important of these cases are *magnitude search*, or location of all items an attribute of which is greater or less than a specified limit, and the search on *Boolean variables*, especially if the keywords have to give a certain value to a complicated combinational function. As for magnitude search, if the ranges of attribute values can be divided into a small number of intervals, it is still possible to provide each of them with a distinct keyword, whereby the searching would be based on identification of the intervals by hash coding. Most often, however, searching on the basis of magnitudes as well as on Boolean arguments is implemented as a batch run, by scanning all entries sequentially and substituting their keyword values to a function which indicates the matching condition, or by more traditional methods discussed below. Thus, before proceeding to the details of hash coding, it may be necessary to mention that the methods for the handling of large files seem to have developed from the beginning in two distinct directions. The hash coding methods constitute one of them, but the other main line in the data management business is the one in which the searching problems are solved by *ordering* or *sorting* the entries and *structuring the files* in accordance with their argument values. One advantage thereby achieved is the possibility of performing magnitude searches directly. The highlights of this latter approach are *search trees* and *multidimensional indexing*. Search trees as well as ordering and indexing of entries, in fact, facilitate the manipulation of data by their contents, but there are some good reasons to leave these topics outside the scope of this book. One of them is that the principles of hash coding, although actually belonging to the area of computer programming, have been included in this kind of book since there has for long existed a debate about the superiority of hardware versus software in content addressing, and the latter is usually identified with hash coding. This book is now intended to provide some documentary material to the reader for settling this question. The second reason for not expounding the other

searching and sorting methods is that there already exist many good textbooks and survey articles on those topics; it seems unnecessary to duplicate
them here. Readers who are interested in data management may find an overview to the material they need, e.g., in the books of KNUTH [2.1], MARTIN
[2.2], FLORES [2.3], DESMONDE [2.4], and LEFKOVITZ [2.5]. Reviews on the
same topics have been published in the articles [2.6-9]. Special questions
partly relevant to content addressing have been dealt with, e.g., in the
papers [2.10-19]. It ought to be emphasized, too, that the searching of
entries by their data contents is one central task in certain business-
oriented high-level computer languages such as COBOL; there was a working
group in 1965, named CODASYL Data Base Task Group, which has published two
reports, [2.20,21] on the development of a generalized data management system.

   To recapitulate, if the problem is to find entries from a data base by
given *identifiers* or *multiple descriptors*, as the case is with documents,
publications, etc., then hash coding usually offers effective means for performing the task. If, on the other hand, the searching conditions are specified by giving the values or ranges of certain *attributes*, especially using
magnitude relations and Boolean functions, then the traditional sorting and
tree searching methods might be preferred. Hash coding has been selected
for presentation in this book since it is what many high-level programming
languages and linguistic data structures make extensive use of, and a view
is generally held that hash coding constitutes one of the most direct implementations of associative and content-addressable processing functions in
computer science.

   The discussion of hash coding to follow is carried out by first intro-
ducing the fundamental ideas and then proceeding to quantitative comparisons.
References to survey articles and other collateral reading are found in
Sect. 2.9.

2.2 Hashing Functions

As already mentioned, the central idea in hash coding is to try to store an
entry at an address which is arithmetically computable from the contents of
its keyword. In principle, only very mild restrictions should be imposed on
the choice of keywords. Natural names or arbitrary number codes should be
eligible, and in particular it should not be necessary to append any error-
checking marks or characters to them. The length of a keyword should also be

arbitrary, although in practice only a number of leading characters might be taken into account. For these reasons, the set or domain of all legitimate words, the name space, becomes very large. For instance, if only six-letter words formed of the English alphabet were used, the number of different possible words would be 26^6, or about 300 million. In practice, the volume of the name space may be much larger.

The first idea that may come into one's mind when thinking of a mapping of the name space into the address space is some form of *compression*: since obviously the name space is very sparsely occupied, a function with the address space as its range ought to be found such that the map of names on it is more densely and evenly distributed. If the keyword values v were scalar numbers, and they had a statistically well-defined distribution, then, indeed, this problem would have a nice solution. Assume that the statistical sum function of v is F(v). If then for the "hash address" of v, the value H · F(v) + B were taken, where H is the number of available addresses and B the first address, the distribution of the hash addresses is expected to be uniform in the interval [B,B+H]. Unfortunately there are some complications with this type of approach: for one thing, the keywords are usually codes which are converted into numerical values using different weights for their elements. It is then not plausible that the numerical values would possess a well-defined distribution. Secondly, the problem of collisions would not be removed since identical hash addresses may result from discretization errors in compression. Since the address mapping can be implemented much more easily without consideration of keyword distributions as discussed in this section, the compression methods will be left without further mention.

The mapping of keywords into the available memory space is in general defined by a *hashing function*. If the content of a key is denoted by K, then h(K) is its *calculated address* or *hash address*. A *collision* is always due when for two different keywords K_1 and K_2 there holds $h(K_1) = h(K_2)$. In this case one of the colliding entries, usually the latter one, must be transferred to a spare address which will be simply derivable from the calculated address. In some cases, the next empty location to the calculated address can be used as the *spare* or *reserve location*. Other methods are reviewed in Sect. 2.3. If the reserve locations are now taken from the address space which is the range of the hashing functions, a collision may occur with the reserve locations, too. In principle the collision is handled in a similar way, regardless of whether its target was a calculated address or a reserve location.

Different hashing functions can be compared according to their ability to minimize the number of collisions in the case of general keywords. Notice that it was not held necessary nor desirable to control the selection of keywords. Although procedures have been devised by which collisions can be totally avoided when the keywords have a certain minimum distance (cf the discussion below), the practical value of such refinements is low. It has turned out that a generally good hashing function is one which spreads the calculated addresses uniformly over the allocated memory area.

There are normally two parts in the definition of a hashing function: 1) Conversion of the keyword into numerical form. 2) Mapping of the domain of numerical values into the set of hash addresses.

Conversion of Keywords into Numerical Form. All information entered into computers or other automatic data handling devices must be in coded form. Punched cards have their own formats for numbers, letters, and other markings, and data are transferred internally and externally in computer systems in the form of special codes, usually in the so called ASCII code where the *characters* (numbers, letters, and control marks) are encoded by eight bits and thus correspond to one *byte* in computers. For the arithmetic computations, numbers are usually represented either as pure binary numbers, or as binary-coded decimal numbers. (Notice that the octal and hexadecimal numbers result only from grouping the bits of binary numbers into subsets of three or four bits, respectively.) In view of the above considerations, since all computational variables are already expressed in numerical form, one might think that further conversion is unnecessary. However, some algorithms for address computations presume that the numerical values must fall upon the legitimate range of integers in machine arithmetics.

The most common type of keyword is a typed string of either numerals, letters, or alphanumeric characters. The alphabet may consist of all the characters of the ASCII code, or it may be restricted to the English alphabet augmented by numerals. If a keyword may consist of several separate words, the blank may be included in the alphabet, too. It will further be emphasized below that it is advantageous to have the number of characters in the alphabet be a prime number; this number occurs as a radix in the numerical conversion. Assume now that for all different characters i in the alphabet, numerical values $d_i \in \{0,1,2,\ldots,w-1\}$ are assigned in some order whereby a string of characters, written as $k = d_N d_{N-1} \ldots d_1 d_0$, may be considered as the representation of an integer in the base of w. The above integer then has the numerical value

$$v = \sum_{i=0}^{N} d_i \, w^i \quad . \tag{2.1}$$

For instance, for the English alphabet, one may choose $A = 0$, $B = 1$, ..., $Z = 25$. The word 'ABE' thus has the numerical value $0 \cdot 26^2 + 1 \cdot 26 + 4 = 30$. It has turned out that this method works rather well in the numerical conversion of English words and names.

Relating to another application of hash coding, the following example shows how *index variables* can be converted into a numerical form suitable for hash coding. This application describes a subtle method for the storage and reading of *sparse matrices* [2.22]. A sparse matrix is one which has only a small fraction of its elements nonzero, and they are distributed in some arbitrary way over the matrix array. In order not to waste memory for the representation of zero values, the nonzero elements can be stored in a hash table whereby the pair of indices (i,j) that indicates the position of this kind of element in the matrix array is now regarded as its "keyword". Assume that $i \in \{1,2,...,p\}$ and $j \in \{1,2,...,q\}$; it is then possible to represent the numerical value of the "keyword" (i,j) in either of the following ways:

$$v = i + q(j-1) - 1 \, , \quad \text{or} \quad v = j + p(i-1) - 1 \quad . \tag{2.2}$$

The address in the hash table at which the matrix element is stored is then calculated as a function of v, in the same way as the hashing function for a string would be calculated, given v.

In the above methods for numerical conversion, a distribution of the v values which is already reasonably uniform is strived for. Usually there exists yet a second step in hashing, however, in which the v values are mapped onto the domain of hash addresses, and where the final randomization is carried out. The quality of the hashing function depends primarily on this latter phase.

Preliminary Discussion of Randomization Methods. Before further discussion of the hashing algorithms, it may be useful to review some methods for the generation of *pseudorandom numbers*. These are numbers generated by numerical algorithms with the purpose of imitating stochastic values. There are plenty of applications in numerical mathematics for which sequences of numerical values that imitate stochastic processes are needed. Incidentally, this need is rather small in statistics, because there the processes are often analyzed using theoretical distributions. Maybe the most important applications of pseudorandom numbers occur in the analysis of complex physical problems, for

47

instance, in the approximation of diffusion and transport integrals which may have to be computed by the so called Monte Carlo methods. Systems theory and systems analysis also use pseudorandom numbers for the imitation of stochastic noise, and there exist electronic devices with built-in hardware circuits for these algorithms.

In hash coding, the pseudorandom numbers serve twofold purposes. For one thing, the *hash addresses* which are functions of the keywords are pseudo-random numbers. Secondly, key-dependent or key-independent random numbers are needed in probing algorithms (cf Sect. 2.3).

The most desirable property of pseudorandom sequences is that the subsequent numbers are as little correlated as possible. This requirement is most severe in analytical problems because the accuracy of results directly depends on it. On the other hand, hash coding by its nature is a trial-and-error method and the quality of pseudorandom numbers thereby used affects only the *number of trials* made. Differences in average speed of a few percent can hardly be noticed, and the most important factor is the speed of operation, which is determined by the number of computational steps. Accordingly, the algorithms applied in hash coding should be computable by as few and as short machine instructions as possible.

An important property of sequences of pseudorandom numbers which is utilized, e.g., in the *probing algorithms* (cf Sect. 2.3) is that when initialized with a particular value, the sequence generated by the pseudorandom number algorithm is completely *deterministic*. This is because most sequences of pseudorandom numbers are computed *recursively*, by using results from earlier computational steps as parameters in the next steps.

It is a known fact that round-off errors of real numbers may often be regarded as random numbers with a uniform distribution. In a similar way, numbers formed of the least significant digits, or from digits drawn from the middle of a numerical representation, usually look as if they had stochastic properties. In general, the so called *residual numbers* in arithmetic normally have very little correlation with the original numbers from which they are formed. All the numbers v-A, v-2A, ... are said to be *residues* of v *in the modulus* A. The least positive residue, denoted v mod A (read: v modulo A) is simply the *remainder* in the division of v by A. A number X is said to be *congruent* to another number Y modulo A, or $X \sim Y$ mod A, if $X = Y + kA$, in which k is an integer. As will be seen, such *congruence relations* are widely utilized in hash coding.

By far the most important generator for pseudorandom numbers is the so called *multiplicative congruence*. The sequence of numbers $\{R^{(k)}\}$ is computed recursively, using the last number obtained as the argument for the next step. In order to decrease correlations in the sequence, it has been found advantageous to modify the algorithm, e.g., into the following form:

$$R^{(k+1)} = (aR^{(k)} + b) \bmod c , \qquad (2.3 \text{ a})$$

or, alternatively,

$$R^{(k+1)} = integer \ [d^{-1}(a \ R^{(k)} + b) \bmod cd] , \qquad (2.3 \text{ b})$$

in which a, b, c, and d are suitably chosen constants; first of all, $0 \le R^{(k)} \le c$. In the latter expression, c is often a power of two and d another small power of two, e.g., d = 4.

A simplified version of the above pseudorandom number formula which may be used, for instance, in the *random probing algorithm* discussed in Sect. 2.3, may be expressed in the form of the following computational steps [2.22] that generally do not require more than five machine instructions. Assume that the hash table size is $H = 2^n$. Whenever the following computational routine is called, an integer I is initialized to be equal to 1. The following steps are then executed:

1) Set I = 5 I.
2) Mask out all but the lowest n + 2 bits of the product and substitute the result in place of I.
3) For the next pseudorandom number, take I/4 and return to 1.

Finally, if special hardware can be developed for hash coding, it may be proper to mention the most commonly applied electronic principle for the generation of pseudorandom numbers. Its fundamentals have been laid down in the theory of *linear sequential circuits* [2.23]. The practical implementation makes use of a binary shift register, the contents of which are understood as a binary number. The register takes a new bit value at its left end through a special feedback of the old contents of the register. Normally this feedback is obtained as the EXCLUSIVE OR function of two specified bits in the register. The positions from which the bits for the EXCLUSIVE OR function are extracted depend on the length of the register. The state sequence thereby obtained looks rather random; a further reduction in the correlation of subsequent numbers can be obtained by picking up the numbers from the state sequence at greater intervals.

If the length of the register is n bits, and the initial state in the register is not zero, it can be proven that the register sequences through

all the states excluding zero, in other words, the length of the sequence
is $2^n - 1$.

An example of linear sequential circuit with the maximum-length sequence
is shown in Fig. 2.2. For details with other lengths of the register, see,
for instance, [2.24].

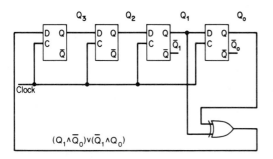

Fig. 2.2. Linear sequential circuit. The rectangles represent D flip-flops
where each Q state attains the logic value prevalent at the D input upon the
Clock signal

Of course, if a computer has a fast random-number algorithm in its systems
program repertoire, especially if it is micro-programmed, then further consideration of these methods in a practical coding application becomes unnecessary.

The Division Method of Hashing. Assume now that the consecutive addresses
in a memory area B...B + H are used as hash addresses. For the hashing function h(v) which defines the hash address it is then possible to select

$$h(v) = v \bmod H + B \tag{2.4}$$

where $v = v(K)$ is the numerical representation of the key argument obtained
in one of the ways described above. Since h(v) is obtained in a division
operation, the above hashing is generally called the *division method*. This
operation is in general found to produce rather uniformly divided hash addresses in most practical applications; moreover it directly spreads the hash
addresses on any chosen size of memory.

Although the division method is often preferred to other alternatives in
the computation of hashing functions, some precautions are necessary when
using it. For uniform spreading of the hash addresses, it will be profitable
to choose the exact hash table size H in a particular way. First of all, H
should be an odd number since if H were even, h(v) would be even when v is

even and odd when v is odd; an odd-even unbalance in the keyword values
would then directly be reflected in an unbalanced occupation between even
and odd hash addresses. Neither should H be a power of the radix used in
the numerical conversion (for instance, in the case of the English alphabet
H should not be of the form 26^p) since v mod H would then always be identical if the p last letters of the keywords were the same. KNUTH [2.1] has
pointed out that if w is the radix used in numerical conversion, then it is
advisable to avoid table sizes the multiples of which are near any of the
values w^k, where k is a small integer. Moreover, it is good to select a
prime number for H.

Special consideration is necessary if the hash table size is a power of
the radix used in the computer arithmetic (e.g., r = 2 or 10). In this case,
namely,

$$v \bmod H = d_{p-1}\, d_{p-2}\, \ldots d_1\, d_0 \tag{2.5}$$

where the d_{p-1} through d_0 are the p least significant digits of v, which
together with B then define the hash address. In other words, no separate
computation for division then need be performed. As the computation of the
hash function is thereby made in a very convenient way, it must be remembered
only that the numerical conversion must not be carried out in a base, the
radix w of which is a power of the radix r of arithmetic.

The Multiplication Method. This method is to be preferred with computers
having a slower algorithm for division than for multiplication. While there
are no great differences in speed between these algorithms in contemporary
computers, there is yet another aspect of this approach; namely, by this
method the normalized hash address in the interval [0,1) is computed first.
After that the result is applicable to a hash table of an arbitrary size,
without any consequences possibly resulting from an improper table size, as
in the division method.

Assume that v is the nonnegative integer obtained in the numerical conversion of the key. It shall fall into the range of representable values. Further
let c be a constant chosen from the range [0,1). The *normalized hash address*
$\varphi(v)$ in the range [0,1) is first defined as

$$\varphi(v) = cv \bmod 1 \,, \tag{2.6}$$

that is, as the fraction part of the product cv, whereafter the final hash
address becomes $h(v) = H\, \varphi(v)$, with H the hash table size. In integer arith-

metic, the practical computation of $\varphi(v)$ proceeds in the following ways. Let
b be the word length in bits, whereby $d = 2^b$ is the upper bound of the representable integers. Consider c as the integer cd, and for $\varphi(x)$, take the
last b digits of the double-length product (cd)v; the radix point must then
be imagined at the left of these digits. Notice further that if the table
size H is a power of 2, e.g., $H = 2^p$, then the leading p digits of the above
b digits constitute the hash address. (With the multiplication method, there
is no harm from the table size being a power of 2.)

The quality of the multiplication method obviously depends on the selection
of the constant c. It has been pointed out [2.1] that if ϕ is the "golden
ratio", then $c = \phi - 1 = 0.618034$ is a rather good choice.

Hash-bit Extraction. Digit Analysis. One of the oldest and by far the simplest
of the hashing functions is named *hash-bit extraction*. When the keyword is
regarded as a binary string, then a suitable number of its bits, taken from
some predetermined positions on the string, are simply concatenated to form
a binary number corresponding to the hash address. The number of such *hash
bits* should be large enough to address the complete hash table. Usually a
special *bit analysis* is accompanied with the selection of hash bits: when a
statistical sample of keywords is available, it is advisable to pick up such
bits for hash bits in which approximately as many 0's and 1's occur.

Instead of bits, it is possible to think of the binary string as a binary-
coded representation of digits by dividing the string into suitable segments.
The statistical distribution of the various digits in each segment may be
studied, and digits corresponding to the best distributions are chosen for
the hash address. This method is named *digit analysis*.

The Mid-Square Method. One method which has very good randomizing properties
and in some comparisons (cf Sect. 2.5.2), with a particular hash table organization, may yield better results than the others, is named the *mid-square
method*. In this algorithm, the keyword value v is squared, after which the
hash bits are extracted from its middle part. This algorithm is computationally very simple. Notice that if the keywords are of different lengths, the
hash bits are not taken from fixed positions, which further increases the
degree of randomization.

A precaution with the mid-square is to check that v does not contain a
lot of zeroes, in which case the middle part of the square may not become
very well randomized.

In the multiplication method, the address bits could be chosen in a similar way from the middle of the product; no experimental evidence is available,

however, about significant benefits achievable over those obtained by taking
the least-significant digits.

The Segmentation Methods. The following methods are useful only if the keywords are considerably long or consist of several words. The binary string
is usually formed of the concatenated character codes, but possibly also
the numerical representation of the keyword may be used. The string is first
divided into segments corresponding to the length of the hash-table address.
These segments can then be combined into a single address in various ways.
First of all, the segments as such may be regarded as binary numbers, whereby
bits in the same position correspond to the same power of two. Alternatively,
every second segment may be reversed so that the bit positions in the segments
are defined in the same way as if the original string were "folded" like a
strip of paper. It seems that the latter method yields a slightly improved
randomization. The segments defined in either of the above ways may then be
combined, whereby several alternatives exist: 1) the sum of the segments
modulo H may be taken, for which H is the hash table size; 2) the respective
bits of the segments may be summed up modulo 2, which is equivalent to taking
the recursive EXCLUSIVE OR function over the segment words; 3) the product
of segments modulo A can be formed. It seems that the EXCLUSIVE OR method
has particular benefits since it is computationally simple and the randomization properties are not deteriorated even though one or more segments consists of mere 0's or 1's; on the other hand, in the multiplication method
special attention must be paid to the possibility that the numerical value
of a segment may be zero, in which case this segment must be rejected.

In particular, in the LEAP language to be described in Chap. 6, the hashing
function for a pair of argument words (X,Y) is formed as the following function of its components:

$$h(X,Y) = (X \oplus Y) \bmod H \qquad\qquad (2.7)$$

where X and Y denote bit strings, and $\oplus$ is the EXCLUSIVE OR operation over
corresponding bit positions.

Radix Conversion. This method is based on the idea that if the same number
is expressed in two digital representations with radices which are relatively
prime to each other, the respective digits generally have very little correlation. Notice that this effect was already utilized in the basic numerical
conversion whereby the number of characters in the alphabet was selected to
be a prime number, and this number was chosen for the radix (2.1). If radix
conversion is taken into consideration in later steps, the keyword is then

regarded as a binary string which is divisible into segments of arbitrary
length. Each segment is regarded as a binary-coded representation of a digit
in some suitable base (which need not be a power of two); for instance, with
four bits it is possible to encode the ten numerals of the decimal system.
Assume that the chosen base is p. The numerical value of the binary string
which is now expressed in a p-nary number system is computed, and the result
is taken modulo Q, where Q is approximately the number of hash table addresses
(preferably Q < A). LIN [2.25] has considered the case that $Q = q^m$ where
p and q are relative primes, and m is a small positive integer.

Algebraic Coding. In many practical applications the keyword values tend to
be "clustered", whereas such agglomerates are usually more or less randomly
distributed. These "clusters" are reflected in hash coding in different ways.
First of all, one might expect that if two or more keywords are very similar,
their hash addresses would be correlated, possibly being near each other.
Fortunately, with a proper choice of hashing function, such as that based on
the division method, the hash addresses will be spread reasonably well, in
spite of apparent similarity in the original keywords. There is another type
of clustering, however, which is more fatal in a practical application, and
it is connected with collisions. Assume that the hash table is rather full
so that collisions occur frequently, or that several identical keywords occur
(in a multi-key hash table). The reserve locations to the calculated addresses
may then become badly clustered unless proper means for the handling of collisions are taken. We shall revert to this problem in Sect. 2.3.

Although, as mentioned above, the danger for the agglomeration of hash
addresses is not particularly great in spite of the original keywords having
similarities, there frequently appears an idea in theoretical discussions of
hash coding that the hash function itself ought to be capable of guaranteeing
different hash addresses for keywords which differ from each other on only a
few character positions. Such a property can indeed be achieved by the application of algebraic coding theory (cf [2.26,27]). The results thereby obtained are generally good, although we have to state that, at least when the hash
table is not very full, the difference with respect to the division method
is minor, and the computational load is heavier. This method is mentioned
here mainly for theoretical interest.

Assume that numbers, arithmetical expressions, and their results that
occur in the following computations are expressed over the Galois field
GF(q) which is a group of numbers with exactly q distinct elements. These
elements form a ring so that, for instance, addition, multiplication, and
division of these numbers is equivalent to the usual arithmetic expressions

modulo q. The key K is now made to correspond to a polynomial $K(x) = k_{n-1} x^{n-1} + \ldots + k_1 x + k_0$, where the k_i's are the digits of K. Another polynomial $P(x) = x^m + p_{m-1} x^{m-1} + \ldots + p_0$ with constant coefficients is chosen properly, as stated below. If the following expression is formed by the application of polynomial arithmetic over GF(q),

$$K(x) \bmod P(x) = h_{m-1} x^{m-1} + \ldots h_1 x + h_0 , \qquad (2.8)$$

then the hash address is obtained as the q-nary number

$$h(K) = (h_{m-1} \ldots h_1 h_0)_q . \qquad (2.9)$$

Obviously the biggest problem in this method concerns the selection of $P(x)$. In the theory of error-correcting codes [2.28] it has been shown that if all polynomials which contain $P(x)$ as a factor have the minimum Hamming distance of d (at least d of their coefficients are different), then no two sets of coefficients $\{k_{n-1}, \ldots, k_1, k_0\}$ which differ in r positions, $1 \le r \le d$, produce the same result h(K). Thus, a heuristic method is to try different $P(x)$ to find a sufficient d.

It has also been demonstrated [2.29] that the algebraic coding method yields reasonably good results even though the coefficients of $P(x)$ were selected quite randomly, with the restriction that p_0 is nonzero.

As stated above, when the polynomial division of $K(x)$ by $P(x)$ is performed, then every elementary step must be computed in modulo q. The value q = 2 is suitable because it is possible to construct simple digital hardware or micro-programs for this arithmetic. KNUTH [2.1] has given the following example: for q = 2, n = 15, m = 10, and $P(x) = x^{10} + x^8 + x^5 + x^4 + x^2 + x^1$, the hash addresses for two distinct keywords will be guaranteed to be unequal when the keys differ in r bit positions, $1 \le r \le 6$.

"Optimal Hashing Functions". In addition to the algebraic coding method which guarantees different hash addresses for distinct keywords if their Hamming distance stays within specified limits, there have been some attempts to generalize this result. SPRUGNOLI [2.30] has recently devised a computational scheme which guarantees different hash addresses for rather general distinct keywords. Unfortunately this scheme is practicable for small tables only, since the computation of the hash address grows quadratically with the number of entries.

Another attempt to devise an optimal hashing function which would minimize probings is due to MITRA [2.31]. This discussion has been carried out
for a hash code of length 3 which is too small for practical purposes.

Reduction of Hash Table Size by a Special Hashing When Errors Are Allowed.
In normal hash coding, collisions between hash addresses are allowed, but
they have to be handled by the methods described in Sect. 2.3 in order to
guarantee error-free operation. The "optimal" hashing functions were intended
to get rid of collisions. The method described in this presentation has quite
different aims; namely, the main objective is ultimately to reduce the hash
table size thereby sacrificing error-free operation. The purpose is to use
fast primary memory for extensive hash tables. Searching tasks with an allowable percentage of errors may occur in many statistical problems. There also
exist applications, for instance, in systems programming where for optimized
overall computational efficiency, a choice between two different procedures
has to be made, depending on the content of variables; mistaken choices are
thereby not fatal, but affect only slightly the theoretically optimal performance. A further example of searching with allowable errors occurs with
the handling of words, e.g., in hyphenation, where a simple rule may be
applied in a majority of cases, whereas for more difficult decisions, a re-
sort to a dictionary has to be made.

All cases discussed above are in effect examples of *membership tests*,
i.e., classification of identifiers (keywords) into two categories, "member"
and "nonmember". A convention can be made that "members" are stored in a
hash table and "nonmembers" left outside of it. Since then no further associated information is defined, the entries in the hash table comprise only
identifiers of the keywords and no associated data.

Assume that a small percentage of errors in the membership test is allowed.
BLOOM [2.32] has discussed two methods for the reduction of information to
be stored in the hash table. In the first method, a function of the identifier,
containing fewer bits than the original one, can be referred to; this method
is based on an idea that if the reduced identifier is stored in the table and
it matches that of the search argument, chances are good that the original
keywords are identical, too. Since information is lost in the reduction of
the identifiers, this method may commit errors whereby an item is falsely
identified as a member. The converse cannot happen, i.e., a true member always gets its reduced identifier correctly identified.

The second method discussed [2.32] assumes that the hash table consists
of N individually addressable bit positions, and a hashing function defines

d pseudorandom addresses a_1, a_2, ..., a_d as a function of the keyword; if all bits in the table were initially 0, the values at locations a_1, a_2, ..., a_d are then set to 1. *With this method, no collision handling is needed.* For every new keyword, a different new set of bit locations is calculated, but when storing 1's in them, it may well happen that earlier some of these bits were already set to 1. When a search argument is tested for membership, and if it is a true member, then the bit locations calculated for it certainly contain 1's; however, there is a nonzero, although generally small, probability that even nonmember search arguments have all bits equal to 1.

The rate of errors committed can be computed by simple statistics [2.32]. If the number of bits defined per keyword is d and they are distributed at random, if H is the hash table size (in bits), and N is the number of keywords stored, then the probability for an error is

$$P = [1 - (1 - d/H)^N]^d \ . \tag{2.10}$$

For instance, for $H = 10^4$ bits, $d = 10$ bits, and $N = 1000$, we have $P \approx 0.01$.

2.3 Handling of Collisions

2.3.1 Some Basic Concepts

In short, the various methods for the handling of collisions can be divided into two main categories: 1) Methods called *open addressing, open hash,* or *open overflow,* which utilize empty locations of the hash table as reserve locations. 2) Methods using a separate *overflow area* for the colliding items.

The techniques applied in open addressing contain many detailed and even complex ideas for the improvement of efficiency in searching. In contrast to this, the methods using an overflow area are relatively simple and straightforward, and if the efficiency were measured in terms of searches to retrieve an entry, the performance of the latter method might seem superior (cf Sect. 2.5.1). Therefore there must be good reasons for the particularly high interest in the open-addressing methods. One of them is that large hash tables are usually held in secondary storages such as disk memories, and the contents are only partly buffered in the fast primary memory in which, however, the searching computations are in effect performed. It will be easy to understand that for efficient buffering, it is advantageous to have all information referred to in a program in a small local memory area, and this is made pos-

sible by open addressing. Another aspect is that there are no unique theoretical criteria by which the optimal performance in multilevel memories
could be defined, first of all because a compromise between speed and demand
of memory space always has to be made. In view of the large differences in
access times between primary, secondary, tertiary, etc. storages, and taking
into account the possibility for transferring data between the memory levels
by blocks in a single access, the performance, thus, cannot simply be measured in terms of memory accesses per retrieval; the overall speed should
actually be evaluated by benchmarking runs and performing a practical cost
analysis.

The procedure by which the reserve locations in the open-addressing method
are found is named *probing*. In principle, probing is a trial-and-error method
of finding an empty location. Whatever method is used in probing, the sequence
followed in it must be *deterministic* because the same sequence is followed
in the storage of entries as well as during retrieval.

In recent years, effective methods for probing and associated organizational procedures have been developed. It has thereby become possible to reduce
the number of accesses per retrieval to a fraction of that achievable by the
simplest methods, especially at a high degree of loading of the hash table.
These methods are here reviewed in a systematic order, starting with the basic
ones. It is believed that the choice of the most suitable variant will be
clear from the context.

Identifiers and Special Markers in the Hash Table. The accesses to the calculated address and its probed reserve locations are made in the same sequences
during storage and retrieval, except for minor deviations from this when
entries are deleted from the hash table, as described below. The probing sequences are thus unique, but there usually exists no provision for the indication of the position of an entry in them. In order to resolve which one
of the reserve locations belongs to a particular entry, a copy of the corresponding keyword may be stored at the calculated or probed address, whichever
is proper. During search, probing is continued until a matching keyword is
found. Instead of the complete keyword, sometimes a simpler *identifier* of it
can be used for the resolution between the colliding entries. For instance,
if the division method is applied in the hashing algorithm, and the separate
chaining method discussed in Sect. 2.3.4 is used for the handling of collisions, the *quotient* of the division (which, in most cases occurring in practice, is representable by a few bits) will do as identifier. (With all the
other collision handling methods, however, complete keywords have to be used
for identification.)

A location can be marked *occupied* by an entry in different ways. For instance, the value zero of a location may indicate that it is empty, and an inserted entry then must always be nonzero. Alternatively, one bit in the location may be reserved for a *usage flag* which has the value zero as long as the location is empty, or after an entry is marked deleted (cf the discussion below); it is set to 1 at the time of storage of an entry.

In case the keyword might occur in several entries, an exhaustive search of the probing sequence must always be carried out. When the entry to be sought does not exist in the table, an exhaustive search must always be performed until it shows unsuccessful. In order to make the search terminate in such cases, the end of a probing sequence must be indicated somehow. Notice that the end of the chain cannot be deduced by examining whether the next location is empty or not; the next probed location may belong to another chain. A special marker, e.g., a *terminal flag* bit may be reserved in every location for this purpose. If the chains of reserve addresses tend to be very long on the average, more space may be saved if an extra location is appended to the chain and it contains a *terminal identifier*, i.e., an otherwise forbidden keyword or data value.

There yet exist some further flags which may be used for operations aiming at an improved efficiency. They will be discussed below in due context.

2.3.2 Open Addressing

Basic Probing Methods. The simplest type of hash table makes use of only one memory area which is shared by the calculated addresses and their reserve locations. This is called *open addressing*. The addresses of the reserve locations are derived from the calculated address, by incrementing the latter according to some rule. A procedure by which the reserve locations are determined in relation to the calculated address was named *probing*. In *linear probing*, the calculated address is incremented or decremented by unity until an empty location is found. The increments can also be computed by various algorithms. The best known procedures of the latter type are *quadratic probing* (often inaccurately named *quadratic hash*) in which the sequence of reserve locations is defined by increments to the calculated address which are computed from a quadratic function, and *random probing* in which the corresponding increments are pseudorandom numbers. It should be noted that the addresses always have to be incremented cyclically: if H is the number of locations in the hash table, and the table starts at B, and if the incremented address is f_i, then as the true address in the probing sequence, $f_i \bmod H + B$

must be used. As mentioned earlier, the algorithm for pseudorandom numbers must produce a deterministic sequence when initialized with a particular parameter.

Deletion of Entries. When a hash table has been used for an appreciable time, there becomes a need to delete useless entries from it in order to make room for new ones as well as to avoid unnecessary delays caused by the processing of irrelevant entries. Notice that it is not allowed to simply mark a location empty since the probing sequence would stop there; instead, the gap caused by deletion must be closed up by moving the entries in the chain one step backwards (towards the calculated address). This is especially cumbersome with the use of random probing since the pseudorandom number algorithms do not work in the reverse direction. Linear probing has some advantages in this respect. There is a need to consider only those entries which are located between the deleted one and the next empty place, because no linear probing sequence can skip an empty location.

It is possible to reclaim the location of a deleted item, to be used for further entries, by using a special marker (e.g. a bit or flag) in the location to indicate this. In probing, the search continues when the marker is encountered. Any new entry can be placed in this kind of location, and the location can also serve as a calculated address for a new chain. In the latter case, the reserve locations are appended to the old chain which passes this location. The number of table look-ups does not decrease because of this arrangement, however, but the time needed to skip a location may be smaller than that necessary to perform a complete searching operation. With linear probing, an additional advantage is gained: when the deletion mark is found, one can look at the next location to find whether it is empty. If this is so, the location corresponding to the deleted entry can be marked empty. Similarly, when scanning the table backwards, all locations corresponding to deleted entries can be marked empty until a location with no deletion marking is encountered.

In place of the flag bit, in order to save space, a special keyword or data value otherwise not allowed to occur can also be applied to indicate a deleted item. This bit code works as a marker as long as the location has not been reclaimed, and after storage of a new entry, the gap will be closed whereby the probing sequence is able to continue without the marker. A disadvantage of the use of bit groups as markers is that restrictions on the allowable keywords or data must be set.

Primary Clustering. Quadratic and Random Probing. The most important motive
for the introduction of quadratic and random probing is to avoid a phenomenon
named *primary clustering*. With linear probing it often occurs that the reserve locations will be agglomerated in long chains; when the hash table
starts filling up, such chains tend to meet frequently and coalesce. As a
result, especially in the case of an exhaustive study of all reserve locations, the number of searches may become extremely large (e.g., some hundreds).
To be more specific, assume that the hash address h(K) and its reserve addresses for a particular key K form a sequence

$$\{f_0(K),\ f_1(K),\ f_2(K),\dots\}\ ,\quad f_0(K) = h(K)\ ,$$

whereby $f_i(K) = [h(K) + g_i] \bmod H$

(i = 0,1,2... and g_0 = 0). $\hspace{4cm}$ (2.11)

Assume that for some reason, for two distinct keywords K_1 and K_2, there now
happens to be

$$f_i(K_1) = f_j(K_2)$$

(i $\neq$ j mod H and $K_1 \neq K_2$). $\hspace{4cm}$ (2.12)

Primary clustering is then said to exist if

$$f_{i+L}(K_1) = f_{j+L}(K_2)\quad \text{for L = 1,2,3,...}\ .\hspace{2cm} (2.13)$$

Since the above condition has to hold when $f_i(K_1)$ and $f_j(K_2)$ are arbitrary
members in two independent probing sequences, obviously primary clustering
is associated with linear probing only.

Quadratic probing, as one of the many possible nonlinear methods, will be
able to eliminate primary clustering. It has a nice feature of being easily
computable. This probing procedure was originally introduced by MAURER [2.33]
in the following form:

$$g_i = a \cdot i + b \cdot i^2\ .\hspace{4cm} (2.14)$$

It has been concluded from practical experiments that if the hash table size
is a power of two, the probing sequences tend to become cyclic, and only a
part of the table is then covered up. For this reason, it was suggested that
the hash table size ought to be a prime, in which case the g_i will be guaranteed to span exactly half the hash table before the sequence becomes cyclic.

The following or nearly equivalent parameter values are commonplace in several third-generation computers in which this algorithm has been used extensively: a = -787, b = 1. For an effective use of arithmetic and index registers in the computation of (2.14) see [2.33]. See also a correction in [2.34].

RADKE [2.35] points out that actually it is sufficient to choose a = 0, b = 1 to get the same result as with (2.14); moreover, in order to guarantee that the entire hash table will be covered in probing, he suggests the following modification. The hash table size is selected to be of the form H = 4n + 3, where n is some integer, and H is further a prime. The increments g_i are alternately picked up from the expressions A and B,

$$A = (h + i^2) \bmod H ,$$

and $$B = [H + 2h - (h + i^2) \bmod H] \bmod H$$

$$(i = 1,2, \ldots, (H - 1)/2) \tag{2.15}$$

for which h = h(K) is the calculated address. It can be shown that addresses obtained from A cover half the table and those generated by B the rest.

The above method has been worked out in an easily computable form by DAY [2.36]. The computational algorithm is given below without proof:

1) Set i to -H.
2) Calculate the hash address h = h(K).
3) If location h is empty or contains K, the search is concluded.
   Otherwise set i = i + 2.
4) Set h = (h + |i|) mod H.
5) If i < H, return to 3; otherwise the table is full.

HOPGOOD and DAVENPORT [2.37] have found that if the coefficients in the quadratic function (2.11) are selected such that b = 1/2, then, with the hash table size a power of two, the length of the probing sequence is H - R + 1 in which R = a + b. With R = 1, the entire table will be searched. The probing algorithm in this case takes on the following form:

1) Calculate the hash address h = h(K). Set j = R.
2) If location h is empty or contains K, the search is concluded.
   Otherwise set h = h + j, j = j + 1 and return to 1.

With this method, the end of the probing sequence has been reached when the ith and (i+1)th probed addresses are the same; this condition can be used to indicate that the entire table has been searched.

ACKERMAN [2.38] has generalized the above maximum length results for the case in which the table size is a power of a prime. BATAGELJ [2.39] has pointed out that the maximum probing period can be reached if the factorization of the table size contains at least one prime power, or is twice the product of distinct odd primes. Some additional theoretical results on the period of search with quadratic probing and related algorithms have been presented by ECKER [2.40].

With *random probing,* two randomizing algorithms are needed: one for the computation of the hashing function h(v) where v is the numerical value of the keyword, and one for the definition of a sequence of pseudorandom numbers $\{d_i\}$. The address of the first probed location is $f_1 = (h + d_1)$ mod H, and further reserve locations have the addresses $f_i = (f_{i-1} + d_i)$ mod H, for i = 2,3,... . Because of the mod H operation, it might seem that no restrictions on the range of the d_i are necessary. Nonetheless, some simple restrictions imposed on the d_i will be shown useful. In order to avoid unnecessary probing operations, no part of the table should be examined twice until all the H locations have been probed. This is so if d_i is relatively prime to H, i.e., d_i and H do not have common divisors other than unity, and if d_i is less than H. For instance, if H is a power of 2, then d_i is allowed to be any odd positive number less than H, and if H is a prime number, then d_i can be any positive number less than H.

Secondary Clustering. Double Hashing. One of the problems associated with the use of open addressing is named *secondary clustering*: whenever two or more collisions occur with the same address, it is then always the same sequence of locations which is probed. Secondary clustering will somewhat deteriorate the performance of quadratic as well as random probing with long chains of reserve locations. If we do not consider multiple-keyword searching applications in this connection, we can assume that the colliding entries have different keywords. It is then possible to make the probing increment depend on the keyword by the application of another hashing algorithm. Thus, in this kind of *double hashing*, two hashing functions are defined:

   h(v) = the hash address corresponding to the numerical value v of the keyword

   i(v) = the "hash increment" corresponding to v.

If H is the number of addresses in the hash table, the ranges of h and i
are [0,H-1]. Moreover, i should be relatively prime to H. If an algorithm
for the hash increment is used by which all of the possible pairs of values
[h(v), i(v)] become equally probable, then this method is named *independent
double hashing*; the term *uniform hashing* is used, too.

The Quadratic Quotient Method. As the computations in random probing tend
to become complicated, it would be advantageous to modify the other probing
methods into a form which avoids secondary clustering. While quadratic probing
was introduced in order to avoid primary clustering, one has tried to modify
it to eliminate secondary clustering, too. The basic idea is to make the parameters in the quadratic function key-dependent; thus, for the probing function one may take

$$f(K) = [h(K) + g_i(K)] \bmod H$$

where

$$g_i(K) = a \cdot i + b(K) \cdot i^2 \ . \tag{2.16}$$

BELL [2.41] has chosen the function b(K) in the following way in order to
make it rapid to compute and yet able to make secondary clustering unlikely.
If the initial hash address is computed by the division method, h(K) =
v(K) mod H + B, there will be available a key-dependent number, namely, the
quotient (denoted Q in the following) at no extra cost of computation. The
quotient may now be chosen for the parameter b(K), whereby the method is
named the *quadratic quotient method*. In order to cover the whole table, the
algorithm might be modified in analogy with (2.15); notice that a table, the
size of which is a power of two, can no longer be searched fully since one
is no longer free to choose the parameter b equal to 1/2. BELL has later
remarked, in response to a comment of LAMPORT [2.42], that the probing function

$$f_i(K) = [h(K) + (Q/2)i + (Q/2)i^2] \bmod H \tag{2.17}$$

is an advantageous choice, provided that Q never becomes congruent to zero.
In this case Q must be replaced, for instance, by Q + 1.

Weighted-Increment Linear Probing. It has further been remarked by BELL and
KAMAN [2.43], LUCCIO [2.44], PAWSON [2.45], as well as BANDYOPADHYAY [2.46],
that linear probing, in which the increment is made a function of the key,
is practically almost as good as the quadratic quotient method. It is to be
noted that usually *primary clustering is also eliminated by this method*, and

so this solution to probing ought to be considered seriously. In the case
that the hash table size is a prime, and when the division method of hashing
is used, BELL and KAMAN suggest that for the probing function one should
select

$$f_i(K) = [f_{i-1}(K) + Q] \bmod H ,\qquad\qquad\qquad (2.18)$$

where Q is the quotient obtained in the division of v(K) by H. This algorithm
is called the *linear quotient method*. Since the probed addresses are computed
recursively, a trivial pitfall is that when Q happens to be congruent to zero,
all $f_i(K)$ become identical. In this case Q should be changed to 1 in order
to eliminate this degeneracy.

With this method, indeed, primary clustering may occur only in the special
case that two keys K_1 and K_2 produce the same quotient Q and their hash codes
differ by nQ (n = 1,2, ...), i.e., $(f_0(K_1) - f_0(K_2)) \bmod Q = 0$.

When the hash table size is a power of two, for the probing function
LUCCIO suggests

$$f_i(K) = [h(K) + ((2 \ h(K) + 1)i) \bmod H] \bmod H .\qquad\qquad (2.19)$$

In a short correspondence, MAURER [2.47] suggests another variant of
linear-quotient probing which also may be an improvement to quadratic probing.
With Q the quotient and R the remainder in division hashing, take h(k) = R.
For the sequence of probed address, take $f_1 = R'$, where R' is the remainder
in the division of Q by H, and continue by taking $f_2 = R' + 29$, $f_3 = R' + 2 \cdot 29$,
$f_4 = R' + 3 \cdot 29$, etc. After the first key-dependent probing, this is then
a constant-increment linear probing method.

*Reduction of Probings in Retrieval by Rearrangement of Old Probing Sequences
During Storage.* A further procedure associated with probing, which yields
significantly shorter probing sequences, especially with nearly full tables,
than the other open-addressing methods, has been suggested by BRENT [2.48].
It can be viewed as a modification of any probing method but it is particularly suitable with the linear quotient method. The leading principle is to
rearrange the hash table when inserting a new key, so that an entry which
resides at a probed address is set aside to another place, to make room for
the new entry. This place is automatically determined so that it would be the
next reserve location if the original probing sequence of the old entry were
continued. The arrangement is made conditionally, if and only if it is possible to deduce that the sum of probings to reach the old entry in its new lo-

cation, and probings to reach the new entry in the place where the old one
was earlier, respectively, are thereby reduced.

The computational load in BRENT's method during storage, because of auxiliary probing operations performed during reorganization, is somewhat heavier
than with the simpler methods. On the other hand, the computations in retrieval are the same as with other methods. As will be pointed out in Sect.
2.5.2, with large hash tables stored in secondary storage it is possible to
compute rather complicated hashing functions in a time comparable with a
single memory access, and if the number of accesses is reduced by the new
method, then the overall searching time is correspondingly reduced, too.

Consider first that the hash table is not full and the probing sequence
relating to the new key K will terminate in s steps when an empty location
is found. Assume that the probed addresses are defined by an algorithm of
the form

$$f_i = (r + q \cdot i) \bmod H, \quad i = 1, 2, 3, \ldots , \tag{2.20}$$

where $r = r(K)$ and $q = q(K)$ are key-dependent parameters, and H is the hash
table size. If $T(f_i)$ is the keyword actually found at address f_i (notice
that in a probing sequence associated with a particular key, the latter is
not stored in the probed locations except in the last one) then another
probing sequence, formally similar to (2.20),

$$f_{ij} = (r_i + q_i \cdot j) \bmod H, \quad j = 1, 2, 3, \ldots , $$

with

$$r_i = r[T(f_i)] \quad \text{and}$$

$$q_i = q[T(f_i)] , \tag{2.21}$$

is defined relating to location f_i. Notice that according to (2.21), f_{ij} is
the next element in the probing sequence relating to key $T(f_i)$ and location
f_i. Notice, too, that i and j are the respective *probing numbers* (i.e., the
numbers of steps in probing) in these two partial sequences, and $i + j$ is
the total number of probings to reach location f_{ij}. A branching to the "side-
track" f_{ij} can be made at any location f_i. There are thus plenty of possibilities to find an empty location f_{ij}, and for a particular key K it is
then possible to minimize $i + j$ under the condition that f_{ij} is an empty
location. In case of a multiple optimum, i is minimized. In this optimization
process, two cases may occur:

1) min $(i + j) \geq s$ in which case it is most advantageous to store K at f_s (f_s is the first empty location in the original sequence).

2) min $(i + j) < s$ whereby the number of probings to find the moved (old) entry would be smaller than the number of probings ($= s$) to find an empty location f_s at the end of the original sequence. In this case it is most advantageous to move the old entry to f_{ij} and to store the new one at f_i.

It is necessary to recall that f_{ij} *is the next reserve location of the old entry*, whatever key was stored with it, since (2.21) defines a probing function which is formally identical with that defined by (2.20) and $T(f_i)$ is the key of the entry which was moved. In this way *the probing calculations during retrieval can always simply be based on (2.20).*

For the functions $r = r(K)$ and $q = q(K)$, BRENT suggests

$$r = K \bmod H, \quad q = K \bmod (H - 2) + 1 \ . \tag{2.22}$$

If this procedure is used with the linear quotient method, then one should select

$$r = K \bmod H, \quad q = Q \ (\text{quotient}) \ . \tag{2.23}$$

The efficiency of this method, when compared with the other methods, has been reported in Sect. 2.5.1.

2.3.3 Chaining (Coalesced)

In most data management applications, the time spent for retrieving operations is very precious, whereas more time is usually available for the entrance of items into the table (initially, or in updating operations). Since the probing calculations have already been made during the storage phase, it seems unnecessary to duplicate them during retrieval, if the results can somehow be stored. The calculated address sequence of the reserve locations can indeed be recorded and utilized in search in a convenient way. Using an additional field in every memory location, the address sequences can be defined by *chaining* or the *linked list* data structure. Starting with the calculated hash address, every memory location is made to contain the address of the next reserve location (which is computed when the items are entered). During retrieval, the address of the next location is then immediately available without arithmetic operations. These addresses are also named *pointers*. This type of data structure is also called *direct chaining*, in order to distinguish it from another alternative discussed in Sect. 2.3.4. It is to be noted that the number of searches at the time of retrieval is significantly reduced, because the reserve locations are found directly, without intervening (unsuccessful) probings.

One particular advantage of chaining is that new items can be joined to the linked list in a very simple way. When a collision occurs and a new reserve location is found, its address is made to replace the previous pointer at the place of collisions, whereas the previous pointer is written into the pointer field of the new reserve location. The rest of the chain need not be altered, and none of the stored entries need be moved to other places (cf Fig. 2.3). It is to be noticed that the order of the colliding items in the chain no longer remains identical with the order in wich the reserve locations were initially probed. There is no harm from this change, however. On the contrary, the most recently entered items are now nearer to their calculated address than otherwise and are retrieved more quickly. This is an advantage since the newer entries are usually more relevant than the older ones.

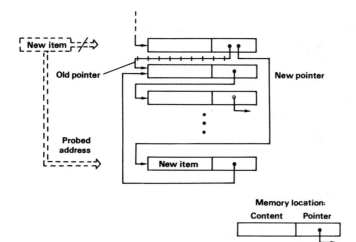

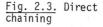

Fig. 2.3. Direct chaining

Because there is a possibility that several chains with different calculated addresses may be joined together, the chaining method discussed in this subsection is also named *coalesced chaining*.

Deletion of items from a chained structure is a particularly simple operation, in contrast to the many data moving operations which are necessary

when data are deleted from an unchained hash table. It is necessary only to
notice that in a hash table, it is completely equivalent which locations are
marked empty. The contents (identifier + entry) of that memory location which
is next in the chain are copied in place of the deleted item, and the location
which was next in the chain is available for new items and is marked empty.
The contents of the rest of the chain need not be altered. (Notice that it
is not practicable to change the pointer of the location which was prior to
the deleted item since in a usual list structure there is no provision for
tracing the chain backwards.)

The end of the chain may be indicated by giving a particular value that
is easy to check to the last pointer: a constant (e.g., zero), the calculated
address, or the address of the location itself will do.

There exists a possibility for compressing the link fields needed to store
the pointers [2.49]. When the probing sequence is defined by a recursive
algorithm which generates the hash increments, then the next reserve location
can be defined by giving the *probe number*, i.e., the number of steps in recursion by which the new location can be found. The probe number, instead of
a pointer, is then stored in the link field. This method may be named *pseudo-
chaining*. It should be noted, however, that if the maximum length of a probing
sequence may equal the hash table size, in the worst case, the probe number
contains as many bits as the hash address; if the link field is then dimensioned according to the maximum length in probing, no space savings are
achieved.

2.3.4 Chaining Through a Separate Overflow Area

Open addressing is particularly advantageous with multilevel storages since
the maintenance of the latter is based on buffering principles, i.e., transfers of blocks of contiguous data between the different levels. It is thereby
of great advantage if related data are kept in the same local memory area.
Nonetheless, since the symbol tables for assemblers and compilers form another major field of application of hash coding, it will be necessary to
review the second main principle for the handling of collisions, which is
more suitable for them. This is the use of a separate memory area, the *overflow area*, from which reserve locations are allocated to colliding items in
consecutive order and according to need. The name *indirect chaining* has,
therefore, been used for this method. Since the chains associated with different calculated addresses are now completely independent of each other,

this method is also called *separate chaining, chaining with separate lists,* or *noncoalesced chaining* (Fig. 2.4).

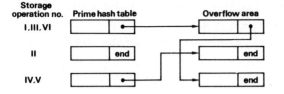

Fig. 2.4. Separate chaining through an overflow area. The results of handling two collisions at the first location and one collision at the third location are shown, whereby the Roman numerals tell in what order the entries have been stored

As the addresses in the overflow area are in no arithmetical way derivable from the calculated addresses, chaining remains the only alternative for their linking into a sequence of reserve locations. No probing arithmetic need be performed, however, since the reserve locations are taken in numerical order. This already speeds up the operation, especially in the storage of items. There are some positive consequences from the use of this method in retrieval, too. First of all, since colliding items are not located in the hash table, the latter will contain more empty locations, with the result that the number of collisions increases more slowly with filling up of the table than with open hash. The demand of memory is also smaller because the overflow area is used up in a regular order. All of the above factors contribute to the fact that the average number of searches (which is not proportional to the average time for retrieval!) remains much lower than with open addressing, especially at a high degree of load factor of the memory.

Contrary to the principle applied in Fig. 2.4 which was taken for a simplified example only, it is more customary to link every new item, when it is stored in the overflow area, immediately behind the calculated address; in this way the chain need not be traced to find its last item whereby plenty of computing time may be saved when new items are inserted.

With the use of separate chaining, there is one additional advantage achievable. As pointed out by FELDMAN and LOW [2.50], the space needed for the keyword in the prime hash table can be compressed to a fraction if in the hashing function the division method is applied and *if the quotient is used instead of the complete keyword* for the identification of the reserve locations. Notice that the remainder already defines the calculated address, and the keyword is uniquely determined by the quotient and the remainder. The quotient is in many practical cases only a few bits long, and the space thereby saved compensates to a great extent the extra space needed for the pointers in the prime hash table.

Finally it may be mentioned that with multilevel storages, most of the benefits of an overflow area can be achieved, and yet one is able to keep the reserve address locally near the calculated address if the overflow area is divided into several parts, each one belonging to a particular range of calculated addresses. Only that part which corresponds to the calculated address needs to be buffered in the primary storage.

2.3.5 Rehashing

With the use of a separate overflow area, no serious problem arises from the prime hash table having a load factor which exceeds unity, possibly even being a multiple of unity. Nonetheless there seems to be a practical limit for a load factor which is many times considered to be around 0.8 to 0.9. With open addressing, the filling up of the hash table is a more serious incident, and there must exist an automatic procedure to handle this case. One possibility is to assign additional addresses to the hash table with a continued range of the original hashing function; since the old entries must stay where they are, the distribution of occupied locations is not expected to become uniform over the combined table. Another, preferable approach devised by HOPGOOD [2.51,52] is to assign a completely new table of a greater size and to *rehash* or *reallocate* all the entries into the new table. Transition to the new table can be made when the old one becomes, say, 80 percent full. The entries can be moved from the original table to the new table serially; notice that in this case it is necessary to buffer complete keywords instead of their identifiers in the hash table locations for the calculation of the new hashing function.

Rehashing techniques have been discussed in detail by BAYS [2.53,54], as well as SEVERANCE and DUHNE [2.55].

2.3.6 Shortcut Methods for Speedup of Searching

Conflict Flag. A simple and effective way to cut down unnecessary searches has been devised by FURUKAWA [2.56]. This method makes use of the fact that in a substantial portion of attempts, a search will show *unsuccessful*, i.e., the entry looked for does not exist in the hash table. It is now possible to reserve a special bit named *conflict flag* at every location to obtain the above answer more quickly. If during the search the search argument or its identifier disagrees with the keyword or its identifier at the calculated address, this may be due to either of the following possibilities. Either an entry with this keyword is in the table but it has been set aside to a reserve location due to a collision, or it does not exist. The conflict flag can now be used to resolve between these cases. If the flag was initially 0 and was set to 1 when the first collision with this location occurred, then, provided that the value 0 for it was found during searching at the calculated address and the keywords disagreed, one can be sure that the keyword no longer can be found in either of the reserve locations, and the search is immediately found unsuccessful.

The usage flag, in connection with the conflict flag, can also be used to mark deleted items. Assume that the conflict flag in any location is set to 1 if it is followed by a reserve location. The value 0 given to the usage flag upon deletion of an item indicates that the location is empty (available), but if the conflict flag has the value 1, this is an indication that the chain of reserve locations continues.

Ordered Hash Tables. The concept of an *ordered hash table* was introduced by AMBLE and KNUTH [2.57], in an attempt to speed up retrievals in case the search is unsuccessful. This method is based on the observation that if the entries in a list are ordered, for instance, alphabetically or by their magnitudes, the missing of an item can be detected by comparing the relative order of the search argument and the list items. As soon as the ordering relation is reversed, one can be sure that the item can no longer exist in the rest of the chain. Fortunately an algorithm for the ordering of items during insertion is very simple, and it is described by the following computational steps.

Assume that the probed locations in the chain are indexed by p, $1 \leq p \leq L$, and $L+1$ is the index of the next reserve location to be assigned. Denote by K_p the keywords existing in the chain, let $K_{L+1} = 0$, and let K be the search argument. The ordered insertion of items, which is exemplified below in the case of magnitude ordering, is defined by the following algorithm:

1) Set $p = 1$.
2) If $K_p = 0$, set $K_p = K$ and stop.
3) If $K_p < K$, interchange the values of K_p and K.
4) Set $p = p + 1$ and return to 2.

This algorithm will sort the items in descending order.

It has been pointed out in [2.57] that for successful search the number of accesses is identical with ordered and nonordered reserve locations, whereas for unsuccessful search, especially if the chains are long, the number of accesses with ordered locations is approximately the same as the number of searches when the search was successful and the chains were unordered.

Virtual Hash Address. One of the tricks in connection with open addressing by which the inspection of reserve locations can be made faster is described in this paragraph. It is to be noted that in open addressing, the probed locations may contain entries which belong to different calculated addresses. The idea applied here is to check as quickly as possible whether the entry belongs to the prevailing calculated address; if the result is negative, further inspection of this location becomes unnecessary. To this end, the hashing function is made to produce a *virtual hash address* which is a few bits longer than the one referring to the hash table. The extra bits from the more significant end are used as identifiers of the calculated address and they are stored together with the keyword. The less significant part of the virtual hash address is then used as the true address in the hash table. Before the comparison of the keywords, the extra bits are checked first. Especially if the keywords are long, comparison of the extra bits can be made significantly faster, with the result that time is generally saved. This speedup is obtained at the cost of space for the extra bits.

2.4   Organizational Features and Formats of Hash Tables

2.4.1 Direct and Indirect Addressing

Hash Index Table. The data items which are stored together with the keywords or their identifiers in the hash table may sometimes demand a lot of space. Their lengths may also vary, but a normal hash table has to be dimensioned according to the longest one. This then normally results in an inefficient use of the memory space. Moreover, it is cumbersome to manipulate with long entries when making the hash table searches. If one can afford an extra memory access (which with secondary storages is necessary anyway), the operation becomes much more flexible when the entries are stored in a separate memory area, and only *pointers* to this area, showing the beginnings of the records, are stored in the prime hash table together with the identifiers. It should

be noticed that only keywords or their identifiers need to be examined when
the prime hash table is searched, and an access to the separate memory area
is made only when the right entries have been found. Another benefit achieved
is that the prime hash table can rather easily be reallocated to another memory area upon need. This construct is named *hash index table* in the following.
(The original name used, e.g., by MORRIS [2.22] was *scatter index table,* but
this term must be related to the name "scatter table" earlier used for the
hash table.) Especially with data bases, the principle of a hash index table
is commonplace. There are few occasions in which it may not be quite profitable, for instance, for a symbol table in which only short names may occur
and an indirect addressing of the above kind brings about an unneccessary
overhead.

If chaining is used in the prime hash table, additional pointers are yet
needed for the linked list structure. A general name *link word* may be used
for any pointers that occur in the hash table.

Direct/Indirect Addressing in the Hash Table (Hash Linking). The idea of a
hash index table can be developed further if it is noticed that this principle may be viewed as a primitive form of *indirect addressing,* the latter
being extensively applied in computer programming. While the indirect addressing mode is predetermined in the hash index table, it is possible to
improve the utilization of the memory space by having two types of addressing,
direct and indirect, in the same table, and the mode used with a particular
location may then be indicated by a special bit named *link flag* reserved in
the location for this purpose. This idea has been worked out by BOBROW [2.58].

In order to realize the possibilities of *direct/indirect hash addressing,*
also named *hash linking,* it may be useful to review briefly similar principles
which have been applied in contemporary computers. As illustrated in Fig.
2.5, a machine instruction usually consists of an *operation code* (opcode)
and an *address field.* It is possible to specify by the opcode whether the
contents of the "address field" represent a constant parameter or an address.
In the latter case, if the special D/I flag takes the value 0, then the address is a *direct* or *absolute* one, and it immediately defines that storage
location in the memory (usually primary memory) in which the due operand can
be found. However, when auxiliary memories are incorporated in the computer
system, the capacity of the address field may be too small for direct addressing; one solution is then to make a convention that the next word is
automatically used for the additional address bits. Another method makes use
of *indirect addressing* which means that when the value 1 is stored in the

D/I flag, the true address is found from an auxiliary table or register array, and the contents of the address field only define the index or address of the corresponding location. Indirect addressing also facilitates reallocation of programs into another memory area without a need to move the data. It is further possible to continue indirect addressing on several levels, every time looking for a D/I flag in the location found in order to deduce whether literal contents or a further address is meant.

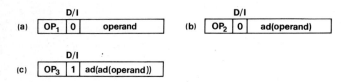

<u>Fig. 2.5a-c.</u> Various types of machine instruction (a) immediate indication of an operand, (b) direct addressing of an operand, (c) indirect addressing of an operand. OP_1, OP_2, OP_3: operation codes; $ad(x)$ = address of x

When this more developed form of direct/indirect addressing is applied to hash tables, the value 0 of the link flag (corresponding to the D/I flag) in the hash table location may then tell that the associated data are found in the hash table location itself. Otherwise the associated field only contains a *hash link* to another, usually much smaller table in which the words, however, are longer. If yet longer entries have to be stored, they can be kept in a third table or memory area accessible through a second hash link, and so on.

One good reason for the introduction of hash linking is that the backup storage area in the secondary storage usually requires a rather long address (typically over 20 bits) with the result that the pointer to be stored in the hash table would become long. In many applications, however, the majority of entries are short and could be stored in a memory area with a small memory capacity and short address, or possibly in the hash table itself. The longer pointers can then be kept in a separate memory accessible by hash linking.

BOBROW [2.58] mentions some examples in which hash linking can be applied advantageously: for instance, in certain bank accounts, 99 percent of the entries may be represented by 16 data bits. The remaining exceptional cases making up only 1 percent of the entries can be kept in a secondary table in which 32 bits are reserved for data. Thus, for the storage of a maximum of

32 bits of information, less than 17 bits per entry have then been used on the average.

In general, advanced data structures based on hash linking might be useful in certain high-level languages developed for artificial intelligence research, list processing, etc.

2.4.2 Basic Formats of Hash Tables

Formats and Contents of Hash Table Locations. The size of the storage location in the hash table is determined primarily by the amount of information to be stored, thereby taking into account possibilities for compression of the keyword identifier (cf the division method of hashing) and efficient utilization of the memory space by direct/indirect addressing as described above. However, the addressing principle used in most computers dictates that the hash table location must be a multiple of a word or byte. Sometimes even one excess bit to be stored may mean an extra word and extra readout operation to be performed when searching a hash table location, and, therefore, the elements of the location must be chosen carefully. This problem is alleviated if there is space left in the last word belonging to the location.

Obviously the indispensable constituents of a hash table location are the *keyword* or its *identifier*, and the associated *data* or their *pointer*. If open addressing is used, a further item to be stored is a flag which indicates the *end of a probing sequence*. On the other hand, this flag can be dispensed with if a dummy location, with a special code as its contents, is appended to the sequence. One might also choose a much longer way in retrieval of probing until an empty location is found, or until the whole table is probed, but this is not very practical. When chaining is used, a special value of the *link word* usually indicates the termination of a chain.

Optional contents of the hash table location are the following one-bit markers: the *usage flag*, the *conflict flag* (cf Sect. 2.5.3), the flag which marks a *deleted item*, and the *link flag* discussed above. The usage flag and the link flag can be replaced by a special value given to the contents of the location, e.g., all bits equal to 0 may mean an empty location, and all bits equal to 1 in a particular field may mean that the rest of the location is a hash link. Consequently, entries with these data values then cannot be stored. The usage flag facilitates a faster operation; the flag indicating a deleted item may be replaced by the conflict flag (Sect. 2.5.3), and it is unneccessary if chaining is used.

Examples of hash table locations are given in Fig. 2.6.

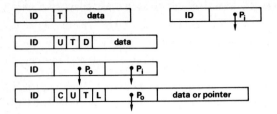

Fig. 2.6. Various types of hash table location. Legend: ID = keyword identifier; P_i = pointer to data area (index); P_0 = pointer to overflow area (or
next entry in direct chaining); T = terminal flag; U = usage flag; D = flag
denoting a deleted entry; C = conflict flag; L = link flag

A further problem occurs in multi-key searching, as discussed in more
detail in Sect. 2.6 and Chap. 6: if several chains pass through the same
entry, in order to define the proper branchings it may be necessary to store
several keyword identifiers in the storage location, each one associated with
a pointer to the next location in the corresponding chain.

Indexing of Words in Storage Locations. In a usual hash table, the length
of each entry must be known beforehand. Assume that this length is k words.
Especially in primary random-access memories, the entries may then be stored
in k contiguous words [see Fig. 2.7 (A)], in which case the calculated addresses must be divisible by k. (This implies that the hash address, calculated by some algorithm, is finally multiplied by k.) In the second method
the table is divided into k sections [see Fig. 2.7 (B)]. The calculated hash

(A) (1,1) (2,1) (3,1) ....
     (1,2) (2,2) (3,2) ....
     (1,3) (2,3) (3,3) ....
     (1,4) (2,4) (3,4) ....

(B) (1,1) (1,2) (1,3) (1,4)
     (2,1) (2,2) (2,3) (2,4)
     (3,1) (3,2) (3,3) (3,4)
     (4,1) (4,2) (4,3) (4,4)
       .   .   .   .
       .   .   .   .
       .   .   .   .

Fig. 2.7. Indexing of words in the hash table (see text). In this example
it is assumed that k = 4

address is used in this case as the index to an element of the entry in each section which results in less computation during searching; especially with symbol tables, this is a rather important benefit.

Buffering of Hash Tables. Bucket Organization. When an auxiliary storage device is used for a large hash table, only part of the table which contains the calculated address may be buffered in the fast primary memory. If linear probing is used for the handling of collisions, it is then likely for one to find a substantial number of the reserve locations, possibly all, in the buffered area. Some systems programs allow flexible swapping of data between primary and secondary storages in quantities named *pages*; with disk memories, a typical page size is 256 words which corresponds to an addressable sector on a track. In order to increase the probability for finding all reserve locations on the same page, the increments in probing algorithms can be taken cyclically, in the modulus of the page size. Only in the case that the page would be filled up completely, the chain of reserve locations will be continued on the next page. Although the distribution of occupied addresses in the hash table thereby becomes less uniform, the overall speed of retrieval is probably increased as the number of slow accesses to the auxiliary storage is thereby reduced.

Another useful principle which is commonplace, especially in connection with auxiliary storages, is named *bucket organization*. In it, several consecutive locations are combined under a common hash address. Such a combination is named *bucket*, and it has a fixed number of "slots" for colliding entries. There is thus no need to look for reserve locations at other addresses as long as all slots of the bucket are not full. In the latter case, another bucket is appended to the chain of reserve locations, and only one pointer per bucket is needed. (Fig. 2.8).

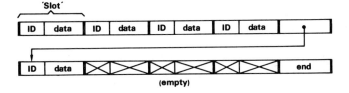

Fig. 2.8. Two buckets with four "slots" each, chained, the latter partly empty

In linear probing, overflows can also be stored in the next bucket which
has empty space for them. There obviously exists an optimum number of slots
per bucket in a particular application. This depends at least on two opposing
factors. With a greater number of slots in a bucket, the common pointer represents a smaller overhead. On the other hand, since the number of hash
addresses then becomes significantly smaller than the number of available
memory locations, it is expected that the distribution of entries in the
hash table becomes uneven, with a result of poorer efficiency in its utilization and an increased number of collisions. The theoretical optimum
seems to be very shallow; we shall revert to the question of bucket size
at the end of Sect. 2.5.

Quite independently of theoretical optimality considerations, one may
easily realize that the reading mechanism of a rotating memory device such
as a disk favors large buckets; if the bucket is made equivalent to a track,
its contents are transferable to the primary storage in one operation.

HIGGINS and SMITH [2.59] refer to a possibility of making the bucket
size grow in a geometric series in a chain; the leading idea is that the
frequency of overflows from the first bucket, especially with multiple keywords, depends on the frequency of usage of that keyword, and the disk space
might thereby be utilized more effectively. This idea which may be justified,
e.g., in library indexing is in contradiction to the ideas of KNUTH [2.1]
and SEVERANCE and DUHNE [2.55] who suggest that the overflow buckets (with
single keywords) should be smaller than the prime buckets.

It seems that the principle of buffering of data was understood already
in the early days of computer technology. Even the bucket organization was
described in the first papers published on hash coding, e.g., those by
PETERSON [2.60] and BUCHHOLZ [2.61]. Incidentally, the first applications
of hash coding were meant for large files!

On the Use of Buckets for Variable-Length Entries. Since the bucket organization is normally connected with the use of secondary storages, even long
entries can be stored in the locations. The searching time depends primarily
on the time needed to transfer a bucket to the primary storage, whereas a
linear examination of the "slots" can be made by fast machine instructions
in the mainframe computer. One possibility with variable-length entries is
now to pack them linearly (contiguously) within the bucket, without consideration of any slot boundaries. It is necessary only to demarcate the different entries using, e.g., special bit groups. For a comparison of various
methods for variable-length entries, see, e.g., the article of McKINNEY [2.62].

2.4.3 An Example of Special Hash Table Organization

Of all hash tables implemented so far, maybe the most complex ones are those designed for various artificial intelligence languages, for instance, SAIL (Stanford Artificial Intelligence Language). A subset of it which was initially developed as an independent version for the storage and manipulation of *associative (relational) data structures* is named LEAP, and it was implemented around 1967 by FELDMAN and ROVNER [2.63-66]. The language processor of LEAP contains many features which are not pertinent to this discussion; however, the special hash table organization of it is reviewed in this subsection. The most important new feature which distinguishes the hash table of LEAP from other forms discussed earlier is that there are several types of memory location in the same table. Their divided roles allow an optimal usage of the data space. This solution is partly due to the fact that entries are encoded by multiple keywords and made retrievable by any combination of them. Although multikey searching procedures will otherwise be presented in Sect. 2.6, this particular case may serve as an introductory example of them. It must be emphasized, however, that the most important reason for the definition of so many types of location has been the necessity of facilitating an automatic generation and manipulation of data structures by the language processor.

Since the data structure of LEAP was designed for a paged, time-shared computer TX2 with a large secondary storage, the memory space could be allocated redundantly, which resulted in a high speed in retrieval. The version of LEAP implemented in SAIL is actually a restricted version, and it is the original form which is discussed here.

In order to facilitate the understanding of data structures in presentation, it will be necessary to describe first the modes of searching expected to occur. The entries in associative languages are constructs named *"associations"* which are represented by ordered triples (A,O,V). The elements A, O, and V have the following roles: O is the name of an *object*, A is an *attribute* of it, and V is some *value* (e.g., name) associated with the pair (A,O). For example, a verbal statement "the color of an apple is red" has the following equivalent representation: (color, apple, red). The value can also be another entry, namely, a triple linked to (A,O). We shall abandon the consideration of data structures thereby formed in this context; let their discussion be postponed to Chap. 6. The aspect in which we shall concentrate here is that a triple is always stored as such in the memory space, but it must be retrievable on the basis of one, two, or all three of its elements A, O, and V. The case in which

the whole triple is specified for retrieval is connected with a   check of
whether the triple exists in the table.

Without having yet discussed various possibilities for multi-key search,
let it be stated that the fastest method for the implementation of all the
above types of query is to have several hash tables, each of them accessible
on the basis of a particular combination of the arguments (*"compound keyword"*).
So, in the most complete case, we should have seven hash tables, namely, one
in which the entries were hash addressed by the A values, and the others in
which the compound keywords were O, V, (A,O), (A,V), (O,V), and (A,O,V), respectively. Notice that, for instance, the compound keyword (A,O) may be
understood simply as a single keyword which results from the concatenation
of the representations of A and O, and the other compound keywords can be
defined accordingly. When a particular combination, say, (A,O) occurs as a
search argument, it is deducible from the form of the query that the corresponding (fourth) hash table has to be consulted.

In LEAP, it was considered sufficient to have only six tables since a
retrieval on the basis of (A,O,V) is implementable as a combination search
(cf Sect. 2.6.1) which may proceed, e.g., along the A chain, thereby only
checking O and V. Three of these tables, namely, those corresponding to
queries on (A,O), (A,V), and (O,V), respectively, are usual (bucketed) hash
tables. As for the remaining three tables, which allow an access on the basis
of A, O, and V alone, the principle of the so-called *inverted list structure*
(cf Sect. 2.6.1) has been applied. In our terminology, an inverted list
structure would be named *hash index table* and its pointers are made to point
at entries stored in one of the above hash tables. This will become more clear
from the example given.

Hashing Function. The three hash tables addressed by (A,O), (A,V), and (O,V),
respectively, are symmetric with respect to A, O, and V, and it will suffice
to exemplify the (A,O) table only. The table is divided into *pages* of variable
size and the *page number* (e.g., the index defining the beginning address) shall
be derivable from A. The numerical value of A, denoted here v(A), is called
internal name, and it was expressed as a decimal number in the original work.
The two higher order digits of v(A) were identified with the page number. Let
the number formed of the remaining digits be denoted v'(A). The *relative
address* of the entry on the page was obtained by a hashing function

$$h = [v'(A) \oplus v(O)] \bmod H \qquad\qquad (2.24)$$

where $\oplus$ denotes the EXCLUSIVE OR operation over respective bits in the binary

strings of v'(A) and v(O), and h corresponds to the binary number thereby
formed, taken modulo H.

A property of the EXCLUSIVE OR function is that $(x \oplus z = y \oplus z)$ implies
$(x = y)$. Accordingly, in order to resolve between the colliding items in the
hash table, it was considered enough to use v(O) as the identifier of the
entry. Notice that the quotient in the mod H division of Eq. (2.24) might
have required yet less space as an identifier (cf Sect. 2.3.1).

Subchain Organization of Reserve Locations. The special subchain organization
of the reserve locations applied in LEAP was intended for speedup of retrieval. Due to multiple keys and unlimited occurrence of the elements A, O, and
V in the triples, a normal case would be that there are two types of collisions
in the hash table; they may result from the mapping of distinct compound
keywords on the same address, or occurrence of identical compound keywords.
The reserve locations were, therefore, structured in two dimensions, shown
horizontally and vertically, respectively, in Fig. 2.9.

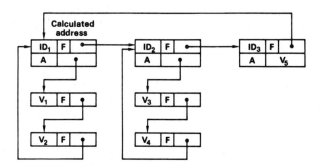

Fig. 2.9. An example of two-way chaining in LEAP, with six colliding entries
(see text). Fields on the top row contain different keyword identifiers ID_1,
ID_2, and ID_3. Field F contains various flags (location type identifiers).
A: see Fig. 2.10. Notice that no keyword identifiers are needed in members of
the subchains except the head; the same goes for the A fields (see Fig. 2.10).
Notice, too, that the circular list structure facilitates a continuous transfer from one subchain to another. V_1 through V_5: values of "association"

The collisions between distinct compound keywords (marked by their identifiers ID_1, ID_2, ... etc. in Fig. 2.9) were handled by linking the corresponding entries in a horizontal chain; colliding entries with identical compound
keywords were organized in vertical subchains appended to the respective
elements of the horizontal chain. This facilitated the fastest possible ac-

cess to all entries with a particular compound keyword. The ends of the
chains are always indicated by pointers back to the calculated address.

The extra blank fields as well as the A fields in the entries shown in
Fig. 2.9 shall be explained below. Observe the number of necessary pointers,
which varies with the role of the entry; it, therefore, seems possible to
optimize the utilization of memory space by using variable-size locations.

The operations associated with the language processors, in fact, require
that yet more types of locations be defined. For an immediate check of whether
a chain of colliding entries was appended to the calculated address, a special
conflict flag can be used in the field on the upper row shown blank. In this
field it is also possible to show whether a subchain has been appended to an
entry. Further specified roles of entries, indicatable by flags stored in
this field are: last member in a horizontal chain, member in the vertical chain,
member which contains its internal name (for language processor operation),
and type of memory location used as independent register.

Inverted List. In order to make the stored entries accessible on the basis of
single elements A, O, and V, respectively, the inverted list (hash index
table) structure is used. A convention is made to store the list accessible
by A (called briefly the A list) on the same page as the hash table accessible by (A,O); accordingly, this page is called A-type page. The O list and
V list are on O-type and V-type pages, respectively. The keyword A is hash
addressed, and in the corresponding location there are stored the identifier
of A, and a pointer to one of the entries (the one with this particular A

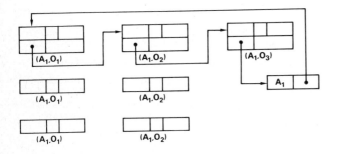

Fig. 2.10. The inverted list structure which facilitates searching on the
basis of A_1 (see text). Notice that all entries shown in this picture are
accessible by A_1 through the subchains. Notice that if there exist other
chains of reserve locations similar to that shown in this picture which
have the same A_1 value, they must be appended to this A chain. The A list is
always confined on the A-type page

value which was stored first in the table). Further entries which contain the same A value are linked to the same list, by using the A pointer fields shown in Fig. 2.9. Let A_1 be a distinct A value. The chained structure used for the inverted list is shown in Fig. 2.10. The entries shall be identical with those occurring in Fig. 2.9. For clarity, the various types of compound keywords (A,O) thought to occur in this example have been written below the storage location in question. As usual, the end of the A chain is indicated by a pointer to the calculated address (A).

2.5 Evaluation of Different Schemes in Hash Coding

Up to this point in this chapter, the emphasis has been on the introduction of various designs for hash tables. For a practical application, one has to choose from many options. Some of these choices are clear, for instance, one may easily deduce whether open addressing or overflow area has to be used, since this depends on the nature of the hash table as well as the type and size of memory device available. For instance, let us recall that symbol tables are usually kept in the primary storage, and chaining through the overflow area is thereby considered most effective. On the other hand, with multilevel memories, it was considered significantly beneficial to store all related data close to each other, whereby open addressing, or partitioned overflow area, is advantageous.

The size of a memory location is also rather obvious from the amount of information to be stored with an entry, and it is normally a multiple of the length of an addressable word. But one of the most important architectural features which is not quite easy to decide is the bucket size, i.e., the number of slots in the bucket; although an addressing mechanism of the secondary storage may make it convenient to use, say, one track for a bucket, nonetheless for the best efficiency it is desirable to learn how the number of memory accesses theoretically depends on the number of stored items with different bucket sizes.

Another option for which plenty of alternatives are available, but only a few of them are to be recommended, is the hashing function. The quality of different functions is relative with respect to the keywords, and cannot be known until benchmarking runs have been carried out in real situations. It seems that the problem of the best hashing function now has a rather generally valid answer, and in spite of the fact that so many alternatives have been introduced, the recommended choice in most cases is the *division method*, possibly preceded by the EXCLUSIVE OR function if several keywords are used. We may also recall that there is a particular benefit obtainable when the division method together with separate chaining is used; the quotient can occur as identifier in the reserve locations whereby plenty of space is saved in the hash table. In the conversion of character strings into numerical values, precautions such as a proper choice of the radix must of course be taken into account, whereas for the size of the hash table (the number used in the division algorithm) there are again several alternatives.

However, there yet remain open questions such as what is the difference between memory accesses when linear, random, quadratic, etc. probing in open addressing is used, what is the corresponding number with the use of an overflow area, what is the effect of ordering the entries in the reserve locations, etc. The purpose of this section is to provide answers to such questions, by representing results obtained in a number of different studies.

It will be necessary to point out that experiments performed with perfectly random keys may in some particular cases yield too optimistic results. Nonetheless, analytical expressions based on such theoretical keywords are useful in showing the dependence of the performance on various parameters. Moreover, in many applications the hashing functions are able to randomize the hash addresses to a degree which in practice may be almost as good as that of theoretical randomization.

2.5.1 Average Length of Search with Different Collision Handling Methods

Average Number of Accesses with Random Probing. With the aid of the following simplified example, it becomes possible to gain an insight into the most important property of hash coding, namely, the average number of accesses as a function of loading of the hash table. This example describes an open addressing situation when random probing with independent double hashing is used, and ideally randomized hash addresses are assumed. In other words, degradation of performance due to primary and secondary clustering is completely ignored. For further simplicity, only the case of *successful search* is considered in this example, i.e., it is assumed that the entry to be located really exists in the table so that the search ends when it has been found. Moreover, all keywords are assumed different (multiple-response cases are not considered). The average number of accesses to the hash table during readout can then be derived theoretically using elementary statistical considerations. It may be clear from the above reservations that an optimistic estimate to the average number of accesses is thereby obtained.

Assume that N entries are stored at random and with a uniform distribution in a hash table containing H locations. The *load factor* of this table is defined as $\alpha = N/H$. The expected or average number of trials to find an empty location is obtained in the following way. The probability for finding the empty location on the first probing is $1 - \alpha$; if probing is assumed completely independent of hash addressing, the probability for finding an empty location on the second but not on the first trial is $[1 - (1 - \alpha)] (1 - \alpha) = \alpha(1 - \alpha)$; the corresponding probability on the third, but not on the first nor the second, trial is $[1 - (1 - \alpha + \alpha(1 - \alpha))](1 - \alpha) = \alpha^2(1 - \alpha)$, etc. The average number of trials T_{av} is obtained when the ordinal number of the trial is weighted by the corresponding probability, i.e.,

$$T_{av} = (1 - \alpha) \sum_{i=1}^{\infty} i \cdot \alpha^i = \frac{1}{1 - \alpha} \quad . \tag{2.25}$$

It is now assumed that the same address sequences are followed during the storage and retrieval of items (which holds true if no items have been deleted during use). As a rule, the length of an address sequence grows with the filling up of the table. It is to be noted, however, that when an entry is retrieved, the corresponding sequence may be new or old. It is noteworthy, too, that the table was initially empty, and when it was filled up so that the final load factor is α, the average number of probings made during this process is smaller than the expected number of probes at the last step. If the number of probes is averaged over the history of the table, the value so obtained then must be the same as the expected number of accesses in the readout process. With random probing, the distribution of entries in the table can be assumed to stay random and uniform at all times, and if the table has a large number of locations, the expected number of searches made during readout may be approximated by the integral

$$E_r = \frac{1}{\alpha} \int_0^{\alpha} \frac{dx}{1 - x} = - (1/\alpha) \ln(1 - \alpha) \quad . \tag{2.26}$$

For instance, when the hash table is half full, $E_r(0.5) = 1.86$. We shall return to the above expression $E_r = E_r(\alpha)$ with the analysis of other probing methods.

The idea of random probing, and the above result were publicly mentioned for the first time by McILROY [2.67], based on an idea of VYSSOTSKY. Methods for derivation of (2.26) that have been presented in the literature are somewhat different.

Average Number of Searches in Linear Probing. Although linear probing is the simplest method for finding the reserve locations, its statistical analysis turns out to be tedious, in view of the fact that the probing sequences may coalesce. ERSHOV, who was one of the first to introduce hash tables and linear probing [2.68], conjectured that if the table is less than 2/3 full, then the average number of searches would be less than two. In the classical work of PETERSON [2.60], the average number of searches was determined by computer simulations, not only for simple hash tables but for bucketed tables, too (cf the discussion later on). These figures are consistently higher by a few percent than those obtained later on. One of the first reasonably accurate theoretical analyses on a probing scheme that is very much similar to the linear probing method described above was made by SCHAY and SPRUTH [2.69] who used a Markov chain model to compute the average length of probing sequences. This analysis, a bit lengthy to be described here, applies to the normal linear probing scheme, too. Thus, if certain statistical assumptions and approximations (such as stationarity of the stochastic processes in the Markov chains, and replacement of the binomial distribution by the Poisson distribution) are made, one arrives at the following approximative expression for the average number of searches:

$$E_\ell(\alpha) = \frac{2 - \alpha}{2 - 2\alpha} = 1 + \frac{\alpha}{2(1 - \alpha)} \ . \tag{2.27}$$

The validity of (2.27) has later been strictly discussed by KRÁL [2.70] as well as KNUTH [2.1] (cf also Table 2.1). It is noteworthy that the above expression describes the number of accesses in the case that the search was *successful*; we shall revert to the case that the entry does not exist in the table later on in Sect. 2.5.3. If the hash table is half full, (2.27) yields the value $E_\ell(0.5) = 1.5$; moreover, $E_\ell(2/3) = 2$.

Comparison of Searches in Random, Linear and Quadratic Probing. One of numerous experimental analyses of probing methods, presented below, has been made by HOPGOOD and DAVENPORT [2.36]. We shall review these results separately since they contain a comparison of linear and quadratic probing referring to the same set of data. As a reference, the theoretical function $E_r(\alpha)$ that describes random probing is used. Moreover, this experiment contains results

obtained with artificial, ideally random keywords as well as with keywords
picked up from real applications.

Table 2.1 reports the average number of searches with artificial keywords
(random hash addresses). In addition to the theoretical values computed from
$E_r(\alpha)$ and $E_\ell(\alpha)$, results from numerical experiments for linear and quadratic
probing are given. In quadratic probing, the increments were computed from
the function $g_i = h + ai + bi^2$, with $R = a + b = 7$ and $Q = 2b = 1$. The hash
table size in the experiments was 2048. It may be necessary to point out that
primary clustering is an effect which occurs with ideally random keys, too.

Table 2.1. Average number of searches with random keywords

Load factor α	$E_r(\alpha)$ (random)	$E_\ell(\alpha)$ (linear)	Searches in linear probing	Searches in quadratic probing
0.1	1.054	1.056	1.076	1.064
0.2	1.116	1.125	1.135	1.123
0.3	1.189	1.214	1.212	1.207
0.4	1.277	1.333	1.312	1.316
0.5	1.386	1.500	1.492	1.441
0.6	1.527	1.750	1.733	1.605
0.7	1.720	2.167	2.127	1.819
0.8	2.012	3.000	2.956	2.187
0.9	2.558	5.500	5.579	2.818

It can be clearly seen that quadratic probing which eliminates this effect
is theoretically almost as good as random probing. Another interesting point
to notice is the validity of $E_\ell(\alpha)$ when compared with numerical results.

In the experiments reported in Table 2.2, the keywords were drawn from
algorithms published in "Communications of the Association for Computing
Machinery", using (presumably) the division method over the table size H =
2048. Only the first three characters of the identifiers were selected for
the keywords. It may be mentioned that the linear probing algorithm was
tested with several *constant integers as the probing increment* (denoted by
a), and the value a = 173 reported in Table 2.2 is an optimum obtained in ten
runs.

Comparison of Quadratic Probing and the Quadratic and Linear Quotient Method.
Those numbers which are available to us about the quadratic and linear quotient
methods are due to BELL [2.41] as well as BELL and KAMAN [2.43]. Unfortunately,

Table 2.2. Average number of accesses with real keywords

Load factor α	Accesses in linear probing $a=1$	Accesses in linear probing $a=173$	Accesses in quadratic probing $R=Q=1$	Accesses in quadratic probing $R=7,\ Q=1$
0.1	0.160	1.165	1.165	1.180
0.2	1.417	1.396	1.400	1.381
0.3	1.727	1.657	1.641	1.597
0.4	2.216	1.981	1.938	1.871
0.5	2.748	2.689	2.322	2.234
0.6	3.762	3.199	2.744	2.648
0.7	5.338	4.529	3.218	3.042
0.8	8.728	6.302	3.917	3.647
0.9	16.674	8.927	4.957	4.542

comparable figures have been reported only for the average number of trials to insert a new entry; in the random probing with independent double hashing, this would be given by the theoretical expression $T_{av} = (1 - \alpha)^{-1}$ [cf (2.25)]. The corresponding values for the other methods are given in Table 2.3. The hash addresses were random.

Table 2.3. Average number of trials to insert a new entry

Load factor α	Random probing	Quadratic probing	Quadratic quotient method	Linear quotient method
0.5	2.000	2.14	2.00	2.002
0.6	2.500	2.72	2.50	2.503
0.7	3.333	3.67	3.33	3.322
0.8	5.000	5.50	5.00	4.992
0.9	10.000	10.79	10.00	9.996

It is obvious that with ideally random keys there is no significant differ-ence between the quadratic and the linear quotient method. In another, more realistic experiment, the quadratic quotient method used in a COBOL compiler was replaced by the linear quotient algorithm; the probings made by the hashing routine in real computational tasks were counted for both versions. The qua-dratic method in a series of runs used 14,896 probings and the linear method

15,320 probings, respectively. The time per probing was 10 percent less with the linear quotient method, however.

Comparable simulation results are available about the average number of searches with quadratic probing and the quadratic quotient method (for random keys) [2.41], and they are given in Table 2.4. In view of the previous numbers (Table 2.3) one can deduce that the last column is approximately valid for the linear quotient method, too.

Table 2.4. Average number of searches in quadratic probing and the quadratic quotient method

Load factor α	Random probing $E_r(\alpha)$	Quadratic probing[a]	Quadratic quotient method
0.5	1.386	1.44	1.38
0.6	1.527	1.61	1.52
0.7	1.720	1.84	1.72
0.8	2.012	2.18	2.01
0.9	2.558	2.79	2.55

[a] cf the results of HOPGOOD and DAVENPORT in Table 2.1.

Average Number of Searches from a Rearranged Hash Table Using the Linear Quotient Method. The method of BRENT (cf Sect. 2.3.2) for the reduction of searches in open addressing by rearrangement of the reserve locations has the best performance so far reported for any open-addressing methods. BRENT has carried out a theoretical analysis as the result of which the following serial expression for the average number of searches $E_B(\alpha)$ is obtained:

$$E_B(\alpha) = 1 + \alpha/2 + \alpha^3/4 + \alpha^4/5 - \alpha^5/18 + 2\alpha^6/15 + 9\alpha^7/80 - 293\alpha^8/5670$$
$$- 319\alpha^9/5600 + \ldots \tag{2.28}$$

If α is very close to 1, the convergence of this series may be poor; theoretically $\lim_{\alpha\to 1} E_B(\alpha) \approx 2.4941$. Values for the function $E_B(\alpha)$ are given in Table 2.5.

Table 2.5. Average number of searches with Brent's improvement of the linear quotient method

Load factor α	0.20	0.40	0.60	0.80	0.90	0.95	0.99
$E_B(\alpha)$	1.1021	1.2178	1.3672	1.5994	1.8023	1.9724	2.2421

Searches with Coalesced Chaining. With open addressing, the number of probings increases progressively as the load factor approaches unity. However, if the reserve locations are defined using direct or coalesced chaining, the tedious probing calculations are made only during insertion of entries into the table, while the unsuccessful probings are not involved in the searches during retrieval. The number of searches from a chained hash table is therefore greatly reduced even if the table is completely full. KNUTH [2.1] has worked out the following approximation for the average number of (successful) searches $E_c(\alpha)$ from a hash table with coalesced chaining (assuming uniform hashing):

$$E_c(\alpha) = 1 + \frac{\alpha}{4} + \frac{1}{8\alpha} (e^{2\alpha} - 1 - 2\alpha) . \tag{2.29}$$

Some figures computed from (2.29) are given in Table 2.6, together with reference values for uniform random probing.

Searches with Separate Chaining (Through the Overflow Area). The load factor in open addressing cannot exceed unity. With the use of a separate overflow area, only the calculated addresses exist in the hash table; accordingly, if no restrictions are imposed on the size of the overflow area, an arbitrary number of entries can be stored in the latter. If the load factor α is again defined as the ratio of the number of stored items to the number of addresses in the hash table, it may exceed the value unity by any amount. The average number of searches vs α may then be evaluated. Actually a comparison with the other methods based on this figure is not quite justified since the demand of memory in the overflow area is not taken into account, and a too optimistic view of separate chaining may thereby be obtained. However, the hash table and the overflow area usually have different organization, and it becomes difficult to define the load factor in any other simple way.

The analysis of searches with separate chaining is very simple and can be delineated in the following way. If N keywords are randomly and uniformly hash addressed over a table of size H, the average number of collisions at a particular address is $N/H = \alpha$. As all collisions will be handled by chaining the overflowing entries separately for every calculated address, the average number of reserve locations will also equal α. If the sought entry really exists in the table, the expected length of search is

$$E_s(\alpha) = 1 + \alpha/2 \quad . \tag{2.30}$$

(One search is always needed for looking at the calculated address.) The same result was obtained by JOHNSON [2.71] on more complicated grounds. Indirect chaining has also been analyzed by HEISING [2.72].

A comparison of searches with random probing and with coalesced and separate chaining, respectively, is given in Table 2.6.

Table 2.6. Average number of searches with random probing and with coalesced and separate chaining (successful search)

Load factor α	$E_r(\alpha)$ random probing	$E_c(\alpha)$ coalesced chaining	$E_s(\alpha)$ separate chaining
0.5	1.386	1.305	1.250
0.6	1.527	1.383	1.300
0.7	1.720	1.471	1.350
0.8	2.012	1.568	1.400
0.9	2.558	1.676	1.450

Effect of Bucket Size on Searches (Simplified Case with Open Addressing).
If stored in a multilevel memory system, the hash table is normally organized in buckets, and a hash address is then made to refer to the bucket as a whole. The data block corresponding to one bucket is transferable to the primary storage in one operation. Naturally the transfer time depends on the total length of records transferred; however, as is the case with many disk memories, the latency time due to arm movement and disk rotation is predominant in the transfer operation. In the first approximation, the number of retrievals of buckets can, therefore, be considered as a basic cost factor, comparable with the average number of accesses to the one-level, nonbucketed hash tables.

92

It is a trivial observation that overflows from a bucket are less frequent the more slots are available in the bucket. The average number of
bucket accesses can quantitatively be determined using the results derived
for linear probing. A thorough discussion can be found in KNUTH [2.1]. The
figures given in Table 2.7 have been published by SEVERANCE and DUHNE [2.55].

Table 2.7. Average number of bucket accesses vs bucket size with linear
probing and assuming successful searches

Bucket size (slots)	Load factor α				
	0.5	0.7	0.8	0.9	0.95
1	1.50	2.17	3.00	5.50	10.50
2	1.18	1.49	1.90	3.15	5.64
5	1.03	1.14	1.29	1.78	2.77
10	1.00	1.04	1.11	1.34	1.84
20	1.00	1.01	1.04	1.14	1.39
50	1.00	1.00	1.01	1.04	1.13
100	1.00	1.00	1.00	1.01	1.05

For all values of load factor, the number of searches decreases monotonically
with increasing bucket size. If the whole hash table formed one bucket, the
number of retrievals would equal unity. It would be absurd to state that this
is the optimum, and an obvious flaw in this analysis is that neither the
effect of bucket size on the transfer time of records to the primary memory,
nor on the time needed for comparisons connected with the bucket access, is
taken into account. Accurate evaluation of the optimal bucket size is very
difficult since the computing costs may be defined in many ways. SEVERANCE
and DUHNE [2.55] have made one analysis in which they arrived at a theoretical
optimum for the bucket size which is between 4 and 18. This question needs
further investigation, however. For instance, independently of optimality
consideration, the reading mechanism of the secondary storage usually favors
a bucket size equivalent to that of one track or an integral number of pages.
    Another figure which may be helpful in the design of bucketed organizations
is the percentage of entries overflowing from the buckets during insertion.
The results given in Table 2.8 contain theoretical results with random keys,
calculated by SEVERANCE and DUHNE, as well as practical results that are
averages from several different practical files (cf [2.29]). The agreement

Table 2.8. Percentage of overflows from buckets

Bucket size		Load factor				
		0.7	0.9	1.0	1.5	2.0
1	T	28.08	34.06	36.79	48.21	56.77
	P	30	37	-	-	-
	S	3	5	-	-	-
2	T	17.03	23.79	27.07	41.63	52.75
	P	18	24	-	-	-
	S	3	4	-	-	-
5	T	7.11	13.78	17.55	36.22	50.43
	P	8	13	-	-	-
	S	2	3	-	-	-
10	T	2.88	8.59	12.51	34.25	50.04
	P	3	10	-	-	-
	S	1	3	-	-	-
20	T	0.81	4.99	8.88	33.50	50.00
	P	2	4	-	-	-
	S	1	1	-	-	-
50	T	0.05	2.04	5.63	33.34	50.00
	P	1	2	-	-	-
	S	1	1	-	-	-

T = theoretical value, P = practical result, S = standard error of the practical result

is remarkably good, showing that statistical calculations can be applied for the estimation of the overflow area needed.

Since the cost analysis of the optimal bucket size is rather involved and depends on the way criteria are set, it is left to the reader to form his own opinion on the basis of different approaches presented so far. Articles relevant to this question are those of BUCHHOLZ [2.61], HEISING [2.72], HIGGINS and SMITH [2.59], KNUTH [2.1], OLSON [2.73], PETERSON [2.60], van der POOL [2.74,75], SEVERANCE and DUHNE [2.55], TAINITER [2.76], and WILLIAMS [2.77].

As stated elsewhere in this book, with multilevel storages it is advantageous to have the calculated address and its reserve locations close to each other. Bucket organization and linear probing obviously both contribute to this property; further advantages are achieved if the probed buckets are chained (directly): the unnecessary probings during retrieval are thereby eliminated, which brings about a significant advantage in view of the relatively slow operation of the secondary memory devices. Reorganization of hash tables after deletions is also made more flexible by chaining: no entries need to be moved when others are deleted, and this is an advantage when the buckets contain a lot of data.

The extra pointers needed in a chained structure represent an overhead in memory resources which, however, is negligible if the bucket size is much greater than unity.

Handling of Overflows from Buckets with Separate Chaining. It has been pointed out by several authors that if one wants to apply chaining with separate lists, it is advantageous to have the reserve buckets on the same memory area, e.g., on the same track as the calculated address. However, in the bucket-per-track case there would be no sense in reserving an overflow area on the same track as the bucket since it would then be more natural just to make the bucket correspondingly bigger. If, on the other hand, there are several buckets on a track, for instance, one corresponding to every addressable sector, the transfer time of a bucket may be significantly shorter than that of a whole track (due to the serial nature of transfer, and some possible restrictions set by the operating system). All buckets can then be made to share a common overflow area at the end of the track, and for a sufficient storage capacity, this could be chained by the entries as explained below.

KNUTH [2.1] has made a further notice, namely, that there is no reason to make the bucket size in the overflow area equal to that in the hash table. In view of the fact that overflows from buckets are relatively rare events, the reserve buckets would be very sparsely occupied and space would be wasted. Instead KNUTH suggests that the overflowing entries as such could form lists, i.e., the bucket size in the overflow area could be unity.

The average number of searches with bucket organization and separate chaining of the overflowing entries, given in Table 2.9, is based on the above suggestion of KNUTH. The figures are from the work of SEVERANCE and DUHNE. These values do not, however, describe the average time of search since, for instance, if the complete overflow area were retrieved in the same memory access as the bucket itself, the number of retrievals would be trivially equal to one.

The increase of accesses with bucket size for $\alpha > 1$ is due to chaining of individual entries in the overflow area: every entry thereby needs one

access, and this effect does not manifest itself until at high load factor
and large bucket size.

Table 2.9. Average number of searches with bucket organization and separate
chaining of overflowing entries (successful search)

Bucket size	Load factor α					
	0.2	0.5	0.7	0.9	1.0	1.5
1	1.10	1.25	1.35	1.45	1.50	1.75
2	1.02	1.13	1.24	1.36	1.43	1.82
5	1.00	1.04	1.12	1.27	1.37	2.07
10	1.00	1.01	1.06	1.21	1.33	2.49
20	1.00	1.00	1.02	1.15	1.31	3.33
50	1.00	1.00	1.00	1.08	1.29	5.83
100	1.00	1.00	1.00	1.04	1.28	10.00

2.5.2 Effect of Hashing Function on the Length of Search

The evaluation of different hashing functions has been postponed to this
subsection because it is related to certain schemes discussed in Sect. 2.5.1.

In systems programming, especially with the symbol tables used in assemblers and compilers, the selection of hashing function is rather clear: the
division method is able to guarantee a rather uniform hashing, and it has
some extra benefits, for instance, short computation time and the possibility for using the quotient as identifier and in the definition of the probing
increments. The considerations are somewhat different with large files in
which secondary storages must be used, because even minor differences in the
searching time are directly manifested in the slow operations; on the other
hand, the need to compress the identifiers stored with the entries is no
longer so stringent, and the time for the computation of the hashing function
is negligible as compared with the access time. Thus, even complicated hashing
algorithms (such as the algebraic coding method) may be tolerated. The comparison of hashing algorithms in this subsection is strongly biased toward
multilevel memory operation.

With large files, one seriously has to take into account chances for
accidental poor operation, even if this would occur only in rare cases. Since
it is difficult to predict the performance in advance, but major changes in
systems design are also tedious to make after the final evaluation, it might

be a good strategy to put more weight on a stable performance than on the
ultimate speed.

A thorough analysis of the different hashing functions in the context of
bucketed organization has been made by LUM et al. [2.29]. Their results
have been selected here for presentation. The authors have evaluated the
stability of hashing functions by benchmarking runs referring to eight files
of different nature and reporting the *standard deviations* in the average
number of searches for different methods. Although in view of the arbitrary
selection of files and the numbers of records in each one, the least-square
method is then theoretically not quite justified, nonetheless this analysis
is very orientative. The types of keyword were: county and state codes,
personnel and applicant names, personnel locations, customer codes, product
codes, library keywords, and, as a reference, random numbers. The keywords
were in five cases numerical codes, in two cases alphabetic, and in one
case alphanumeric. The number of characters was typically six in the number
codes, and 10 to 12 with letters. The alphanumeric codes were seven characters long.

The average number of accesses and its standard deviations for different
hashing functions are plotted in Fig. 2.11 as a function of load factor, for
four sizes of buckets. The results are reported separately for open addressing
and separate chaining. More detailed results are found in the original article
of LUM et al.

Some general conclusions can be drawn from these results. It seems that
with open addressing, and with the load factor less than 0.75, the best results are given by the mid-square method. With separate chaining, the best
results are given by algebraic coding methods. However, the best results generally are due to the division method. Varying degrees of performance, generally worse, are obtained by the other methods. An effect which cannot be
seen from the reported results is that if a few bad results were excluded
from all experiments, the remaining numbers would agree much more closely.
As pointed out above, however, chances for poor operation even in rare cases
are generally not allowed because one cannot predict what kind of data will
be received during operation, and even one bad data set will then affect the
future operation.

Some corrections to the above results have later been published by LUM
and YUEN [2.78]. LUM [2.79] has also suggested an algorithmic method for the
generation of test keywords to be used in comparison criteria. A cost analysis
of the efficiency of hashing functions has been given by ULLMAN [2.80].

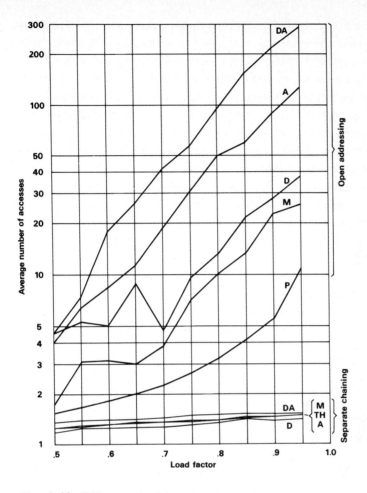

Fig. 2.11. Effect of hashing function on the average number of accesses.
DA = digit analysis method, A = algebraic method (Galois field 2), D =
division method, M = mid-square method, P = Peterson's theoretical results,
TH = theoretical results

2.5.3 Special Considerations for the Case in Which the Search Is Unsuccessful

In most applications, in a substantial fraction of queries the sought entry
may not exist in memory. The analysis of the retrieving costs must take into
account this possibility. One might conjecture that on the average an unsuccessful search is longer than a successful one. Take, for example, a list of
length L; if an item does not exist in it, the number of searches is L, whereas if the item can occur in any position with the same probability, the aver-

age length of successful search is L/2. In hash tables, the situation is not that simple and needs more careful analysis. For instance, with linear probing, the average spacing of reserve locations in a chain grows with the filling up of the table. If an entry exists in the chain, it will be found when, on the average, half of the chain is retrieved. On the other hand, the number of probings on the first half of the chain is smaller since the average spacing is smaller, and retrieval of the rest of the chain (for unsuccessful search) takes progressively more probings. The analysis of an unsuccessful search should actually be carried out for every collision handling method separately. This would extend the material presented in this chapter considerably, and we shall refer to a couple of the simplest cases only. However, if unsuccessful searches would play an important role in practical applications, the conflict flag method or the ordered hash table described earlier can be used.

In the simplest case, the average number of unsuccessful searches is actually the same as the number of trials for inserting a new item. The average number of unsuccessful searches in the cases of linear and random probing are approximately given by the following theoretical expressions [2.1]

Linear probing: $\frac{1}{2} [1 + (1 - \alpha)^{-2}]$ $\qquad\qquad\qquad\qquad$ (2.31)

Random probing: $(1 - \alpha)^{-1}$. $\qquad\qquad\qquad\qquad\qquad$ (2.32)

Numerical values computed from these formulas, together with figures referring to successful searches, are given in Table 2.10.

Table 2.10. Average number of searches

		Load factor α				
		0.5	0.6	0.7	0.8	0.9
Linear probing	S	1.500	1.750	2.167	3.000	5.500
	U	2.500	3.625	6.056	13.000	50.500
Random probing	S	1.386	1.527	1.720	2.012	2.558
	U	2.000	2.500	3.333	5.000	10.000

S = successful search, U = unsuccessful search

2.6 Multi-Key Search

In symbol tables, there is usually a one-to-one correspondence between a
keyword and its associated value, the latter, for instance, being the absolute or internal address of a computational variable used within a program.
Contrary to this, the normal situation with documentary files, customer records, and other nonmathematical data is that every entry to be stored is
provided with several keywords or descriptors. The same keyword may occur
in an arbitrary number of entries; for instance, in personal files, the
binary-valued attribute 'sex' takes on the same value in at least half of
the records.

If there often occurs a need to carry out an exhaustive retrieval of all
entries matching given keywords, then *sequential search methods* may be considered first. This is the case, for instance, in the *selective dissemination
of information (SDI)* that is one form of the services offered by contemporary
libraries. Every subscriber to this service defines his profile of interest
in the form of a logical expression which contains a set of relevant descriptors. In one computer run, a great number of subscribers can be served simultaneously, by comparing their profiles with the descriptors stored with
the documents and using various functions which indicate the degree of matching. However, the SDI application, in spite of its great importance, is not
a particularly suitable candidate for implementation by hash-coding. One of
the reasons for this is that for the records, magnetic tape is normally used
on account of its flexibility and low cost; but magnetic tape is not directly
addressable. Secondly, even if disks or other comparable media are applied,
the great number of simultaneous queries makes batch processing most economical anyway.

If, on the other hand, queries to a data base are made irregularly and
one at a time, then a scanning-type retrieval is not at all advantageous,
but content-addressing would be preferred. There are plenty of examples of
such applications: transactions in the banking business, automated distribution of messages in big organizations, retrieval of data from documentary
archives, data management in computer-aided design (CAD) applications, research on artificial intelligence, etc. With the advent of remote computer
terminals in ordinary households, one has to face the situation that central
data banks have to be reorganized using content addressing in order to make
on-line queries possible. It is obvious that the storage devices, whatever
capacity they may have, then have to be addressable. For the present, one

possible choice for such mass memories would be magnetic disk; special devices such as optic ones may be used in extremely large archives, too.

In view of the above considerations, it seems necessary, within the scope of this book, to have a special discussion of methods connected with the searching of entries on the basis of multiple keywords, or the *multi-key searching*. The problematics of this field are discussed in this section in a rather general form, without reference to particular applications. Some standard principles of multi-key searching are presented first. After that, special solutions implementable only by hash coding are introduced.

It ought to be emphasized that definition of a searching task is often made by giving a set of *names* and their *attributes*. In this case the semantic differentiation of keywords enters into the addressing problem. Searching from semantic structures will be discussed in Chap. 6.

2.6.1 Lists and List Structures

Any ordered set of data items may be named a *list*. A *list structure* results when an element of a list is another list, and if such a nested organization is continued, the list structure may grow very complex.

A list or list structure itself is an entry in memory which must be accessible in a convenient way. There is a particular point in the list, usually its first item, which is called the *head* or *header*, and which is identifiable by the *name* of the list. The names can be kept sorted in which case usual searching methods (described in numerous textbooks) can be applied. A convenient method for the location of list heads is the application of hash coding on the names, whereby the latter need not be sorted. The other items may be linked to the list in a number of ways; the simplest way is to use consecutive locations.

Examples of lists the items of which are not stored consecutively but scattered over the memory area are the chains of reserve locations in hash coding; the calculated address then comprises the list head. The sequence of addresses can be computed by a probing algorithm at every retrieval, or directly defined with the aid of pointers. A *linked list* is any collection of items which is chained using pointers. A *one-way (linked) list* allows tracing of its items in one direction only. In a *two-way list*, also named *symmetric* or *doubly-linked list*, every item contains two pointers, one pointing forward and the other backward, with the result that the list can easily be traced in both directions. A *circular* or *threaded list*, also named a *ring*, is a one-way list the last pointer of which points back to the list head. Various types of linked list are shown in Fig. 2.12.

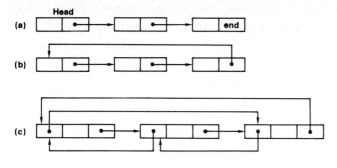

Fig. 2.12. Various types of list; (a) one-way list, (b) circular list, (c) two-way (double-linked) circular list

For implementing a *list structure*, some items must have at least two pointers thus defining a *node*. A *tree* is a list structure with branchings starting from a node named *root*, and it has no recurrent paths to the same node. A *binary tree*, an example of which is given in Fig. 2.13, always has two branches at every node, except for the last items. A *graph* is a general list structure which may have recurrent paths.

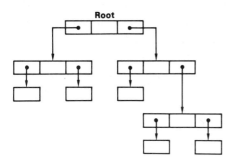

Fig. 2.13. An example of binary tree

There are various methods for the retrieval of information from graphs, as well as means to reduce graphs in simpler forms. We shall abandon such discussions here since the list structures are considered only in a few instances in this book. The principal applications of list structures will be discussed in Sect. 2.8 and Chap. 6. For references of lists and related data structures, see, e.g. [2.81-86].

Multilists. Structures which are formed when the same item may occur in many lists, i.e., when several independent chains pass through the same set of items, are named *multilists*. They differ from meshy graphs by the fact that the common items do not behave like nodes, but every partial list, starting from its head, can be followed in a unique way up to its end. For this purpose, the same principle as that used in the hash table may be applied: the identifiers of the individual list heads must be stored together with the respective pointers, for the resolution between different chains (Fig. 2.14).

Flags	Data
ID 1	Pointer 1
ID 2	Pointer 2
ID 3	Pointer 3

Fig. 2.14. An example of storage location in a multilist

The concept of multilist (also written Multi-list or Multilist) was originally introduced within the context of associative memories and associative computing by PRYWES and GRAY [2.87-91]. The multilist may also be a pure logical structure such that if an entry is common to several lists, in reality there would exist identical copies of the same entry at physically distinct memory locations, each one representing the entry in one of the lists. One example of this is involved in the example presented in Sect. 2.6.2.

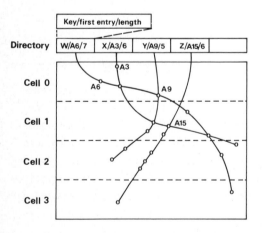

Fig. 2.15. Usual multilist

The principle of a usual multilist is shown in Fig. 2.15. It contains a *directory* in which all keywords, the pointers to the first items, as well as the lengths of the lists are given. The directory may occupy a separate memory area. The keywords in it may be stored in a sorted table, or hash coding may be applied for their location. In Sect. 2.6.2, another possible organization is described: the keywords as well as the first pointers are stored in the same hash table. In this case, the format of every storage location ought to be identical, and indication of the chain length may be abandoned in order to save space. Alternatively, the location at the calculated address might be modified to store the list length instead of an entry, in which case, the first stored entry is placed into the first reserve location.

In Fig. 2.15, the entries are indicated by small circles, and the memory area is shown two-dimensionally in order to illustrate the various chains. If the maximum number of keywords in an entry is m, then the location must be able to hold, in addition to data and control bits, m identifiers and pointers, the latter corresponding to all possible successors in different chains.

The multilist structure is usually constructed at the same time as the entries are stored. The list heads corresponding to the keywords of an entry are first located, and the chains are traced until the last items are found. The address of the storage location of this entry is then inserted as a pointer to every last item of the chains existing so far, and the new entry becomes the new end of these chains.

The most common mode of multi-key search is the *combination search*, or searching on the conjunction of two or more keywords. Assume that in the example of Fig. 2.15 all entries corresponding to the keyword combination 'X and W' have to be found. It is then preferable to start with the shorter chain, corresponding to keyword X, and it remains only necessary to check whether W exists in an entry or not. For this purpose each entry must contain a copy of all its keywords or their identifiers.

With the usual multilist structure, the lists may grow arbitrarily long. The time for the addition of a new entry may become intolerably long, since the complete chains must thereby always be traced. Deletion of items is a similar tedious operation. On the other hand, due to the inherent nature of data, one cannot prevent the searches from becoming long. There may be still other reasons, however, on account of which *limitation of the list length* might be desirable, for instance, when only a restricted amount of buffer space is available in the primary storage. Referring to Fig. 2.15, one can then segment the list into parts of a fixed maximum length; in this case the dictionary must contain pointers to all partial lists. Insertion of a new entry is made to the last partial list which is normally much shorter than the complete chain of a usual multilist (Fig. 2.16).

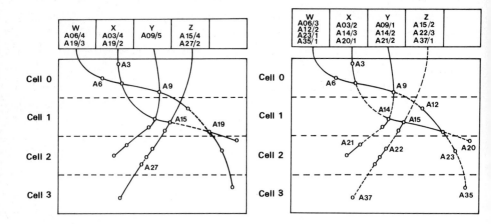

Fig. 2.16. Multilist with limited list length

Fig. 2.17. Cellular multilist

In the so called *cellular multilist*, the lengths of the lists are not limited; instead, a chain is not allowed to exceed the boundaries of a physical memory block, for instance, a track (cell, bucket) (Fig. 2.17).

The advantages of this organization are that all data in a partial list are always transferred to the primary storage whenever one memory block is buffered in it, and furthermore, it is usually deducible from the directory that only a few blocks need to be examined. For instance, if in the example of Fig. 2.17 a combination search on the condition 'X and Y' were to be performed, examination of cell 3 is unnecessary since Y does not have entries in it. Moreover it is possible to carry out the searches in different cells along the shortest chains, for instance, on cells 0 and 1 along the Y chain, and on cell 2 along the X chain.

The *cellular serial organization* (Fig. 2.18) has some features from the cellular multilist and the bucketed organization. In the directory, only indices of those memory cells are indicated in which entries with a particular keyword occur. A complete linear search, however, has to be performed through the corresponding cells. The updating of this organization is simple, and if one cell holds only a few records, the operation is relatively fast.

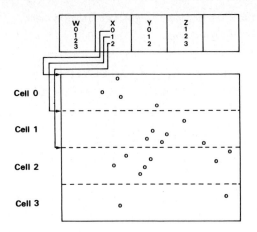

Fig. 2.18. Cellular serial organization

Inverted Lists. As the last example of file organization for multi-key search, the *inverted list* or *inverted file* structure is presented. The name of this method comes from the fact that while in a usual (serial) search operation the entries have to be retrieved before their keywords can be examined, in the inverted method the addresses of all entries associated with a given keyword are made available in the directory (see Fig. 2.19). In a way the inverted list is identical to a cellular serial organization in which there is only one location per cell.

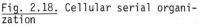

Fig. 2.19. Inverted list

Relating to the hash coding principles earlier introduced in this book, one can derive the basic idea of an inverted list from them. One may simply think that one way for the organization of the directory is a *hash index table* in which the keyword corresponds to the calculated address, and the addresses of the entries are equivalent to pointers to the secondary storage, as explained in Sect. 2.4. These pointers are placed into probed reserve locations. The practical implementation of such a principle is described in Sect. 2.6.2. However, if the keywords in the inverted list are kept sorted, then any standard searching method can be used for their location, and the pointers can simply be tabulated in consecutive locations.

The most marked advantage of inverted lists is a possibility of compressing the information utilized in combination search into the minimum space whence, at least if the file is not very large, the whole directory can be kept in a fast primary storage. Unnecessary mechanical operations such as arm movements in disk memories can be eliminated in advance by the computation of search conditions in the mainframe unit.

2.6.2 An Example of Implementation of Multi-Key Search by Hash Index Tables

The leading idea in multi-key hash coding is to compute a separate hash address for every keyword that occurs in the entries, and to store it in all of these places. Similar keywords in different entries then always hash to the same calculated address, which will add up to an appreciable number of collisions. For this reason it is advisable to keep the load factor rather low, and if the secondary storage were used for the hash table as in this example, open addressing then would seem advantageous.

Instead of storing the same entry in many places in the hash table, it will now be most economical, even for secondary storages, to apply the principle of hash index table; only pointers to a separate memory area where the original entries are stored need to be duplicated. Notice that in multi-key search, plenty of computations are needed before the final entries are defined, and, therefore, one may tolerate one extra indirect addressing step even though the hash table is in the secondary storage.

The above arrangement would already be sufficient for a search on the basis of a single keyword. When it comes to the retrieval of entries which must have two or more specified search arguments among their keywords (combination search), a more sophisticated form of organization will be necessary. One suggestion for it will be discussed in the context of the following example.

Consider the memory organization depicted in Fig. 2.20. It consists of
the following memory areas: the *hash index table (A)*, two *sorting tables
(B and C)*, and a *document area (D)*, in which the complete entries together
with all of their descriptors are stored. An auxiliary list of sifted pointers
may still have to be constructed during the search.

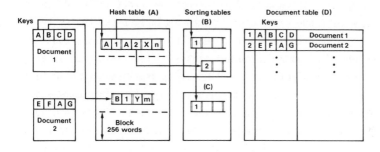

Fig. 2.20. Memory areas in a multiple-keyword search

Most of the following considerations have to be devoted to the hash index
table, which requires plenty of space in view of the fact that every keyword
of every entry must occupy a location in it. The load factor of this table
should remain well below unity. For these reasons, in a real application
(from which this example is taken) one small magnetic disk was reserved for
the hash table. Those parts of the disk which are referred to will be transferred to the primary storage. A suitable size for buffer storage was considered to be 256 words, named a *page*, since the minicomputer in question
had ready provisions for block transfers of this size in its operating system.
Every hash-addressable location of the hash index table was able to hold
a keyword (or its identifier) and one pointer. Every word on the disk was
made hash-addressable; in this example the number of locations was 2^{20} =
1,048,576. Another alternative would have been to apply buckets with linked-
list organization. In this example it would have made the discussion more
complicated. For simplicity, linear probing was used, although random probing
with independent double hashing or the quadratic or linear quotient method
might have yielded a somewhat better performance at a high load factor. As
the hashing function, division method was preferred. It must be pointed out
that many alternative details would have been possible with this application,
and the purpose here is only to convey the basic idea.

There are some opposing opinions about one detail of the probing algorithm, namely, whether the reserve addresses should be on the same *page* as the calculated address and thus be computed modulo 256, or whether they should continue on the next page and be defined modulo 2^{20}. The choice is largely dictated by how many times the same keyword may occur in different entries on the average. Especially if the load factor is relatively small, modulo 256 probing seems profitable since this then reduces the number of disk accesses.

The sorting tables B and C are used in the combination search in the following way. They are hash-addressed with respect to the *pointers* found in the hash index table. The search argument presented first picks up a set of pointers in table A. These pointers (their copies) are then stored in table B at addresses which are hashing functions of the pointers themselves. No other associated information is stored in these locations. The second keyword picks up another set of pointers independently of those found in the first step. However, when the second pointers are hash addressed, it directly becomes possible to check whether they have already been recorded in table B. If this is the case, one can deduce that both search arguments must occur in these entries. In the opposite case the pointers found first may be forgotten. If no further search arguments have been given, the pointers for which a coincidence was verified are entered into a list according to which the final look-up in the document area is made. If, on the other hand, a third search argument is given, those pointers for which a coincidence was due are first copied into table C, again using hash addressing. When the pointers picked up by the third search argument are hash addressed for table C, it then becomes possible to check whether they coincide with the pointers found in the first two passes. If the result is positive, one can reason that all the three search arguments presented so far are contained in entries corresponding to these pointers. If the number of search arguments was three, the pointers can now be listed. With a still greater number of search arguments, one first copies the last pointers in table B, after its previous contents have been cleared up. In this way tables B and C are used in turn until all search arguments are exhausted. The "sifting" procedure described above continuously reduces the number of possible candidates until only those having keywords in agreement with all the specified search arguments are left.

By virtue of hash coding applied at all steps of retrieval, the complete search can be performed extremely quickly. In a real application described in this example, the sifting time was much less than that needed to type in the keywords.

2.6.3 The Use of Compound Keywords in Hashing

The multi-key searching methods discussed above were based on the determination of intersections of lists associated with individual keywords. This
subsection approaches the problem in a different way, by regarding all the
keywords associated with an entry as a single compound keyword. As long as
all keywords are involved in a search, the implementation is trivial, similar
to that of a single-keyword case. A problem arises when arbitrary keywords
are left unspecified.

Hashing of All Keyword Combinations. A solution for combination search which
is fastest of all, but also demands the maximum amount of memory space, is
to hash the entries for all different combinations of their keywords. Thus,
for n keywords in an entry, there are $\sum_{k=1}^{n} \binom{n}{k}$ combinations for which copies
of the entry have to be stored in the hash table.

A standard procedure for the definition of a hashing function for compound
keywords is the following. Assume that $B(K_1)$, $B(K_2)$, ..., $B(K_k)$ are the
binary bit strings corresponding to the keywords K_1, K_2, ..., K_k, respectively,
and they all have been defined by some hashing algorithm, having the same
length as the hash address. For the bit string B corresponding to the hashing
function, then, the following expression is used:

$$B = B(K_1) \oplus B(K_2) \oplus \ldots \oplus B(K_k) \tag{2.33}$$

where $\oplus$ is the EXCLUSIVE OR operation, or modulo 2 summation, over respective
bit positions.

With such a *combination hashing* as described above, the efficiency of the
hash table can be improved if the keywords can be divided into categories
according to their known role. Assume, for example, that x_1, x_2, and x_3 are
the values of three different *attributes* A_1, A_2, and A_3, respectively, for
instance, 'nationality', 'age group', and 'residence' of a citizen. When a
keyword value x_i is given, one immediately knows from its nature whether
i = 1, 2, or 3. Thus the different types of keyword compound can always be
identified, and a *separate hash table* then can be reserved for each one. In
the above example, separate hash tables might be constructed for all the
compounds A_1, A_2, A_3, (A_1, A_2), (A_1, A_3), (A_2, A_3), and (A_1, A_2, A_3), respectively. Operation of such partitioned hash tables is simplified, since the
chains of reserve locations become much shorter than if all compound keywords were hashed into the same table.

Exhaustive Study of "Don't Care" Cases. At the cost of increasing number of searches, the hashing can be limited to a few types of keyword combination that are most likely to occur, and for the rest of the cases, some more tedious searching procedure can be applied. Alternatively, one may construct the hash tables for some of the biggest compounds only, and if some keywords are left unspecified, they are regarded to attain the attribute value "don't care". An exhaustive study of all "don't care" cases must then be made, by giving all possible values to these attributes in turn, and carrying out separate searches for all alternatives.

In some cases the above idea works very well, for instance, if the un-specified attributes can take on only a few values. Consider, for instance, the compound attribute ('sex', A_2, A_3). If the entries were stored at ad-dresses obtained by hashing these triples, then, with the attribute 'sex' unspecified, only two hash coded searches would be necessary, namely one for the compound (male, A_2, A_3), and the second for (female, A_2, A_3), respectively.

In a general case, with unspecified keywords that can take on an almost arbitrary number of values, the above procedure would imply an immense number of "don't care" cases, however. There are some possibilities to overcome this difficulty; one of them is discussed below.

Concatenated Hash Address. The hash address, being a bit string, can be de-fined in such a way that each of the keywords in a compound defines its own section in it, and these sections are concatenated. For instance, if every keyword would define exactly two bits, then the hashing function of every separate keyword would have a range with just four values (00, 01, 10, and 11, respectively). If k keywords are left unspecified, there would be then a need to carry out a searching for 4^k bit value combinations.

Various hashing functions for concatenated hash addresses have been dis-cussed by RIVEST [2.92]. One might expect that the uniformity of the distri-bution of concatenated addresses would not be as good as for other types of hashing. The practical reduction in performance is not too bad, however, in view of the extra benefit of multi-key search thereby attainable.

A combinatorial addressing technique, related to hashing of multiple keywords, has been represented by GUSTAFSON [2.93].

2.7 Implementation of Proximity Search by Hash Coding

Let us once again recall that the usual objective in content addressing has
been to locate all items which match the given search arguments in their
specified parts. This is obviously not the way in which biological associative memory operates; for instance, for a human being it is possible to recall memorized occurrences on the basis of rather vague key information.
On the other hand, a noteworthy feature in the operation of human associative memory, when compared with the content-addressable memories, is that
it does not list all matching recollections but usually concentrates on that
one which has the highest degree of matching. Associative recall from a biological memory is probably more akin to the operation of *pattern recognition*
devices. In the latter, the representation of an object is given in the form
of a set of simultaneous or subsequent signals. After an analysis of some
characteristics (often named *features*) of this set of values, a result is
obtained which usually defines the identity of the object, or alternatively,
triggers a response. The classical approach to artificial pattern recognition is based on operational units named *Perceptrons*, the mathematical equivalent of which is a *discriminant function*. One discriminant function is defined for every category of the patterns to be classified, and it attains a
scalar value when the representative signal values are substituted into it.
The classification decision for a particular pattern is based on comparison
of the relative magnitudes of the discriminant functions.

In this context we shall not be involved further in the problematics of
pattern recognition or biological associative memory; it may suffice to refer
to a recent discourse by the author [2.94]. Instead, since the implementation
of large parallel networks of Perceptrons has technically been shown very
tedious, it may be more interesting in this context to look for the most
effective software implementations of this task. If the features may take on
only *discrete* values, hash coding provides a very efficient solution, presented in this section. The procedure described here was recently developed by
the author in cooperation with REUHKALA [2.95,96] and it directly implements
a mode of retrieval called *proximity search*, or identification of one or a
few patterns which have the highest degree of matching with the search argument given. The first of the central ideas applied in this method is to
represent a keyword or a relevant portion of an item by *multiple attributes*
or *features* derived from it. These features are then regarded as multiple
keywords for the item by which the item is addressable in a hash table. The
second of the central ideas in this method is a procedure for the collection

of statistical results from successful searches relating to the above features. It is believed that the principle described in this section is new, and it will be discussed within the context of *word identification* (and thus, correction of misspelt words) when the search argument is a randomly garbled version of one of the reference words.

The example discussed in this section is primarily intended to demonstrate that, contrary to what is generally believed, *it is possible to implement direct content-addressable searching on the basis of approximately presented descriptions, too*, and that software principles essentially based on hash coding can thereby be applied effectively. The missing, erroneous or extraneous parts of information may be present in unspecified positions of the description, contrary to the conventional content-addressable memories.

Extraction of Features from Words. In the following example selected for the illustration of proximity search, a keyword is represented by a number of features which indicate its *local characteristics*. One feature is a small group of letters, and the groups are allowed to overlap. In order to minimize the effect of local errors on features, the letters of a group are selected consecutively. Groups may also be formed of cyclically (end-around) contiguous letters. Special consideration is necessary, however, concerning the order in which the characters are taken to an end-around group. Since it would be desirable to discriminate between versions of a word that are obtained by limited rotation from each other, (e.g., 'table' and 'etabl') the original order of the characters should be retained. Thus, if the keyword is 'table', and its local features consist of three letters, they should be selected as 'tab', 'abl', 'ble', 'tle', and 'tae', respectively.

When the features are formed of n cyclically consecutive characters, and their order is defined as in the above example, they are hereupon named *n-grams* (in particular, *monograms*, *bigrams*, *trigrams*, and *quadgrams* for the values n = 1, 2, 3, and 4, respectively).

The next problem is associated with the number of characters to be included in a feature. The optimal number is apparently a function of the contents of the keywords as well as of their number. Assume that only letters from the English alphabet are used. The numbers of different monograms and bigrams are 26 and 26^2 = 676, respectively, and these values are impractically small since hash addresses have to be computed for them. The next choice would fall upon trigrams, and their number is 26^3 = 17,576. In natural words, only part of these may occur. However, this number seems large enough for the range of a hashing function.

If the features would consist only of trigrams, the minimum length of words in which at least a single feature would be saved in spite of any single error occurring is four characters; if there is a need to allow errors in shorter words, too, a mixture of trigrams, bigrams, and monograms might be used for the features, at the cost of an increased number of "keywords" and a corresponding demand of memory space. A compromise might be to select for features all trigrams, the three bigrams formed of the first and last two characters, as well as the pair formed of the first and last characters, respectively. It must be pointed out that the type of features selected should not be made to depend strictly on word length since, in view of the fact that insertion and deletion errors change the word length, it is not possible to deduce the accurate lengths of the original words from the search arguments.

Hash Coding for the Features. Before definition of a hashing function for features, one has to consider what happens to the original features when the string to be encoded contains errors. The different types of error are 1) *replacement error* or change of a letter into another one, 2) *insertion error* or occurrence of an extra letter, and 3) *deletion error* or dropping of a letter. In the case of an insertion or deletion error, the positions of all letters to the right of the error will be changed, except when different types of error compensate the effect of each other. Accordingly, in order that as many hash codes as possible would be saved in the search argument, in spite of errors occurring, it might seem profitable to define the hash codes without taking the position of a feature in the string in any way into account. This, however, is not a sufficient solution, in view of the fact that similar features may occur in many places of the words, with the result of poor discrimination between different words. A compromise is to allow a limited amount of translation of a feature. Such a limited translational invariance is very hard to take into account in a hashing function, however. Therefore, a procedure which has an equivalent effect is to compute first the hashing functions for local features (n-grams) without consideration of their position in the word. A multiple search with all such features used as keywords then picks up a number of candidate words, or their associated entries. It is then possible a posteriori to increase resolution by screening out all searches in which the relative displacement between the features of the original keyword and the search argument was more than, say, two character positions. The number two may be justified on the following grounds. The lengths of keywords are seldom much greater than ten letters, and it is then highly improbable that three insertion or deletion errors, all of the same type, would occur close to each other. On the other hand, we would like to allow double errors in arbitrary combinations.

There is yet another aspect which ought to be mentioned in this connection. We shall set no restrictions on the words to be handled, and so the strings of letters can be considered as being stochastically (not necessarily uniformly) distributed. Errors are generated in stochastic processes, too, but it is more justified to assume that their type and occurrence in the strings has a uniform distribution. Now, contrary to the objectives in error-tolerant coding, there exists no possibility of designing an identification procedure for arbitrary erroneous words which would guarantee their identification to 100 percent accuracy. One should realize that errors are *inherently undetectable* if they change a legitimate word into another legitimate word, and so the identifiability depends on the contents of the keywords, although to a smaller extent.

Assume now for simplicity that only trigrams are used for features. Their address in a hash index table may be computed in the following way. Assume that the alphabet consists of 26 letters. If the letters in a trigram are denoted by x_{i-1}, x_i, and x_{i+1}, respectively, the trigram shall first attain the numerical value

$$v_i = 26^2 \cdot x_{i-1} + 26 \cdot x_i + x_{i+1} \ . \tag{2.34}$$

If H is the size of the hash index table and the table begins at address B, the division method gives for the hashing function

$$h(v_i) = v_i \bmod H + B \ . \tag{2.35}$$

If the hash table size, especially when using disk memories as auxiliary storages, is selected to be a prime, the radix 26 used in numerical conversion seems profitable, whereby the division method is expected to guarantee a rather uniform distribution of hash addresses.

Organization of the Hash Table. With multiple keywords, the chains of reserve addresses tend to become long; multilevel storages ought to be preferred, therefore. It is advisable to make the hash table rather large in order to keep its load factor low. Open addressing with linear probing (modulo page size) is thereby applicable. The bucket organization is also natural with secondary storages. It is noteworthy that an additional advantage is provided by the adoption of buckets: in this example the number of hash addresses was limited by the number of different triples, being only 17,576. When using bucket organization, the hash table size equals this number multiplied by the number of slots per bucket.

The linked list structure is also natural in connection with bucket organization, whereby the time of retrieval, which in this method tends to be significantly longer than in simple hash coding, can be kept reasonable.

If the division method is chosen for the hashing algorithm, the extra advantage thereby gained is a minimal need of space for the keyword identifier; the significance of this benefit will be appreciated with large tables.

The overall organization is delineated in Fig. 2.21. Every location in the hash index table contains an identifier for a trigram, and a pointer to a document area where the original entries can be found. The bucket organi-

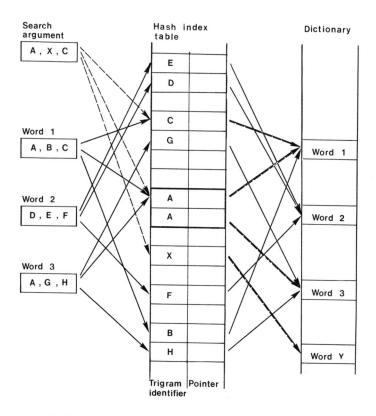

Fig. 2.21. Illustration of the redundant hash addressing method. The letters A, B, ..., X stand for trigram features. Notice that there is a collision of feature A, and the next empty address in the hash index table is used as reserve location

zation is not shown explicitly; as a reserve location, the next empty slot
to the calculated address is used in the illustration. (In the example of
Fig. 2.21, there is a collision of feature 'A'; the colliding item is placed
in the next location.)

A search argument will pick up as many pointers from the hash index table
as there are features in it. As long as at least one of the original features
from the correct keyword exists in the search argument, there will be at least
one pointer to the correct entry. By the nature of this method, however, an
erroneous search argument generates wrong pointers, too. In any case, however, the maximum number of different pointers thus produced is the same as
the number of features, or, which is equivalent, the number of letters in the
search argument. All of the candidates indicated by these pointers can now be
studied more carefully. As mentioned earlier, the position of a trigram in
a candidate entry will be compared with that of the search argument, and only
if the relative displacement is not more than, say, two positions, a match is
said to be true.

Alternative Modes of Search. In the continuation, the redundant hash addressing system may be developed in either of the following directions.

For one thing, a combination search with several, possibly erroneous,
keywords can be performed. The procedure is essentially the same as that
discussed in Sect. 2.6.2. If there are several search arguments associated
with the same entry, during storage their features are entered into the hash
index table independently, as described above. The retrieving operation consists of several passes, one for each search argument, and every pass produces
a number of pointers in the way described for the principle of redundant hash
addressing. By the application of another hashing algorithm to all of these
pointers, they are entered into the sorting tables B and C (cf Fig. 2.20),
or compared with pointers entered at earlier passes. When all the search
arguments have been exhausted, the number of remaining candidate pointers has
been reduced and among the pointers finally left there are probably no further
ones than those which identify the sought entries.

Another possible continuation for redundant hash addressing is a statistical analysis of the features, and the identification of a single entry which
is most "similar" to the search argument. This mode of operation is termed
proximity search. One of its applications exists in the correction of misspelt
words: the search argument has to be compared with reference words stored in
a dictionary (equivalent to the document area described above) and that one
which agrees best with the search argument is picked up. The latter mode of
identification will be expounded below in more detail.

Statistical Analysis in the Proximity Search. If the search argument has
several features in common with the keyword of a particular entry, then
several pointers converge at that entry, and the reverse also holds true.
By counting the pointers of all responding entries (the number of which, as
stated above, is not greater than the number of features extracted from a
search argument) it then becomes possible to deduce which one of the candidates has the highest degree of matching with the keyword.

As mentioned above, a pointer to an entry shall be accepted if and only
if the relative displacement of the same feature in the search argument and
the stored keyword stays within specified limits; in this example, two character positions. However, it will not be quite enough to base the comparison
of words on accepted pointers solely; since the keywords may be of different
length, disagreement of the latter should somehow be taken into account. In
a number of experiments it has been shown [2.88] that if these errors are
generated in a stochastic process, the following measure of dissimilarity
will lead to a good recognition of the original keywords: the *feature distance* (FD) between words X and Y is defined as

$$FD(X,Y) = \max (n_x, n_y) - n_e ,$$ (2.36)

in which n_x and n_y are the numbers of features extracted from X and Y, respectively, and n_e is the number of matching features. In the redundant hash
coding method, n_e is the number of accepted pointers converging to an entry
in the document area.

When counting the pointers, it will be expedient to construct a match
table at the time of each search. The pointers picked up from the hash index
table by the features are entered into this match table if the corresponding
features in the search argument and the stored keyword match. If the pointer
already existed in the match table, only a count index of this pointer is
incremented. If the number of search arguments is small, as is the case in
general, the inquiry for pointers stored in this table can be made by linear
search. For ultimum speed, this table may also be made hash addressable.

The efficiency of this method has been tested with two fairly large dictionaries of keywords. One of them consisted of 1021 most common English
words [2.97]; the second dictionary had only surnames in it, picked up from
World Who's Who in Science [2.98]. In the latter case, some transliteration
was performed to fit the English alphabet. All possible types of single and
double errors were investigated; the single errors were made to occur in each
character position of each word, but the double errors were generated at

random with a probability which is uniform for every possible erroneous letter and for every position. The statistics used in these experiments was rather large: approximately 50,000 realizations for erroneous search arguments were applied for each of the dictionaries. The results, expressed in percentages, are shown in Tables 2.11 and 2.12.

Table 2.11. Correction accuracies of misspelt words in an ensemble of 1021 most common *English words*, for various word lengths (rows) and combinations of errors (columns). The numbers are percentages (rounded); the values between 99 and 100 have been truncated without rounding. Rows labeled Depth 1 correspond to results in which the correct word had the smallest FD number among the candidates; rows labeled Depth 2 show results in which the correct word had the smallest or the next to the smallest FD number among the candidates

Word length			Deletions: 1, Replacements: 0, Insertions: 0	Deletions: 0, Replacements: 1, Insertions: 0	Deletions: 0, Replacements: 0, Insertions: 1	Deletions: 1, Replacements: 1, Insertions: 0	Deletions: 1, Replacements: 0, Insertions: 1	Deletions: 0, Replacements: 1, Insertions: 1	Deletions: 2, Replacements: 0, Insertions: 0	Deletions: 0, Replacements: 2, Insertions: 0	Deletions: 0, Replacements: 0, Insertions: 2
3	Depth	1	-	-	96	-	-	-	-	-	39
	Depth	2	-	-	97	-	-	-	-	-	42
4	Depth	1	89	93	98	-	83	47	-	-	75
	Depth	2	93	95	98	-	84	49	-	-	80
5	Depth	1	88	98	99	23	64	74	19	46	93
	Depth	2	96	98	99	37	71	80	26	51	97
6	Depth	1	95	99	99	46	90	93	25	75	97
	Depth	2	96	99	99	65	95	97	53	79	97
7	Depth	1	96	99	99	57	94	98	26	91	99
	Depth	2	96	99	99	77	95	98	55	97	99
8	Depth	1	98	100	99	76	98	99	36	97	99
	Depth	2	98	100	99	92	98	99	64	98	99
9	Depth	1	98	100	99	92	98	100	67	99	100
	Depth	2	98	100	99	95	98	100	87	99	100
10	Depth	1	98	99	100	93	98	100	81	98	100
	Depth	2	98	99	100	93	98	100	84	98	100

Table 2.12. Correction accuracies of misspelt words in an ensemble of *surnames*, for various word lengths (rows) and combinations of errors (columns). The numbers are percentages (rounded); the values between 99 and 100 have been truncated without rounding. The percentages that might be rounded to 100 have been indicated by the superscript 1. Rows labeled Depth 1 correspond to results in which the correct word had the smallest FD number among the candidates; rows labeled Depth 2 show results in which the correct word had the smallest or the next to the smallest FD number among the candidates

Word length		Deletions	1	0	0	1	1	0	2	0	0
		Replacements	0	1	0	1	0	1	0	2	0
		Insertions	0	0	1	0	1	1	0	0	2
3	Depth 1		-	-	95	-	-	-	-	-	36
	Depth 2		-	-	95	-	-	-	-	-	36
4	Depth 1		98	96	99	-	83	50	-	-	80
	Depth 2		99	98	99	-	84	50	-	-	83
5	Depth 1		95	99	99^1	38	70	75	33	51	95
	Depth 2		97	99	99^1	45	72	80	35	54	98
6	Depth 1		99	99^1	100	63	95	92	52	78	98
	Depth 2		99	99^1	100	73	98	98	67	81	99
7	Depth 1		99^1	99^1	100	69	98	98	50	93	99^1
	Depth 2		99^1	99^1	100	85	99	99	76	98	99^1
8	Depth 1		99^1	100	100	83	99	99_1	55	98	100
	Depth 2		99^1	100	100	96	99	99^1	99	99	100
9	Depth 1		99^1	100	100	96	99^1	100	79	99	100
	Depth 2		99^1	100	100	98	99^1	100	96	99	100
10	Depth 1		100	100	100	99	100	99^1	94	99^1	100
	Depth 2		100	100	100	99	100	99^1	97	99^1	100

Research Related to Proximity Searching. HARRISON [2.99] and GOBLE [2.100] have introduced ideas on how to locate character strings from continuous text on the basis of groups of characters; the hash coding principle thereby used is related to that applied in the above proximity searching implementation. Other works on the matching and correction of words or strings have been published by AHO and CORASICK [2.101], ALBERGA [2.102], DAMERAU [2.103], DOSTER [2.104], LEVENSHTEIN [2.105], MORGAN [2.106], OKUDA et al. [2.107], RISEMAN and HANSON [2.108], as well as ULLMAN [2.109].

2.8 The TRIE Memory

Another content-addressable scheme which is different from hash coding has been selected for presentation in this section. This scheme, dubbed TRIE, was introduced by FREDKIN in 1960 [2.110] and its basic idea is similar to that of *search trees*. A feature which distinguishes the TRIE from hash coding is that the problem of collisions does not exist with it. Comprehensive discussions of searching trees, which form another major field in computerized searching techniques apart from hash coding, can be found in numerous textbooks on data management, searching, and sorting (to mention only [2.1-3], and they fall outside the scope of this book. The name TRIE presumably comes from the word "retrieval", but in order to distinguish it from trees, it is pronounced in the same way as "try".

The basic principle of a TRIE is that the keywords, being strings of characters of variable length, are described by a tree structure in which the characters form the nodes. The number of possible branchings from a node must be at least the number of letters in the alphabet, and one extra branch is needed to link the associated entry to the keyword. With 26 letters in the alphabet, the number of pointers at every node is thus 27. The 26 pointers, however, are used *not* to indicate directly what is the next letter to a node, but information is organized in a more subtle way, explained below with the aid of an example. In accordance with the original article of FREDKIN, the example is presented with a five-letter alphabet consisting of the letters A, B, C, D, and E only. While most of the emphasis in [2.110] was put to devising a method to monitor the occurrence of a keyword only, we shall modify, and also somewhat simplify, the discussion to better comply with the presentation of hash coding. The following discussion refers to Fig. 2.22.

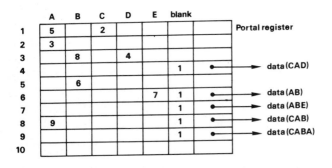

Fig. 2.22. An example illustrating the TRIE (see text)

The TRIE system consists of a number of registers, each one capable of holding 27 pointers. The value of every pointer is 0 before the pointer is set during storage operations. The number of registers depends on the number of entries to be stored. The first of the registers, indexed by 1, is the *portal register* which corresponds to the first letter of every keyword. Assume that the word 'CAD' has to be stored. The pointer in the portal register corresponding to 'C' is set to point at register 2, which happens to be the first empty one. Into its 'A' position, a pointer to register 3 is set. Assume now that the end of a keyword is always indicated by typing a blank. If the next character position is always studied when scanning the letters, and if a blank is found, the letter in question is known to be the last one. Therefore we write a pointer 4 (next empty location) into the 'D' position of register 3, but in register 4, the pointer in the 'blank' field contains the terminal marker, in FREDKIN's scheme pointer 1. The seventh field in every register may be used to hold a pointer to an associated entry in a separate document area. This completes the writing of the first entry.

The storage of further items in the TRIE now divides into two different procedures depending on whether the new keyword has a beginning which has occurred earlier or not. Consider first the latter case, and let the new keyword to be stored be 'ABE'. The fact that 'A' has not occurred before can be deduced from the 'A' position in the portal register, which up to this point was 0. This pointer is now set to point at the next empty register which is 5. The pointer in the 'B' position of register 5 is set to 6, and the 'E' position in register 6 is set to 7. Again, a pointer to the portal register and to the document area, corresponding to the entry associated with the keyword 'ABE', are set in register 7. All operations needed to store the second item have now been carried out.

Let the next case to be considered have the first two letters in common with an earlier keyword, e.g., let the new keyword be 'CAB'. An inspection of the portal register shows that 'C' has occurred before, and following the pointer found at 'C', register 2 is examined next. As 'A' is also found, together with a pointer to register 3, then register 3 is subjected to examination. The letter 'B', however, was not found and therefore, the 'B' location can be used to store a pointer to a new empty register which is 8. The pointer in the 'blank' location of register 8 is set to 1, and a pointer to the entry associated with 'CAB' is simultaneously set.

The measures for the storage of an entry with a keyword which totally contains an old keyword and has extra letters in addition, as well as the case when the new keyword is identical with the front part of an old keyword,

are completely deducible from the foregoing. It is left to the reader to
verify that the pointers corresponding to the keywords 'CABA' and 'AB' are
those written in Fig. 2.22.

The *retrieval* of information from the TRIE must be discussed separately
in the case that a keyword identical with the search argument has occurred
and been encoded in the above way, and when the search argument is completely
new. Consider, for example, searching on the basis of 'ABE'; when the pointers
are followed in the same way as during storage, and since 'E' is known to
be the last letter, then we uniquely end up at the 'E' position in register 5.
We find the end marker 1 there, and the correct pointer to the document area
is found from register 5, too.

Consider now the case that the search argument is 'CA'. When the TRIE is
traced by following the pointers, we end up in register 2 which, however, does
not contain the terminal marker 1. Accordingly the search can be found unsuccessful. The reader may construct other examples with arbitrary new keywords
to convince himself that the search is always found unsuccessful.

Comparison of the TRIE with Hash Coding. Although the TRIE method was discussed above in its simplest form and further developments suggested by
FREDKIN are not considered, one may already be able to draw some general
conclusions from the basic scheme. As mentioned in the beginning of this
section, no collisions occur with this method. In return, there is an appreciable number of comparison operations to be performed which, with general
purpose computers, makes the operation somewhat slower when compared with
hash coding. The worst handicap is that the degree of utilization of memory
in the TRIE method is not particularly high. With small TRIEs the registers
seem to be sparsely occupied. FREDKIN claims that the efficiency will exceed
50 percent in large memories if a special multidimensional organization suggested by him is used. We have to realize, however, that every letter needs
a separate pointer field; the compression scheme proposed by MALY [2.111],
however, brings about some improvement in efficiency.

There have been some attempts to develop TRIEs, to apply them to hash
coding, and to combine hash-coded data structures with TRIEs. Let it suffice
to mention the following articles: BURKHARD [2.112], COFFMAN and EVE [2.113],
MALY [2.111], and SEVERANCE [2.114]. Recently, MALLACH [2.115] has made an
attempt to unify the concepts of hash coding within an abstract framework
based on binary trees.

2.9 Survey of Literature on Hash Coding and Related Topics

Historical. It seems that the possibilities for hash coding were realized soon after the introduction of the first commercial digital computers. In 1953, the basic idea appeared in IBM internal reports: Luhn was the first to suggest a hashing method which resolved the collisions using buckets and chaining. An algorithmic method for the definition of reserve locations is due to Lin. The idea of linear probing was introduced by Amdahl in 1954 during a project in which an assembler was developed for the IBM 701 computer. Other members of the team were Boehme, Rochester, and Samuel. Independent work was done in Russia where ERSHOV [2.68] published the open addressing, linear probing method in 1957.

A few extensive review articles can be found from the early years. DUMEY [2.116] in 1956, while discussing indexing methods, also introduced the division method of hashing. An extensive article, describing the application of hash coding in large files occurring in practice, was published by PETERSON in 1957 [2.60]. This work contained many extensive analyses of the effect of bucket size on the number of searches. Hashing functions were studied by BUCHHOLZ in 1963 [2.61].

Books. The principles of hash coding can be found in some books written either on systems programming or on data management. The monograph on compiling techniques by HOPGOOD [2.52] contains a short section on the principles of hash coding. There is a somewhat longer review which also contains many statistical analyses in the book of KNUTH [2.1]. In the book of MARTIN [2.2], topics of hash coding have also been discussed.

Review Articles. In addition to those writings mentioned above, there are a few review articles that can be mentioned as a reference. One of them was written in 1968 by MORRIS [2.22], and besides an introduction to the basic principles, it also contained answers to many detailed questions concerning the relative merits of different solutions.

More recent review articles have been written by SEVERANCE [2.114], SEVE- RANCE and DUHNE [2.55], MAURER and LEWIS [2.117], as well as SORENSON et al. [2.118]. A thorough comparison of the hashing functions can be found in the articles of LUM et al. [2.29] and KNOTT [2.119].

Developments in Hash-Coding Methods. A few recent studies of the hashing methods may be mentioned. The influence of the statistical properties of keywords on the hashing method has been considered by BOOKSTEIN [2.120], DOSTER [2.121], and SAMSON and DAVIS [2.122]. An analysis routine for search

methods has been designed by SEVERANCE and CARLIS [2.123]; optimal or improved table arrangements have been discussed by YUBA and HOSHI [2.124],
RIVEST [2.125], and LITWIN [2.126,127]; the qualities of some hashing algorithms have been considered by AJTAI et al. [2.128], GUIBAS [2.129], GUIBAS
and SZEMEREDI [2.130], LYON [2.131], KRAMLI and PERGEL [2.132], FORTUNE and
HOPCROFT [2.133], as well as THARP [2.134]. A special *dynamic hashing* method
has been developed by LARSON [2.135]. So called *extendible hashing* is suggested by FAGIN et al. [2.136] . The partial-match problem has been handled using
projection functions by BURKHARD [2.137]. In multikey searching, the statistical principal component analysis has been applied by LEE et al. [2.138].
Various applications are described by HILL [2.139], GRISS [2.140], IMAI et al.
[2.141], WIPKE et al. [2.142], HODES and FELDMAN [2.143], LEWIS [2.144], and
NACHMENS and BERILD [2.145]. Differences between files have been discussed by
HECKEL [2.146], and concurrency in access by GRONER and GOEL [2.147].

Special Hardware for Hash Coding. The worst handicap of usual hash coding
methods, namely, the sequential access to memory can to a great extent be
relieved by special hardware. It is possible to control a multi-bank memory
by a set of hash address generators and some auxiliary logic circuits thereby
facilitating parallel access to many locations. Such systems have been proposed by GOTO et al. [2.148] as well as IDA and GOTO [2.149]. Fast hashing
operations for this hardware are described by FABRY [2.150].

Chapter 3   Logic Principles of Content-Addressable Memories

This chapter contains a survey of various operations, algorithms, and system organizations by which content addressability can be embodied in memory hardware. It includes logic descriptions of searching operations, their circuit implementations, and organizational solutions for content-addressable memory (CAM) arrays that are utilized in searching operations. It may be useful to remember that these principles and solutions aim at the implementation of the following three partial functions: 1) There must exist a rapid method for broadcasting a search argument to all memory locations. 2) An individual comparison of the search argument with every stored keyword or a set of attributes must be performed. 3) All matching data and other associated information must be recollected and read out in succession.

3.1   Present-Day Needs for Hardware CAMs

The motives which originally evoked the interest in associative or content-addressable memories were mentioned in Sect. 1.1.1. This section attempts to specify more closely those application areas for which the hardware CAMs in their basic forms have been generally adopted.

The computational power achieved in the present-day general-purpose computers may primarily be attributed to the high switching speed and small signal delays of the semiconductor microcircuits, whereas not so much has happened in the development of their basic architectures. It is rather clear that the inherent speed of elementary operations cannot be increased very much beyond the present one, even though some radically new physical principles (cf Sect. 4.1.3) were applied. While it might seem that the speed of operation is enough for most purposes and there still exists unutilized computational potential, nonetheless there exist problems of mathematical physics, meteorology, etc., for which more powerful equipment is needed. The conventional or von Neumann-type computer suffers from a serious fundamental handicap which stems from the principle of addressing its memories: 1) The

operands must be read and written serially, i.e., one at a time, which con-
sumes a great deal of time; many large problems could better be handled in
parallel, i.e., simultaneously over a set of variables. 2) With increasing
memory capacity, the length of address code increases, with a result that
more space must be reserved in each memory location which contains an address reference. For this reason, instead of addressing the whole memory
system directly, it is more customary to divide the memory space of large
computers into several smaller banks which are accessed individually, thereby using indexing of the banks in machine instructions, as well as relative
and indirect addressing for the reduction of the address range handled within a program. Then, however, for the determination of the absolute addresses,
several auxiliary arithmetic operations must be performed with each memory
reference.

In order to operate on many memory locations simultaneously, and to simplify the searching of the operands, it would be very desirable to base the computations on *content-addressable memories*, and during more than 20 years,
many attempts have been made for their development. Content-addressable memories would be especially advantageous from the point of view of high-level
programming languages which refer to their operands and procedures by symbolic names and other identifiers. The computer architectures would probably
be quite different from those currently applied if highly parallel computations, distributed all over the data files were simply and directly implementable by hardware. Unfortunately, large content-addressable memories have not
been realized; when it comes to direct access to data on the basis of their
symbolic representation, i.e., by names or alphanumeric identifiers, it seems
that software methods as those discussed in Chap.2 have mostly been resorted
to. No parallel processing, however, is involved in software content addressing.

Nonetheless, hardware content-addressable memories have, in fact, been
used as special parts in computing systems, in certain organizational solutions whereby the CAMs can effectively perform fast *buffering, checking,* and
bookkeeping operations needed in the moving and rearrangement of data. In this
way, although the memory system as a whole is not content addressable, the
CAM devices can effectively contribute to the speed of usual operations by
making the operands more directly available to the processing circuits. Two
such applications, the *virtual memory* and the *dynamic memory allocation* will
be discussed later in Sects. 5.1,2, respectively.

One has to realize further that a substantial part of automatic computations, and the majority of all installed computing devices are concerned with
pure *searching* and *sorting* tasks which in fact comprise the main bulk in ad-

ministrative data processing. With increasing sizes of files, the times
needed to scan the records sequentially would become intolerably long, and,
therefore, all kinds of organizational solutions for data management have
been suggested and developed to alleviate this deficiency. Hash coding is
one of them but as already mentioned earlier, it does not solve many problems
connected with numerical variables. The ways in which data are specified in
a content-addressable search, namely, differ from one application to another:
for instance, in symbolic computations it is only necessary to locate all
entries which *exactly match* a given search argument, for instance, an identifier in its special portions. In the retrieval of numerical data from large
files, however, there frequently occurs a need to locate all entries which
satisfy specified *magnitude relations*, for instance, having one or several
of their attributes above, below, or between specified limits, or absolutely
greatest or smallest. This chapter contains many examples of magnitude comparison algorithms which could effectively be solved by hardware CAM systems.
Much effort has thus been devoted to the development of special CAM devices
with an extremely high storage capacity. However, although progress in this
latter respect has not been particularly fast, even smaller-size CAMs can
be helpful in the following way: although the available content-addressable
memory capacity as such would be too small to hold all files, it can be used
in a *buffering mode* for the execution of searches by large chunks of data,
whereby the processing is parallel over the chunk. In this way, although
many batches of data must be processed sequentially, the overall searching
time is shortened by several orders of magnitude. This application is discussed in Sect. 5.3.

A further area of application of the CAMs and their derivatives is in the
programming of logic operations in the central control circuits of computers
and also in other equipment where many Boolean conditions have to be evaluated,
or table look-ups be performed. It seems that a special form of content-addressable memory, the Functional Memory as discussed in Sect. 3.7 might be
applicable. The general problematics of programmable logic are discussed in
Sect. 5.4.

The CAMs have already had a significant thrust on certain parallel computer architectures. There exist a few types of *content-addressable parallel
processors* which have been developed for certain special applications, e.g.,
air traffic control, numerical transformation, and digital filtering in which
bulk computations on similar pieces of data have to be carried out. A number
of them will be discussed in Chap. 6.

3.2 The Logic of Comparison Operations

The comparison operations of binary words can be described in a general way, using Boolean algebra for the definition of the logic functions thereby applied. These operations are then implementable by electronic, magnetic, optic, etc. devices; sometimes, however, a physically simpler solution may replace a standard implementation of a logic operation (cf, for instance, the special implementations of the OR function with a great number of input variables, as described in this chapter).

Equality Comparison. The basic operation in hardware content-addressable memory (CAM) structures is *bit match*. If the Boolean values of two binary variables are denoted by x and y, respectively, then the Boolean function which has the value 1 (*true*) for *logical equivalence* (match) of x and y, and 0 (*false*) otherwise is

$$(x \equiv y) = (x \wedge y) \vee (\overline{x} \wedge \overline{y}) \ . \tag{3.1}$$

Alternatively, the *mismatch* of the values x and y is indicated by the EXCLUSIVE OR-function,

$$x \oplus y = (x \wedge \overline{y}) \vee (\overline{x} \wedge y) \ . \tag{3.2}$$

In *masked search,* only a subset of the bits of a search-argument word is compared with respective bits of all memory words. (Masking in the CAM usually means masking out, or disablement of certain bits.) Those stored words which agree in the specified (unmasked) bit positions with the search argument are read out. For that purpose, each bit position in the memory is equipped with circuit logic which implements either (3.1) or (3.2), together with an indication whether this bit is involved in the comparison or not. If the j:th memory word is $S_j = (s_{jn}, s_{j,n-1}, \ldots s_{j0})$, with the s_{ji} the Boolean values of its respective bits, and the Boolean search argument is $A = (a_n, a_{n-1}, \ldots, a_0)$, and if further the *mask word* $C = (c_n, c_{n-1}, \ldots, c_0)$ has the Boolean value 1 in all masked (disabled) bits and 0 otherwise, the masked match of respective bits is indicated by the function

$$m_{ji} = (a_i \equiv s_{ji}) \vee c_i \ . \tag{3.3}$$

Note that when the j:th bit is masked, then m_{ji} is identically 1 because $c_i = 1$. The masked mismatch in the i:th bit position is indicated by the

Boolean function

$$\overline{m}_{ji} = (a_i \oplus s_{ji}) \wedge \overline{c}_i \quad . \tag{3.4}$$

The matching of memory word S_j with search argument A in all unmasked positions is defined by requiring a match condition at all bits,

$$m_j = \bigwedge_{i=0}^{n} m_{ji} \quad , \tag{3.5}$$

or, alternatively, a mismatch in at least one bit position indicates the inequality of S_j and A:

$$\overline{m}_j = \bigvee_{i=0}^{n} \overline{m}_{ji} \quad . \tag{3.6}$$

The equality of S_j and A may also be computed recursively, as would be done in the word-parallel, bit-serial search (Sect. 3.4): let

$$e_{ji} = e_{j,i-1} \wedge m_{ji} \quad , \quad i = 1,2,\ldots,n, \; e_{j0} = 1 \; ;$$

$$\text{then } m_j = e_{jn} \quad . \tag{3.7}$$

Alternatively, the following expressions may be used:

$$\overline{e}_{ji} = \overline{e}_{j,i-1} \vee \overline{m}_{ji} \quad , \quad i = 1,2,\ldots,n, \; \overline{e}_{j0} = 0 \; ; \quad \overline{m}_j = \overline{e}_{jn} \quad . \tag{3.8}$$

Magnitude Comparison. Another important operation in content-addressable search is the identification of every memory word the numerical value of which is greater (or less) than that of a given search argument. (The between-limits comparisons, as well as the searches for extremal values, can be executed by combined operations in a memory system, as discussed later on.) For simplicity, it is assumed in the following that all binary words represent nonnegative integers in the binary number system.

Although the CAMs in general have circuit logic to perform the equality comparisons in parallel, magnitude comparison is more often implemented com-

putationally in several search operations by bits, with intermediate results
stored and utilized in later steps. We shall later introduce the circuit principles for this. Let the recursions here formally be defined in terms of sequences of mathematical operations.

One algorithm for magnitude comparison of binary numbers starts at the
most significant (left) end of them. In addition to the search argument $A = (a_n, a_{n-1}, \ldots, a_0)$ and the memory word $S_j = (s_{jn}, s_{j,n-1}, \ldots, s_{j0})$, two
sequences of values for auxiliary variables are defined: $G_j = (g_{j,n+1}, g_{jn},$
$\ldots, g_{j0})$ and $L_j = (\ell_{j,n+1}, \ell_{jn}, \ldots, \ell_{j0})$ (the letters are mnemonics for
greater than and *less than*, respectively). It may be pointed out that if the
comparisons are carried out serially (cf Sect. 3.4), then the sets G_j and
S_j are represented by two machine variables g_{ji} and ℓ_{ji}, respectively, which
are updated during the process of recursion. Consider the following recursive,
simultaneous expressions:

$$g_{ji} = g_{j,i+1} \vee (\overline{a}_i \wedge s_{ji} \wedge \overline{\ell}_{j,i+1}) \; ,$$

$$\ell_{ji} = \ell_{j,i+1} \vee (a_i \wedge \overline{s}_{ji} \wedge \overline{g}_{j,i+1}) \; , \qquad\qquad (3.9)$$

with the initial values $g_{j,n+1} = \ell_{j,n+1} = 0$. If now $s_{jn} = 1$ and $a_n = 0$, then
it is obvious that S_j is greater than A, and it can be computed from (3.9)
that for all $i < n$, $g_{ji} = 1$, $\ell_{ji} = 0$. If, on the other hand, $s_{jn} = 0$ and
$a_n = 1$, then A was greater, and for all $i < n$, $g_{ji} = 0$ and $\ell_{ji} = 1$. But if
$s_{jn} = a_n$, then $g_{jn} = \ell_{jn} = 0$, and the magnitude relations must be resolved
on the basis of less significant bits. In the latter case, the same deductions
are carried out on the bits $s_{j,n-1}$ and a_{n-1}, and so on, proceeding until the
rightmost position which yields the final output signals g_{j0} and ℓ_{j0}. If now
$g_{j0} = 1$, $\ell_{j0} = 0$, then S_j is greater than A; if $g_{j0} = 0$, $\ell_{j0} = 1$, then A is
greater. If $g_{j0} = \ell_{j0} = 0$, the numbers are equal. The combination $g_{j0} = \ell_{j0} = 1$
cannot occur.

Magnitude comparison is implementable by another algorithm, too, in which
the Boolean expressions are slightly more complex than before. A significant
advantage attainable by this method, however, is that since the recursive
comparison is now started at the least significant (right) end, the result
is known sooner and the process can be stopped in case the numbers contain
leading zeroes.

Using the same notation as in the previous method, but with new meanings
given to the recursive functions g_{ji} and ℓ_{ji}, the following expressions

are defined:

$$g_{j,i+1} = \left(\overline{a}_i \wedge s_{ji}\right) \vee \left(\overline{a_i \wedge \overline{s}_{ji}} \wedge g_{ji}\right)$$

$$\ell_{j,i+1} = \left(a_i \wedge \overline{s}_{ji}\right) \vee \left(\overline{\overline{a}_i \wedge s_{ji}} \wedge \ell_{ji}\right) \tag{3.10}$$

with $g_{j0} = \ell_{j0} = 0$. As before, when all digits have been exhausted, the last values of g_{ji} and ℓ_{ji} indicate the magnitude relation. This result may be justified in the following way. If the bits a_i and s_{ji} agree, then $g_{j,i+1} = g_{ji}$ and $\ell_{j,i+1} = \ell_{ji}$. However, if they disagree, $g_{j,i+1}$ becomes 1 if $s_{ji} = 1$ and $a_i = 0$, but zero if $s_{ji} = 0$ and $a_i = 1$. Similarly $\ell_{j,i+1}$ becomes 1 if $s_{ji} = 0$ and $a_j = 1$, but zero if $s_{ji} = 1$ and $a_i = 0$. After recursion, S_j can be found greater than A if the last values are $g_{j,n+1} = 1$, $\ell_{j,n+1} = 0$ and less than A if $g_{j,n+1} = 0$, $\ell_{j,n+1} = 1$. Equality of S_j and A is due if $g_{j,n+1} = \ell_{j,n+1} = 0$.

A paradigm of logic circuits, the so-called *iterative circuit* may be used to implement (3.9) by hardware; its block scheme is shown in Fig. 3.1. The bit values to be compared together with their inversions are available as the contents of two registers, and the rectangular blocks situated between the respective bit positions contain identical logic circuits, defined by (3.9). Results from the recursive operation are then represented by signals propagated to the right, and the outputs of the last stage indicate the final result of comparison.

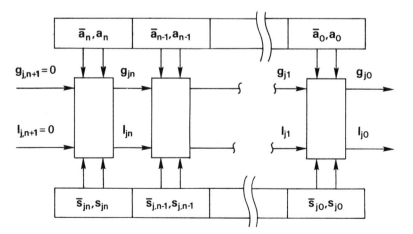

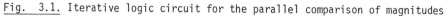

Fig. 3.1. Iterative logic circuit for the parallel comparison of magnitudes

Magnitude Comparison by Arithmetic Operation. While (3.10) are readily implementable by special logic circuits and microprogram control (cf, e.g.,
Subsect. 6.5.1), a more straightforward, although possibly somewhat more
expensive solution is to perform magnitude comparison by arithmetic subtraction. A *serial subtractor circuit* for positive binary numbers is shown
in Fig. 3.2; the operands A and S_j to be compared are held in two shift
registers in which they are synchronously shifted end-around to the right.
The *full subtractor circuit* (FS) generates the difference and borrow bits
d_i and b_{i+1}, respectively; the borrow flip-flop buffers the previous borrow
bit b_i.

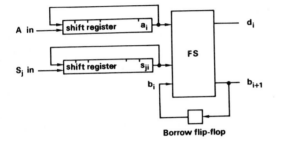

Fig. 3.2. Serial subtractor circuit

The FS circuit is a combination logic circuit with the following truth tables
(Table 3.1).

Table 3.1. Truth tables of the full subtractor

Difference bit (d_i)			Borrow bit (b_{i+1})		
		b_i			b_i
a_i	s_{ji}	0 1	a_i	s_{ji}	0 1
0	0	0 1	0	0	0 1
0	1	1 0	0	1	1 1
1	0	1 0	1	0	0 0
1	1	0 1	1	1	0 1

The result of comparison will be indicated by the last borrow bit b_n; if it is 1, then S_j was greater than A.

A typical implementation of Table 3.1 requires 7 logic NAND gates; this should be contrasted with the two AND plus two NAND gates needed for equations (3.9).

3.3 The All-Parallel CAM

The memory structures presented in this section are direct implementations of equations (3.3,5), or (3.4,6) by hardware circuit logic. When compared with some other CAM structures (cf Sects. 3.4,5), the number of logic gates per bit cell in a commercially available electronic all-parallel CAM is found to be appreciable. Some simpler principles have been suggested, but not yet demonstrated to be practicable. The direct logic-circuit implementations are extremely fast, with access times only a few nanoseconds, and the all-parallel CAMs are, therefore, typically used in fast but small *buffer memories*, as well as in *programmed logic circuits* accociated with central control operations of computers and their peripheral devices. On the other hand, the price per bit cell is much higher than for usual (addressed) memories, and when it comes to applications requiring complex searching or processing operations, the solutions presented in Sect. 3.4 are more preferable, at least for the time being.

3.3.1 Circuit Logic of a CAM Bit Cell

The comparison of the search argument with all stored words is in principle based on either the equivalence function (3.1), or the mismatch (3.2). In practice, it is most economical to provide every bit of the search argument with its logical negation, and these values are then broadcast in parallel along pairs of lines to respective bit positions of all storage locations. The double lines can also be controlled in a way which directly implements the masking function in (3.3) or (3.4), as described below. The results of bit comparisons in a word have to be collected by an AND or NOR circuit with very high fan-in (number of inputs). Implementation of this function by a logic gate is not a practicable solution. One way to circumvent this problem would be to compute the NOR function recursively as shown in Fig. 3.3a, whereby an arbitrary number of bit cells can be chained. A drawback to this solution is a delay in the cascaded stages which drastically increases the access

134

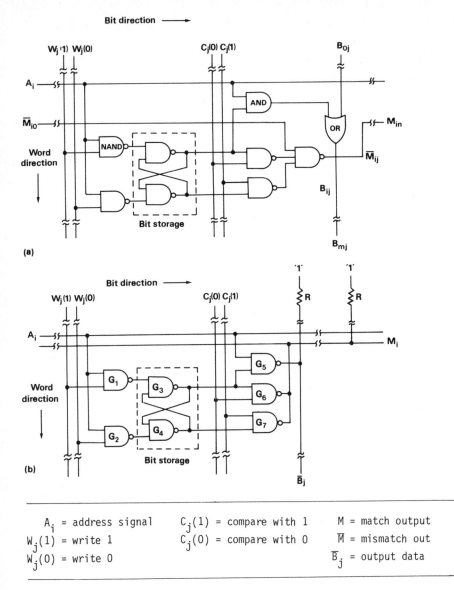

(a)

(b)

A_i = address signal $C_j(1)$ = compare with 1   M = match output
$W_j(1)$ = write 1 $C_j(0)$ = compare with 0 $\overline{M}$ = mismatch out
$W_j(0)$ = write 0 $\overline{B}_j$ = output data

<u>Fig. 3.3.</u> (a) Bit cell of an all-parallel CAM, with iterative circuit imple-
mentation of word comparison (mismatch, $\overline{M}_{ij}$) and addressed readout (B_{ij}).

(b) Bit cell of an all-parallel CAM, with Wired-AND implementation of word
comparison (match, M_i) and addressed readout ($\overline{B}_j$)

time. A much better alternative, whereby high speed and also rather good electrical characteristics can be achieved, is the so-called *Wired-AND* function wich means that the output circuit of the bit logic is equivalent to a switch connected between the word output line and the ground potential. If, and only if, all the parallel switches are open, the word line potential is able to rise high, whereas a single closed switch is able to clamp the word line to the ground potential.

Figure 3.3b shows the complete circuit of a standard commercial CAM bit cell. Besides the comparison logic, every bit cell contains provisions for addressed reading and writing of data.

All of the logic gates shown in Fig. 3.3b are standard NAND gates except those which are connected to the M_i and $\overline{B}_j$ lines which in addition have to perform Wired-AND logic. For this purpose they must have a suitable output circuit, e.g., the *open-collector circuit* if bipolar transistors are used.

Writing. Gates G_3 and G_4 form the bit-storage flip-flop which is set and reset by the signals $W_j(1)$ and $W_j(0)$, through gates G_1 and G_2, respectively, under the condition that the address line signal A_i is high. The output of G_3 represents the stored bit value, and it will be set to 1 if $W_j(0) = 0$, $W_j(1) = 1$, and $A_i = 1$. The value 0 is set at the signal values $W_j(0) = 1$, $W_j(1) = 0$, and $A_i = 1$. If in a writing operation a bit value of the addressed word shall be left unaltered, the writing can be *masked out* by holding $W_j(1) = W_j(0) = 0$.

Reading. For *addressed reading* of the bit storage it is only necessary to hold the A_i line high, whereby G_5 directly mediates the inverted bit value to the $\overline{B}_j$ line, which is common to this bit in all words. *Content-addressable reading,* taking into account *masking,* is done using the lines $C_j(0)$ and $C_j(1)$ which pass all words at the same bit position. Then it is set $C_j(0) = 0$, $C_j(1) = 1$. If now the stored bit was 1, the output of neither G_6 nor G_7 is low (both output switches are open) with the result that these gates, when connected to the M_i line, allow it to rise high. A similar condition is expected if the search argument has the value 0, whereby $C_j(0) = 1$, $C_j(1) = 0$, and if the stored bit is 0. On the other hand, if the search argument bit disagrees with the stored bit, then the reader may convince himself that either G_6 or G_7 will switch down. This is sufficient to keep the potential of the M_i line low, and it is seen that a single *mismatch* in one of the bit positions is able to give the value 0 to the M_i signal. When this bit position is *masked* in the comparison operation, it will be sufficient to hold

$C_j(0) = C_j(1) = 0$, whereby G_6 and G_7 will represent an open switch, irrespective of the stored bit value.

3.3.2 Handling of Responses from the CAM Array

Consider a rectangular m-word, n-bit CAM array built of bit cells of the type shown in Fig. 3.3b. The basic searching task is to locate all stored words in the array which match with the search argument in its unmasked portions, whereby the $C_j(0)$ and $C_j(1)$ lines are controlled in the way which defines the masked search argument. A normal situation is that several M_i lines yield a response. The contents of these words must then be read out in full, one at a time. It is for this purpose that the CAM has a provision for addressed reading. For the sequential readout of all responding words, however, some problems discussed below must be solved. First of all there must exist some sort of *queue serving* organization which is able to pick up only one response at a time, usually in the top-down order, and to subsequently reactivate the corresponding word location, this time using the address line. The selected word then appears at the $\overline{B}_j$ lines, albeit with every bit complemented. The addressed readout operation will be repeated for all responding words.

There exist two principally different strategies for queue serving. For both of them it is necessary to provide every M_i line of the CAM array with a *buffer flip-flop*. Another necessary circuit is some kind of *priority resolver* which shall single out one of the multiple responses. The flip-flops together form the *response store*, and the latter combined with the priority resolver is named *multiple-response resolver*.

In the first queue serving strategy, the parallel equality comparison in the CAM array is performed only once. The priority resolver displays the uppermost response at the respective output line, whereby the corresponding word can be read out. After that, the uppermost active buffer flip-flop is reset whereby the response which is next in order appears uppermost in the response store and can be served, and so on. In the second strategy, all flip-flops except the uppermost "responding" one are reset to leave only one response, whereby information about the lower responses is lost. The corresponding word can then be read out immediately. This location in the memory must now be marked in some way to show that it has been processed, for instance, using a "validity bit" in the word reserved for this purpose. To continue, new content-addressable reading operations must be performed to find the rest of the matching words which are yet unprocessed.

Cascaded Priority Resolvers. A simple iterative network which displays only the uppermost '1' of its inputs at the respective output line is the "chain bit" circuit [3.1,2]. Its idea is to let every active input inhibit all lower outputs, whereby only the uppermost active input will be propagated to the output. This circuit can be built in the response store in the way shown in Fig. 3.4a. Notice that clocked buffer flip-flops must be used in this solution because resetting of the uppermost response would otherwise immediately induce resetting of the next flip-flop, etc.

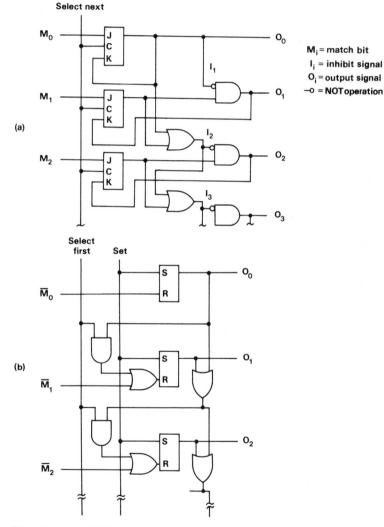

Fig. 3.4a,b. Multiple-response resolver: a) with nondestructive "select first/next" control with JK flip-flops, b) with destructive "select first" control

In view of the fact that plenty of buffer flip-flops are usually needed, the solution shown in Fig. 3.4b, which implements the second strategy, might look more attractive because simple bistable circuits can be used whereby the total costs are reduced. Then one can permit the repeated content-addressable readout operations necessary in the second strategy, especially since they are executable almost as fast as the resetting operations in the response store. Notice that the iterative OR network at the output produces only an output which is called *inhibit vector*: all of its output signals below the uppermost response are 1 and they are used for resetting the flip-flops.

It should be realized that the sequencing control is different in these two cases. In Fig. 3.4a it is very simple: the first response is automatically displayed at the outputs, and the "Select next" control is used only to reset the uppermost "responding" flip-flop. The "Select first" control in Fig. 3.4b, on the other hand, is applied after content-addressable reading, to single out the uppermost response and to reset all the lower flip-flops. The subsequent marking of the validity bit, resetting of the response store, and repetition of content-addressable reading operations need some central control not shown in the diagram.

In both of the above methods, the last stage in the cascade produces an extra signal which indicates that there is at least one response. This information is necessary in some extremal-searching algorithms (cf Sect. 3.4.5).

A drawback of the cascaded networks is that with a large number of storage locations it takes an appreciable time for the iterative logic to settle down to its final state, because of the propagation of signals through all the gates. For example, if the delay per stage were 10 ns, it would initially take 10 μs to search through a 1000-word store although this time decreases with lower bits. For this reason, a circuit, the upper part of which is shown in Fig. 3.5, was suggested by FOSTER [3.2] for the generation of the inhibit vector to speed up priority resolution. This circuit must be used in connection with the principle of Fig. 3.4b. It uses the idea of a binary tree, with the root on the right; for more details, the reader should consult the original article. It has been shown by FOSTER that if there are 2^n words in the memory, the worst-case delay is only $2n-1$ gate delays. On the other hand, the number of gates in the circuit is only $3 \cdot 2^{n-1} - 1$, or approximately 1 1/2 gates per word. In a one-million word memory with 10 ns gate delay, the worst-case total delay would be only 400 ns.

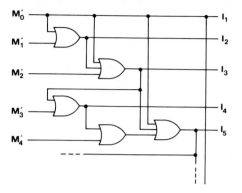

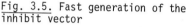

Fig. 3.5. Fast generation of the inhibit vector

Modular Tree Organization for the Priority Resolver. The structure of the circuit of Fig. 3.5 is irregular and thus not very suitable for fabrication by the LSI technology. Another method which conveys a similar idea but is implementable by standard modules is due to ANDERSON [3.3].

Consider the block depicted in Fig. 3.6a. Without the I_A ("inhibit all") signal it could, in fact, be a simple generator of the inhibit vector, and being a one-level circuit, it is yet faster than the construct shown in Fig. 3.5. Its drawback is that the number of inputs to the OR circuits increases linearly with the number of words, which makes this solution impractical for large memories. For CAMs with a small number of locations this might be applicable as such, however (cf, e.g., the buffer memory organization discussed in Sect. 5.1).

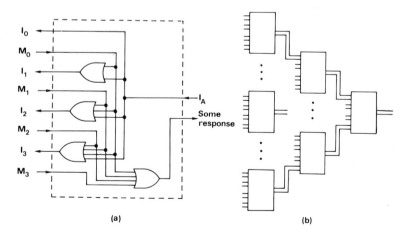

Fig. 3.6a,b. Modular-tree implementation of the inhibit vector: a) one module, b) modular tree

In a hierarchical organization of blocks, the I_A ("inhibit all") signal
is used to generate the value 1 at all outputs I_i of the block, i.e., to
"inhibit" the block. (Inhibition actually means propagation of a resetting
signal to the buffer flip-flops through the network.) In principle, it would
then be possible to apply the "chain bit" method in a cascade of blocks, in
analogy with Fig. 3.4b, to implement a priority resolver with a higher number of inputs. However, a more efficient solution with the minimum of gate
delays is obtained in another way. Assume that a tree structure, delineated
in Fig. 3.6b, is built of these blocks. All output ports of the blocks, let
alone the *"Some response"* signal, are drawn to the left. When the "Some response" signals from the lower levels converge to blocks of higher levels,
an active input at any block will induce the value 1 at all the lower I_i
outputs of that block. Since the I_i outputs act as I_A inputs to the next
lower level, the action of the I_A will be propagated left until the I_i outputs of the first level where they generate the *inhibit vector* to be used
in the resetting of the buffer flip-flops. By tracing the branches of the
tree it will be easy to see that the inhibition vector is obtained correctly for all response patterns.

Notice that if there is at least one response at the first level, it will
yield a response at the "Some response" terminal of the last stage (root of
the tree).

Anderson's tree structure can easily be modified to produce the priority-
resolver circuit for the principle demonstrated in Fig. 3.4a [3.3]. This
circuit will be described below in connection with address generation.

Generation of the Address of the Uppermost Response. Most LSI CAM arrays are
provided with an address decoder for the addressed selection of word locations; this has been found necessary on account of the limited number of pins
in the integrated circuit packages. As shown in Fig. 3.9, the outputs of the
multiple-match resolver then ought to be encoded to produce the address code
of the first response. However, it might be more effective to derive the address code within the multiple-match resolver circuit. Such an operation, by
a slight modification of the tree circuit is shown in Fig. 3.7. (Cf also a
similar circuit of KOO [3.4]). The cascaded gate structure in Fig. 3.7a is
intended to display only the *uppermost response* at its left-side outputs.
By adding two gates with Wired-OR output to the basic block, as shown in
Fig. 3.7a and interconnecting the blocks as in Fig. 3.7b, the circuit will
yield the correct address of the uppermost response, as will be found upon
inspection of the diagram.

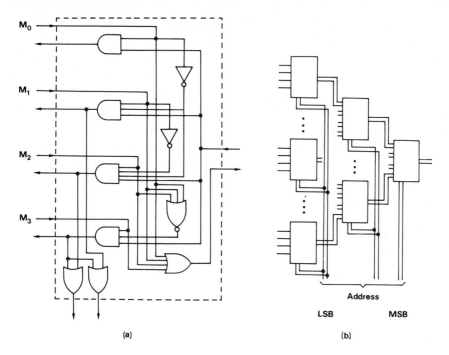

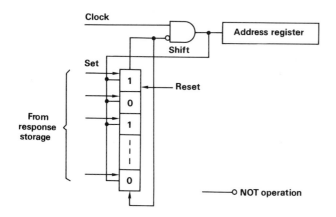

Fig. 3.7a,b. Generation of the address of the uppermost response:
a) one module, b) modular tree (three levels)

Fig. 3.8. Shift-register implementation of the multiple-response resolver

An extremely simple although a bit slow circuit for the generation of
the address of the uppermost response is shown in Fig. 3.8. It is implemented by an end-around shift register in combination with an address counter.
All the responses from matching storage locations, indicated by the bit value
'1', are first copied into the shift register. Periodic clock pulses are
used to advance the counter as well as to shift simultaneously the contents of
the shift register upwards, until the first '1' appears at the uppermost
position of the latter; the clock pulses will then be automatically blocked
out from the counter as well as from the shift register. The contents of the
counter, if it was initialized by 0, now directly indicate the address of
the first matching word, and after this number is transferred to the address
register, the complete matching word can be read out. After resetting the
uppermost 1 in the shift register, the shifting automatically continues until
the next '1' appears at the uppermost position, whereafter the reading process is repeated, and so on, until all responses have been served. With this
solution, no further buffer flip-flops at the M_i lines are needed.

Related multiple-response resolvers have been described by BEHNKE [3.5],
HILBERG [3.6,7], RUDAKOV and IL'YASHENKO [3.8], CAMPI and GRAY [3.9], and
DAVIDSON [3.10].

It is to be noted that the resolvers usually have an output to indicate
that "some" responses (at least one) have been obtained (cf also [3.11]).
A more complex problem is to *count* the exact number of responses. Such solutions are given, e.g., by FAVOR [3.12], HILL [3.13], and FOSTER and STOCKTON
[3.14].

3.3.3 The Complete CAM Organization

The block scheme of a complete CAM system is shown in Fig. 3.9. Its central
parts are the *CAM array*, the *search argument register*, the *mask register*,
the *response store*, which consists of a flip-flop for every M_i line, the
multiple match resolver, and some auxiliary circuits needed for interconnection and control.

The mask bits explained in Sect. 3.2 are stored in the mask register which
together with the search argument forms the $C_j(1)$ and $C_j(0)$ signals. During
writing this unit forms the $W_j(1)$ and $W_j(0)$ signals, respectively, whereby
the (masked) search argument will be written into a word location defined by
the address code given at the input of the address decoder.

The output lines of the multiple-match resolver could in principle be
connected directly to the address lines of the associative memory, whereby

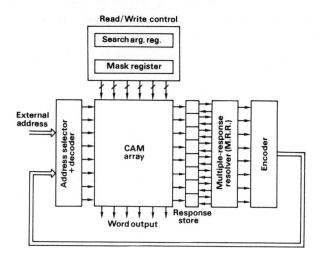

Fig. 3.9. The complete CAM organization

the complete contents of all matching words can be read one item at a time. In practice, CAM arrays are normally made by integrated circuit technology, and for the addressed reading and writing there is always a built-in address decoder. For this reason the output from the multiple-match resolver must be *encoded*, whereby only a small number of feedback lines is needed, as shown in Fig. 3.9.

It is also possible to have both content-addressable and usual memory arrays in the same memory system. This type of memory might then be used as a simple *Catalog Memory* the main features of which were shown in Fig. 1.1. The left-hand part is a normal all-parallel CAM intended to hold keywords, and the right-hand subsystem is an addressable (linear-select) memory for the associated items. If the catalog memory is used to represent symbol tables, then it may be assumed that all keywords in its content-addressable part are different. During search, all entered keywords are compared in parallel with the search argument. If all bits of the stored keyword match with those of the search argument, then the corresponding output line is activated. This line again acts as the address line in the linear-select memory. Notice that entries can be stored at arbitrary locations, for instance, in the order of their declaration.

Various Uses of Masking. The provisions for masking the search argument can be applied in several operations. The most important of these may be in the searching on selected attributes which are parts of the search argument. The words in the content-addressable memory may be composed of several fields describing different attributes, and a subset of these can be addressed by unmasking the search argument correspondingly.

Another important application of masking is the *writing of new data into memory locations which happen to be vacant*. Because a list of the occupied positions usually cannot be maintained, the empty places must be found automatically. For this purpose there is an additional *vacancy indicator bit* in each word which is comparable to a data bit. In an empty location this bit is initially 0, and it is marked 1 when an item is entered. After deletion of the entry this bit is again reset to 0. The vacant places can now be found by masking all other bits except the vacancy indicator bit, and performing an equality search. The vacant location into which a word is written is then determined using, e.g., the multiple-match resolver.

The third possible use of masking is in the *magnitude search* on numerical entries explained in the next subsection.

3.3.4 Magnitude Search with the All-Parallel CAM

The all-parallel CAMs usually do not have logic circuitry which would directly implement the magnitude comparison, e.g., by the algorithm (3.9) or (3.10). As a search on the basis of magnitude comparisons, or *magnitude search* as it is usually called, may become necessary in some systems which have an all-parallel CAM, it may be useful to show how the organization can be modified simply for this mode of searching.

The *response store* which is included in the organization shown in Fig. 3.9 can now be utilized effectively. Let us recall that its contents are normally reset to 0 before usual searches. Since the following searching algorithm is made in several passes, it is necessary only to leave the values obtained in each pass in the response store where the final search results are accumulated. Initially all bits of the response store are set to 0, and every search operation changes the value to 1 if a match is found. If the bit value is 1 after the previous pass, a value 0 (mismatch) obtained on the match line in a new searching operation cannot reset the value to 0, however.

It is further assumed that the CAM is associated with, say, a computing device such that the search argument bits can be scanned and changed automatically, and the mask can be set depending on the bit values.

The following algorithm is based on an idea of CHU [3.15].

Algorithm Which Finds all Words Greater than the Search Argument:

1) Set all bits in the response store to 0.
2) Scan the search argument from left to right and choose the first 0 for a *target bit*.
3) Change the target bit to 1 and mask away all bits to the right of it, thereby forming a new search argument.
4) Perform a search on the equality match condition relating to the new search argument.
5) Set those bits in the response store to 1 which correspond to matching words.
   Comment: It is easy to deduce that those words which correspond to 1's in the response store are certainly greater than the original search argument, whereas nothing can be said of the other words until this algorithm stops.
6) Change the target bit to 0, retain all bit values to the left of it in the search argument, and scan the search argument further to the right. If no 0's are found, or the bits of the search argument have been exhausted, stop. Otherwise choose the next 0 for the new target bit and repeat from step 3.

This algorithm leaves 1's in the response store at all words which are greater than the search argument.

Example:

Below are shown a search argument, the contents of a CAM, and the contents of the response store after the two searching passes corresponding to the two target bits.

		T_1			T_2	Response store	
						Pass 1	Pass 2
Search argument	1	0	1	1	0		
Word 1	1	0	1	1	1	0	1
Word 2	0	1	0	1	0	0	0
Word 3	1	1	0	1	0	1	1
Word 4	1	0	1	0	0	0	0
Word 5	1	1	0	0	0	1	1
Word 6	1	0	1	1	0	0	0

T_1, T_2 = target bits

In order to find all words which are *less than* the search argument, the above algorithm is modified by changing the bit values 0, 1, and 0 mentioned at steps 2, 3, and 6 into 1, 0, and 1, respectively.

Because magnitude search and searching on more complex matching conditions are normally made using special hardware provisions not provided in the above design, the discussion of other searching modes is postponed to Sect. 3.4 where it is carried out naturally in the context of content-addressable processing with word-parallel, bit-serial CAMs.

3.4 The Word-Parallel, Bit-Serial CAM

The content-addressable memory principles discussed in this section, although in many respects simpler than those of the all-parallel CAM of Sect. 3.3, have a significant advantage; since the memory array itself is inexpensive, complexity can be added to the processing circuits, each of which is common to a word location. In this way the processing power is increased, and this memory architecture has in fact been adopted for certain large and complex special computers. Thus, in order to set the constructs discussed in this section in their right perspective, and to give simultaneously a view of the operation of complete CAM systems, some auxiliary features included in content-addressable processors such as the *results storage* with recursive bit operations performed in it, and the *mixed addressed and content-addressable reading and writing principles* are presented. Further implications of content-addressable structures to parallel processing will be given in Chap. 6, especially, Sect. 6.5.

3.4.1 Implementation of the CAM by the Linear-Select Memory Principle

Consider a usual linear-select memory module depicted in Fig.3.10a; it consists of an m by n array of bit storages, functionally organized in m *words* with n bits each, an *address decoder* which activates the bit storages of one word for reading or writing, the *address code input* to the decoder, the *word input* and *word output* common to all word locations, and some amplifiers and general control for writing or reading.

A *linear search* would normally be performed word-serially by the application of successive address codes to the decoder and giving a series of reading commands, whereby the words of the array subsequently appear at the output port of the memory where they can be compared with external data. This takes m reading operations, and in general, m >> n.

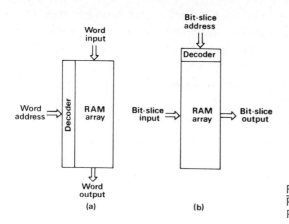

Fig. 3.10. a) Linear-select
RAM module. b) Application of the
RAM module to bit-slice addressing

Content-Addressable Parallel Search. If one would like to speed up the searching which in a simple case means comparison of every stored word with an external *search argument*, a simple and straightforward solution, at the cost of an increased amount of hardware in the reading circuits, but at the same time corresponding reduction in the address decoding, is to take an array with n words, m bits each, and to orient it in such a way that the roles of the words and bits, when compared with the addressed memory, are exchanged (Fig. 3.10b). If memory modules with such a word length are not available, it is possible to connect the address decoder input of several units in parallel, whereby the sets of their word inputs and outputs appear as words of extended length.

If now the rows in the array of Fig. 3.10b are redefined as words, then the decoder selects a *bit slice*, or the set of the same bit positions from all words. This time the searching would be implemented simultaneously and in parallel over all words, by reading the bit slices, one at a time, into the output port where a comparison *with the respective bit of the external search argument* can be made. By the application of a sequence of bit-slice address codes, obtainable from a *bit slice counter* (cf Fig. 3.11), at the decoder inputs, all bit slices can be made to appear at the memory output port where they can be compared with the corresponding argument bits.

It is a simple task to pick up the argument bits, corresponding to the bit slices, from a *search argument register* for comparison: this can be implemented, e.g., by a *multiplexer* controlled by the bit slice address code (Fig. 3.11). A more delicate operation is the combination of intermediate results from the various bit slices. If a comparison for *equality* were the only operation to be performed, then it would be necessary only to form somehow the logical products of bit comparisons as defined by (3.3); results from the successive bits can be collected recursively as indicated by (3.7).

The comparison of words with the search argument is most often done in word-parallel, bit-serial CAMs with the aid of a *results storage* which in this case is very simple. (More complex recursive operations are described in Sect. 3.4.5.) The results storage consists of a set of flip-flops, one for every word location, which is set to 1 from a common line before searching, and which is conditionally reset by an EXOR gate taking one input from the memory output, and the other from the search argument register as shown in Fig. 3.11.

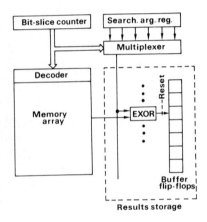

Fig. 3.11. Control organization of the results storage

The idea is to reset the flip-flops in the results store whenever a bit value read from the memory array disagrees with that of the search argument. The schematic principle shown in Fig. 3.11 does not take into account the asynchronism of the bit values to be compared; for a practical solution, some extra circuits for the resetting operation are necessary. When all bit slices have been read, there are ones left only in such flip-flops which correspond to completely matching words.

For a complete search in the array, only n reading cycles have to be performed. This is usually considerably fewer than the m cycles needed in linear search.

It is to be noted that the word bits can be read for comparison in any order, by the application of a corresponding address sequence at the decoder and multiplexer inputs; the bit counter is then replaced by a sequential control. Furthermore, it is possible to select any subset of bit positions for comparison. This is then equivalent to the *masking* operation.

Although the above principle is a rather trivial one, nonetheless it has been shown to be influential to many further designs, and its derivatives will be discussed below.

The Problems of Addressed Writing and Reading. There are a couple of characteristic problems and drawbacks in the above design. The *writing* of data into the array must be done by bit slices, and in order to do this, the complete contents of the array must have been buffered elsewhere, e.g., in a memory of corresponding size. There is an important application for this type of memory operation, namely the *buffered content-addressable search memory* as described in Sect. 5.3.

Updating of data, especially alteration of a single word, is not possible in a memory of the above kind without rewriting of the whole array. *Addressed reading* of a word location, in order to find the whole contents of it when part of the word matches with the search argument, is likewise an impossible operation in the basic design. As such, this principle is, therefore, suitable only for simple table look-up or *Catalog Memory* operations as presented in Fig. 1.1, and not for computing or control operations.

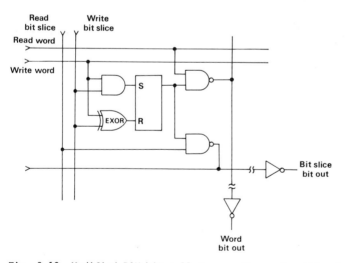

Fig. 3.12. Modified RAM bit cell for reading and writing by words as well as by bit slices. One of the write control lines is the address selector, another the data bit to be written, respectively [3.15]

For the implementation of the word-addressed reading and writing operations, it is certainly possible to modify the bit cell circuits to incorporate the addressed reading and writing operations by words as well as by bit
slices; although this means only a few extra gates (Fig. 3.12), the design
is then no more standard and not available off the shelf. If the bit storages
were implemented by a simple ferrite-core technology, the two-way addressing
could be made by threading more signal lines, as shown in Sect. 4.3.2, through
the cores. All such arrangements are unnecessary, however, if the special
organization described next is applied.

3.4.2 Skew Addressing

There exist a couple of ingenious variants of the basic principle described
in Sect. 3.4.1 which are also based on standard memory modules but in which
the addressed reading and writing *by words as well as by bit slices* is directly rendered possible, using a special method of permutation of the memory array, and some extra circuitry. These solutions have, in fact, been
shown to be so effective that they were selected for the basic design in the
memories of some large content-addressable processors (see Sect. 6.5).

The standard memory module needed is a 1 by m array where m is the number
of words, and for example, 1-bit by 256-word semiconductor memory chips, with
a built-in address decoder (bit selector), have been available for a long
time. In the following basic examples *the memory array is square*, e.g., there
are 256 words with 256 bits in the memory. If a higher memory capacity is
needed, the memory is expandable by adding more arrays. The bit-slice control
can then be common to all arrays. In more advanced design it may be necessary
to search the memory by masking off some selected arrays.

Fig. 3.13. RAM module for skew-addressed memory

When compared with the principle of Sect. 3.4.1, the CAM arrays described
here need somewhat more hardware; they contain m decoders, and each of them
must receive a different address code computed by some functional circuits.

The searching principle applied in the two designs described below is in
general termed *skew addressing*, or originally, *skewed storage techniques*
[3.16,17]. The corresponding hardware may be named *skew network*. The meaning

of these terms becomes obvious if it is stated first that the original or
logical data array is permuted before storage so that the original bit slices
are in a skewed and permuted order in the memory. The mapping of the original
array into the contents of the memory in the case of the *adder skew network*
(STONE [3.16]) is exemplified by a 256 by 256 bit arrray in Fig. 3.14. The
skewed storage, with a separate bit selection on every row, will facilitate
reading and writing by words and bit slices as explained below.

Fig. 3.14. Mapping of stored data in the adder
skew network memory

Adder Skew Network. Every address decoder in this design (Fig. 3.15) is provided with an *arithmetical addition circuit* (full adder), capable of forming
the sum of the externally given address code and a constant (wired-in) number
which corresponds to the number of the row. This sum is taken modulo m where
m is the number of bits. By a proper control, the adders can be disabled,
i.e., a replica of the bit-slice address is then obtained at their outputs.
Data read out and written into the memory is buffered by a *memory register,*
capable of performing end-around shifting operations by a number of steps
corresponding to the bit slice address. Shifting can be in the upward or
downward direction depending on a particular access operation.

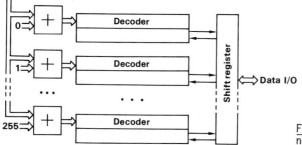

Fig. 3.15. Adder skew
network

Consider first *writing of a bit slice* into the memory. The *two's com-plement of the bit-slice address* is switched to the adder inputs, the addition circuits are *enabled*, and the data to be written are transferred into the memory register where they are shifted *downwards* end-around by an amount corre-sponding to the bit-slice address. By a writing command, this slice is trans-ferred into proper bit storages of the memory array; for instance, if the bit position was 1, the map of the slice would be that indicated by the heavy borderlines in Fig. 3.14.

Reading of a bit slice is done by giving the *two's complement of the bit-slice address*, *enabling* the adders, and performing a reading operation whereby the slice is transferred into the memory register. The data obtained, however, are still in a permuted order, and a number of *upwards* shifting operations (end-around) must be performed to align the contents of the slice.

Writing of a word proceeds by first giving the *word address* to the adder inputs, *disabling* the adders, and transferring the data into the memory re-gister where they are shifted *downwards* by the amount given in the word ad-dress. A writing command then transfers the data into proper bit storages in the memory array. It should be noted that *word i will thereby be written into column i of the memory array*, with its contents rotated by an amount corre-sponding to the word address.

Reading of a word is an operation inverse to the previous one. The *word address* is presented at adder inputs, the adders are *disabled*, and a reading command is given. Contents of the memory array corresponding to one original word appear in the memory register in which it is shifted *upwards* by an amount corresponding to the word address.

EXOR Skew Network. Another skew addressing principle due to BATCHER [3.17] in which the adders are replaced by simpler EXOR gates is presented in Fig. 3.16. The main idea is that when the address mode register contains mere *zeroes*, the writing and reading occurs by *bit slices*, and when the register is full of *ones*, writing and reading by *words* is done. The memory mapping is less lucid than in the adder skew network, but can be followed easily with examples. It can be shown that for any contents of the address mode register the memory mapping is one-to-one and thus reversible. In the EXOR skew network, it is further possible to define other slices than by bits or words. This becomes possible by having special codes in the *address mode register*. For instance, a k-bit field in every kth word (with k even) may be addressed.

A special permutation operation at the I/O port of this network is needed: if w is the index of word to be written and b its bit position, then the per-muted bit position (row in the memory) is w $\oplus$ b.

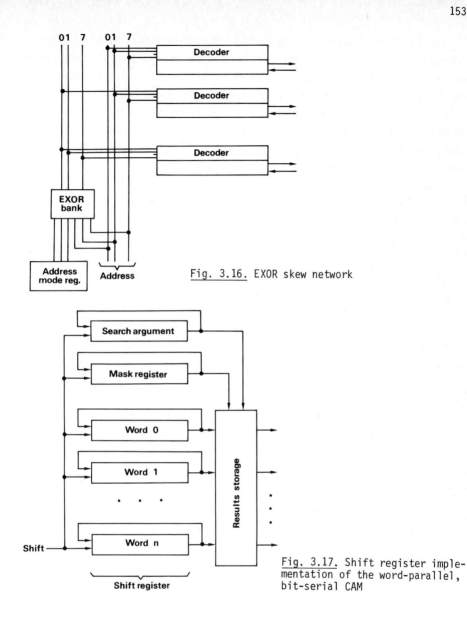

Fig. 3.16. EXOR skew network

Fig. 3.17. Shift register implementation of the word-parallel,
bit-serial CAM

3.4.3 Shift Register Implementation

Besides the above methods which are based on standard random-access memories,
there is another principle according to which a word-parallel, bit-serial
content-addressable memory can be built of standard components. In this solution, delineated in Fig. 3.17, all memory locations are either *shift re-*

gisters or cheap *dynamic memories* such as the CCD discussed in Sect. 4.3.4. The search argument and the mask word are stored in separate registers similar to the storage locations. The contents of every location, including those of the search argument and mask word registers, are made to recirculate in synchronism through the results store, in which the masked search argument is compared in parallel with all storage locations, one bit position at a time. In this solution, too, a results store is needed the contents of which are evaluated recursively during the recirculation.

In this latter type of memory, the words may be stored with either the most significant or the least significant bits to the right, depending on the comparison algorithm applied (cf Sect. 3.2). However, there is no possibility of skipping any bit positions as might be done in the first version by special bit slice addressing.

3.4.4 The Results Storage

In the parallel searching of words by bits, the serial or *recursive* method of processing thereby applied makes it possible to implement several interesting comparison and computing functions in addition to the simple equality method described before. Since the words in word-parallel, bit-serial CAMs are generally long, for instance, 256 or 512 bits, an appreciable amount of complexity can be added to the serial comparison circuit which is common to the word location without significantly increasing the total system cost.

It may be useful to point out first that there are two modes of operation for which the CAM systems are used: *searching* and *computation*. In *searching*, the purpose is usually to locate and **read** out all words which satisfy a given criterium, e.g., being equal to, greater than, or less than a given search argument, lying between two specified limits, or being the absolutely greatest or smallest of all stored numbers. While the contents of the CAM array are usually left unaltered in searching, the purpose in *content-addressable computation* is to operate on the stored words and to leave the results in the CAM array, whereby they frequently replace the previous contents or a particular field in the words.

The various modes of operation are implemented by the logical design of the *results storage* which consists of sequentially operating logic circuits, one per word location. Development of the results storage here is made in three steps. First, a simple design is presented which is capable of searching words which are equal to, greater than, or less than the search argument, respectively. Next, the design of the results storage adopted to a large commercial computer is presented, and finally, an example of content-address-able additions is given.

Combined search operations using magnitude comparisons as well as other criteria will further be described in Sect. 3.4.5.

A Results Storage for Magnitude Comparisons. It is a straightforward task to implement the algorithms (3.9,10) by sequential logic circuits; the former is selected here for illustration. The two categories of circuit implementa-tions to be considered are the *asynchronous* and *synchronous* sequential cir-cuits, respectively. In the former ones, logic gates are directly intercon-nected to implement logic equations derived, say, from (3.9) by the addition of a timing signal for bit control. The pitfalls in asynchronous circuits are all sorts of *races* and *hazards* (cf, e.g., [2.24]). In order to avoid their discussion which might obscure the main goal, and because the example given here is only a tutorial one, not necessarily very economical, the results store described below is a synchronous circuit, implementable by *clocked flip-flops.* For the latter one, the D and the JK flip-flops are available. In this design, for simplicity of discussion, an arbitrary choice for the D flip-flops is made; the reader may try to find out whether the logic design with JK flip-flops is simpler, and also he might attempt to find out the corresponding asynchronous implementation. It may be mentioned that clocked flip-flops are generally used in arithmetic circuits because of their easy control, as well as their high stability against various disturbances.

Consider Fig. 3.18 which shall stand for a simple response-store flip-flop of Fig. 3.9. There are now two response-storing flip-flops g_j and ℓ_j, at every word j, here named *intermediate results stores* with successive bit values g_{ji} and ℓ_{ji}, respectively, corresponding to the variables occurring in (3.9). The logic circuitry around the g_j and ℓ_j flip-flops is shown in detail in Fig. 3.18.

When the search argument will be compared with the contents of a particular storage location, the respective bits a_i and s_{ji} are made to appear serially and in synchronism at the inputs of the intermediate results stores. In Fig. 3.18, the two D flip-flops are used to buffer these bit values, and they sequentially take on new values for every bit position. (The D flip-flop is a device which copies and buffers in it the binary value present on its D input at every clock pulse CP, and displays this value together with its Boolean complement at the output ports up to the next clock pulse.) After the complete serial operation, the last values that remain in these flip-flops indicate the result of comparison.

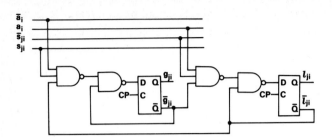

<u>Fig. 3.18.</u> Intermediate-results store for magnitude comparison

It should be noted that all words which exactly match the search argument
are indicated by the values $g_{j0} = \ell_{j0} = 0$ after searching.

It is a straightforward task to design a switching circuit which connects
relevant signals from the results store into a multiple-match resolver to
carry out the readout operations as shown in Fig. 3.8.

The Results Storage of STARAN. One of the largest content-addressable parallel
computers implemented is the STARAN of Goodyear Aerospace Corporation. The
central memory in it is based on the EXOR skew network principle (cf Sect.
3.4.2). The processor system will further be described in Sect. 6.5.1; its
basic CAM is expandable in blocks of 256-word by 256-bit arrays up to a capacity of 8 K words. Let the memory here appear as a normal word-parallel, bit-
serial CAM.

In the results storage of STARAN there are two sequentially operating intermediate results stores, with bits named X_i and Y_i, respectively. There is a
common register F with flip-flops F_i which may appear as a search argument
register (cf Sect.6.5.1). Further there exists a set of flip-flops M_i which
can be used, e.g., for *masking out words* during writing; in this way it is
possible to select a subset of locations which become subject to parallel
computation.

The two intermediate results stores X_i and Y_i have the following specified
roles: X_i can be switched to many modes of sequential operation and it is the
primary results store. Y_i is an auxiliary intermediate results store which
has a simpler sequential control. It should be pointed out that although
STARAN operates on bit slices, these can be picked up in any order from the
CAM array by stored-program control (machine instructions). Accordingly, the
recursive operations described below are not synchronous on pairs of bit values as in the magnitude search example given above; it may be more proper to
imagine that the results stores acquire new values in synchronism with machine

instructions, whereby a bit position in serial arithmetic operation, as described below, usually requires the execution of several machine instructions.

The operation of X_i and Y_i is defined by the following substitutions:

$$X_i \leftarrow (X_i \wedge \overline{Y}_i) \vee [Y_i \wedge (F_i \; op_1 \; X_i)]$$

$$Y_i \leftarrow F_i \; op_2 \; Y_i \tag{3.11}$$

where op_1 and op_2 are any Boolean operations referring to two variables. It may be generally known that there exist 16 different truth tables for two Boolean variables, and any of these can be defined in the machine instructions by which X_i and Y_i are controlled.

An Example of the Use of Results Storage for the Implementation of Addition over Specified Fields. Assume a content-addressable array with four fields, corresponding to variables V_i, A_i, B_i, and S_i in every word, with i the word index. The purpose in this example is to form the sums $A_i + B_i$ and to store the results in the S_i fields. This addition shall be conditional on sums being formed only in words which have a V_i value matching with the search argument in the corresponding field.

In this particular example [3.18], the function $F_i \; op_1 \; X_i$ is defined to be identical to F_i, and for $F_i \; op_2 \; Y_i$, $F_i \oplus Y_i$ (EXCLUSIVE OR) is taken.

Denote the operands $A_i = (a_{i,n-1}, a_{i,n-2}, \ldots, a_{i0})$, $B_i = (b_{i,n-1}, b_{i,n-2}, \ldots, b_{i0})$, and $S_i = (s_{i,n-1}, s_{i,n-2}, \ldots, s_{i0})$. The addition of A_i and B_i commences with the least significant bits. During a bit cycle, Y_i shall form the sum bit and X_i the carry bit, respectively.

The first step is to load the M_i flip-flops, by performing a content-addressable search over the V_i fields (with A_i, B_i, and S_i fields masked off). The M_i are thereby set according to the responses. For the initial values, $X_i = Y_i = 0$ are set. The first bit slice, the set of all a_{i0} is read into the F_i flip-flops by having 1 in the search argument at this position and masking off all other bits. According to (3.11), X_i then attains the value 0 and $Y_i \leftarrow a_{i0}$. Next the bit slice b_{i0} is read into the F_i flip-flops whereby $X_i \leftarrow a_{i0} \wedge b_{i0}$, $Y_i \leftarrow a_{i0} \oplus b_{i0}$. Because there was no previous carry, a new carry ($X_i = 1$) is correctly obtained if, and only if, $a_{i0} = b_{i0} = 1$, and Y_i is formed as the mod 2 addition of a_{i0} and b_{i0} which is correct, too. This value is written into the bit slices s_{i0} on the condition that M_i is 1. Finally, $Y_i \leftarrow X_i$ is set.

The next bit positions, j = 1 through n-1, are operated in the same way. In the first step the bit slice a_{ij} is read into F_i, and in the second step

b_{ij} replaces a_{ij} in F_i. Note that X_i and Y_i are always changed sequentially at both steps. The first step is equivalent with the mappings

$$X_i \leftarrow (X_i \wedge \overline{Y}_i) \vee (Y_i \wedge a_{ij}) \, ,$$

$$Y_i \leftarrow Y_i \oplus a_{ij} \quad . \tag{3.12}$$

At the second step one obtains

$$X_i \leftarrow (X_i \wedge \overline{Y}_i) \vee (Y_i \wedge b_{ij}) \, ,$$

$$Y_i \leftarrow Y_i \oplus b_{ij} \quad . \tag{3.13}$$

The writing of Y_i back into the bit slice s_{ij} proceeds next conditionally on the value $M_i = 1$. As the third mapping,

$$Y_i \leftarrow X_i \tag{3.14}$$

is simply formed thereafter.

A proof that (3.12-14) represent the arithmetic addition is made inductively. If it is tentatively assumed that after the previous bit position (j-1) the carry (c_{j-1}) was obtained correctly and it was left in the X_i as well as Y_i, a substitution and simplification shows that (3.12,13) are equivalent to a single pair of mappings

$$X_i \leftarrow (\overline{a}_{ij} \wedge b_{ij} \wedge c_{j-1}) \vee (a_{ij} \wedge \overline{b}_{ij} \wedge c_{j-1}) \vee$$

$$\vee \ (a_{ij} \wedge b_{ij} \wedge \overline{c}_{j-1}) \vee (a_{ij} \wedge b_{ij} \wedge c_{j-1}) \, ,$$

$$Y_i \leftarrow a_{ij} \oplus b_{ij} \oplus c_{j-1} \quad . \tag{3.15}$$

These equations are known as the equations of a *full adder*, i.e., X_i represents the new carry and Y_i the sum digit, respectively. Since the recursion started correctly at the least significant bit, a formal proof of the algorithm by complete induction has been given Q.E.D.

3.4.5 Searching on More Complex Specifications

If content-addressable memories are used in parallel processors, and this is the area in which the word-parallel, bit-serial memories have primarily been applied (cf Chap. 6), it will become necessary to locate and identify entries

on the basis of more complex searching conditions, too, not only on the basis
of equality or magnitude. For instance, logic functions defined over specified bits can be used to indicate matching. Moreover it will often be necessary to locate entries the bit pattern of which is closest to that of the
search argument, or which have a value next greater or smaller to that of the
entries found in the first search. The following review covers some such
operations that occur in parallel processing. Although these procedures are
discussed in the context of word-parallel, bit-serial memories, nonetheless
many of these methods are in principle applicable to all-parallel memories
as well. However, for the verification of complex matching conditions such
as those based on Boolean variables, it may sometimes be necessary to add
new functions into the comparison logic, and the extra expenditures are then
tolerable only if the logic circuit is common to the memory location, as is
the case with this type of memory.

Between-Limits Search. In this, as well as in most of the other modes of
search discussed, the results storage must contain a further auxiliary storage
flip-flop in addition to the g and ℓ flip-flops already involved. The purpose of this third flip-flop, briefly called *results store* below, is to form
the intersection of those sets of responses which are found in partial searches
on one matching criterium only. The idea is to initially load the results
store with ones, of which every partial search resets a subset to zero upon
mismatch. In this way ones are left only in such results stores for which all
matching conditions are satisfied.

The between-limits search, or retrieval of all words which have their
numerical value between two specified limits, is simply done by first searching for all numbers greater than the lower limit, and then performing the
next search for numbers less than the upper limit. If the bit-cancellation
method described above is used, only the between-limits matches are left in
the results store.

Search for Maximum (Minimum). On account of the special nature of binary
numbers, a search for *maximum* among the stored values can be made in several
passes in the following way. No external search argument is used with this
method. Starting with the most significant bits (MSBs), values '1' are set
in the search argument register into its highest, next to highest, etc. bit
position in succession. By masked equality match, all words are first searched which have the value '1' in the highest bit position, and proceeding then
to lower positions. If, when proceeding from the MSB towards lower bits, a
search does not pick up any candidates (all words have '0' in this bit posi-

tion), then this occurrence must be detected in the output circuits, and no cancellations must be performed in the results store during this pass. The words found in the first passes are possible candidates for further searches because all the other numbers must necessarily be smaller. This process is repeated until all bits of the search argument have been exhausted. It is possible that there exist several identical greatest numbers which all are found in this way.

The search for minimum is obviously the dual of the search for maximum. Starting with the MSBs, the highest zeros are checked in similar passes as before.

Search for the Nearest-Below (Nearest-Above). By combining the successive searches for numbers less than a given value, and searching for the maximum in this subset, the nearest-below to the given value is found. Similarly, a combination of the greater-than and minimum searches yields the number nearest above the given value.

Search for the Absolute-Nearest. After the nearest-below and the nearest-above to a given value are found, the number that is absolutely nearest to the given value is determined by comparing the differences from the given value.

Ordered Retrieval (Ordering). This is nothing else than a successive search for maxima or minima in the remaining subset from which the maximum or minimum, respectively, has been deleted at the preceding step. It can be implemented as a combination of previous algorithms. However, ordered retrieval of items from a file, in numerical or lexicographical order, has sometimes been regarded so important a task that special hardware-supported algorithms for it have been designed; one has to mention the works of LEWIN [3.19], SEEBER and LINDQVIST [3.20], WEINSTEIN [3.21], JOHNSON and McANDREW [3.22], MILLER [3.23], CHLOUBA [3.24], SEEBER [3.25], WOLINSKY [3.26], and RAMAMOORTHY et al. [3.27]. Hardware sorting networks have been invented by BATCHER [3.28].

Proximity Search. There are several occasions on which the maximum number of bits matching with those of the search argument is a useful criterion for location of the entry. It is now necessary to augment the memory system, especially the results store, by bit counters which for each word count the number of bits agreeing with the bits of the search argument. If these bit counters are also made content-addressable, a maximum search over them yields words that have their value in proximity to that of the search argument.

Search on Boolean Functions. By adding circuit logic to the comparison logic,
a search on any Boolean function can be performed. In bulk processing, it
may be necessary to find words which, together with the search argument, give
a particular value to a Boolean function defined over specified bit positions.
Typical binary functions are the logic equivalence, EXCLUSIVE OR (EXOR), OR,
and AND.

Survey of Literature on Complex Algorithms. In addition to the methods for
ordered retrieval, the following references which describe magnitude search
and related algorithms shall be mentioned: GAUSS [3.29], ANDERSON [3.30],
ESTRIN and FULLER [3.31], KAPLAN [3.32], WOLINSKY [3.33,34], LINDQVIST [3.35],
BIRD et al. [3.36,37], SEEBER and LINDQVIST [3.38], FENG [3.39], FOSTER [3.40],
and AGRAWAL [3.41].

3.5 The Word-Serial, Bit-Parallel CAM

The purpose of this section is to show that an associative or content-addressable search need not be performed simultaneously or in parallel with respect
to all memory locations, as long as the responses are obtainable in a time
which is compatible with the other computing operations. In principle, the
method described below involves little more than a linear search in which all
the stored items are scanned sequentially. The main special feature is an
automatic operation for which no extra programming is needed.

It is perhaps easiest to explain the *word-serial, bit-parallel search*
using the system description of CROFUT and SOTTILE [3.42]. It is delineated
in Fig. 3.19 in a somewhat simplified form. The basic storage of information
is a set of *delay lines* in which, by means of transmitting and receiving
circuits applied at their ends, trains of synchronous pulses can be made to
circulate. Such devices were used in early computers and special digital devices as their main memories, and one possible medium is a magnetostrictive
wire, where the propagation of impulses is based on a magneto-acoustic effect.
Certain solid media such as fused quartz, glass, and sapphire are also able
to transmit ultrasonic acoustic waves at hundreds of MHz with quite small
attenuation. The solid is usually fabricated in a form of a polygonal prism
in which the waves are reflected via many paths. (Cf also [3.43-48]).

Delay lines based on simple physical effects can also be replaced by some
other devices for which the technology is still in a progress of development,
for instance the *magnetic bubble memory* (MBM), and the *charge-coupled device*

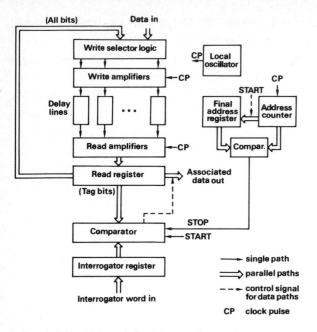

Fig. 3.19. Word-serial, bit-parallel CAM

(CCD). Their physical principles will be reviewed in Sect. 4.3. It must be emphasized, however, that both of these devices are at least two decades slower than the best delay lines.

Finally it may be mentioned that usual, preferably fast integrated-circuit shift registers are a possible, although expensive solution for the dynamic memory of the above kind (cf Sect. 4.4.3).

When electrical pulses in synchronism with a clock frequency of, say, 100 MHz are transduced into longitudinal acoustical signals at one end of a quartz delay line, and an inverse transducer is placed at the other end, the solid may be put to work as a dynamic storage for thousands of bits of information: an occurring pulse represents a bit value 1, and a missing pulse in a synchronous train, a binary 0. When a number of such lines are connected in parallel and driven by a common clock, and every receiver feeds back its signals into the corresponding transmitter, the system may seem like a continuously running drum or disk memory, but having a much higher clock frequency. If now the bits of a word are stored in parallel on the different lines, the consecutive words appear serially, in a bit-parallel fashion.

Comparison of a word with a fixed search argument buffered in the respective register can be made by special fast logic circuits which implement, e.g., the comparison function as discussed in Sect. 3.2. The fastest standard logic circuit family is the emitter-coupled logic (ECL), but special electronic switches, e.g., using tunnel diodes or usual fast diodes, can be applied, too.

The synchronizing marks for the bits are provided by a stable local oscillator the cycles of which are counted by an *address counter*. The contents of the address counter thus directly indicate the instant of occurrence of a particular word at the read amplifiers, i.e., the address of the memory location. The contents of this memory, i.e., the pulse patterns of the delay lines are repeated periodically. If the maximum capacity is 1000 words, the cycling period is 10 μs. Thus the average access time to an entry is 5 μs. In order to start the searching at the earliest possible moment and to continue it for one complete cycle of repetition only, a START command pulse is applied which copies the current contents of the address counter into the so called *final address register*. At the same time, the *comparison circuits* (or their outputs) are activated. Whenever the word appearing in the read register agrees with the search argument in all unmasked portions, the outputs of the comparison circuit become active and the complete contents of the read register are read out. If it is necessary to know the address of the current storage location, the contents of the address counter may be strobed at the same comparison signal. After one full cycle, the contents of the address counter again agree with the contents of the final address register, and this is indicated by another comparison circuit that gives the STOP signal.

Writing of data into the delay line memory occurs in the addressed mode: when the address counter is properly initialized, new data can be copied into consecutive locations by the application of corresponding data signals at the write amplifier inputs through a write selector logic, in synchronism with the memory operation. External control circuits are thereby needed.

It is also possible to apply in this system magnitude comparison operations using, e.g., the iterative circuit depicted in Fig. 3.1.

3.6 Byte-Serial Content-Addressable Search

3.6.1 Coding by the Characters

The standard coding of information on magnetic tape and other media used
for archival storage is based on *bytes*, eight bits wide. One byte could thus
in principle encode up to 256 different symbols or *characters* which may represent letters, numbers, as well as control and signaling marks that occur
in a data stream. Most common of all coding systems is the ASCII (American
Standard Code for Information Interchange) which is used in practically every
teletypewriter and other terminal. A few characters of the ASCII code are
represented in Table 3.2.

Table 3.2. Examples from the ASCII code

Character	Code	Character	Code
(blank)	(00 000 000)	0	10 110 000
leader/trailer	10 000 000	1	10 110 001
line feed	10 001 010	2	10 110 010
carriage return	10 001 101	3	10 110 011
space	10 100 000	...	
+	10 101 011	A	11 000 001
-	10 101 101	B	11 000 010
∘ (period)	10 101 110	C	11 000 011
		...	

The above coding is often used within central processing units, too, especially if the computer system is oriented towards administrative EDP in which the
main bulk of data may consist of numbers as well as text. Many large-scale
computers such as the IBM Systems 360 and 370 have a standard option in their
instruction repertoire of addressing their memory systems by byte locations.
The transfer of information between the CPU and peripherals, as well as the
remote transmission of data is usually made by bytes, very often through
eight parallel lines or channels on which the bits of a character appear
simultaneously.

One further remark ought to be made at this point. While the access to the
storage location of main memories, disks, and drums can be based on address,
usually no such possibility exists for magnetic tape. This is due to the tape
transport mechanism which is based on pulleys and does not allow accurate
stopping of the tape. For this reason, the files of contiguous data must be

written as variable-length blocks of characters, separated by gaps, say, half an inch long at which the tape can be stopped. The whole block must be read and written as one unit. In order to identify a block, it must begin with a unique combination of characters, the *identifier* which has to be found and decoded. Normally this is done by the CPU or a special subcomputer, the *channel controller* which scans the transmitted characters by the execution of a small program loop, possibly using special hardware for it.

3.6.2 Specifications Used in Document Retrieval

Scanning of data by bytes is the normal mode of search in information retrieval, and document retrieval in particular. From the stream of characters it is then necessary to spot a certain character string or pattern which represents an identifier which matches with the given search argument. Sometimes the identifier is made to occur in a field which has a fixed position with respect to the beginning in every record. A more general identification must allow an arbitrary position of the identifier within specified limits, and the length of the identifier is usually variable. Documents normally contain many, say, a dozen different identifiers named *descriptors*; if the occurrence of a particular descriptor in the document is regarded as a logic proposition, or a Boolean variable which has either of the values {*true, false*}, then the condition for searching for a relevant subset of documents may be defined as some Boolean function over these variables.

Some of the identifiers characterizing a document may be quantitative, named *attributes* the *value* of which must satisfy certain magnitude relations in a search. For example, one may look for all documents published between certain years, or the length of which exceeds a certain limit. Such quantitative values of attributes must have been recorded in conjunction with the identifiers, and a more complex search process then must be able to evaluate the attribute and to assign a Boolean value to the corresponding proposition.

One further feature would be desirable in document search. The descriptors may occur in many equivalent forms the *root* of which is common. For instance, 'magnetic', 'magnetizing', and 'magnetization' might alternatively be selected to a set of descriptors. In order to take all of these into account, and yet to exclude other versions like 'magnet' and 'magnetoacoustic', one might want to specify the proposition used in search such that the root must be 'magnet' followed by 2 to 6 nonblank characters.

3.6.3 A Record-Parallel, Byte-Serial CAM for Document Retrieval

This subsection describes a CAM system suggested by PARHAMI [3.49,50]. Although it has not materialized as such, the logic of this design reflects most of those features which would be desirable to have in a CAM intended for document retrieval. Especially in the SDI (selective dissemination of information by libraries) where searching is not made as a batch run but upon request, a principle such as that discussed below might prove useful.

The following requirements which are not met by the other CAM principles discussed in this chapter, were stipulated for this design:

1) Large storage capacity with modular growth.
2) Variable-length records consisting of character strings.
3) Search on various specifications and their combinations as discussed above.

Notice that the items to be identified in document search are not words but variable-length *records* with a number of identifiers for the representation of which the regular memory array is not most suitable. When striving for maximum parallelism it is, therefore, possible to make the retrieval parallel with respect to a large number of records as well as to all bits in a byte. The string matching, however, usually contains so many individual comparisons that the retrieval has to proceed serially by characters (bytes).

For a cheap and relatively fast memory device, a fixed-head (head-per-track) magnetic disk or drum memory can be used. The recirculating dynamic memory operation of the bit-serial CAM is thereby replaced by mechanical rotation. This means a somewhat increased cycle time but much larger capacity. For instance, the contemporary and future disks may have bit densities of the order of 20 000 bits per inch which roughly corresponds at least to some 500 000 bits per track. The number of tracks and thus reading circuits is a question of cost but as many as 100 tracks with fixed reading heads are not uncommon. One surface of a special disk may thus store 50 million bits.

For every record, nine parallel tracks are reserved (eight for the character bits and one for a special marker bit needed in search). Since the records may be significantly shorter than the length of tracks, several of them can be packed in a row. The records can thus occupy sectors of arbitrary width.

In this application, it is desirable to define the matching criteria for character strings in a rather complex way. On the other hand, since the disk speed may be of the order of 100 rotations/s corresponding to a cycle time of 10 ms, there will be ample margin for real-time searching and it is possible to use many revolutions for one searching task, every time performing the same elementary processing operation on all records stored on the disk. In most operations processing is done in several sequential steps.

In the operations described below, an arbitrary location of the record along the track is assumed. The discussion is then independent of whether all records are scanned in parallel, or all or part of them consecutively. For parallel operation it is only necessary to have multiple processing circuits, one for every nine tracks. The following discussion relates to one record, shown in Fig. 3.20. This record is demarcated by the *start* and *end markers*, and special delimiters are used to indicate the boundaries between subfields of the record.

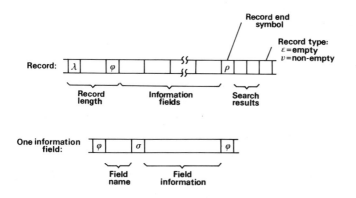

Fig. 3.20. Format of record in the byte-serial CAM

Reserved Characters. Every position in the string shown in Fig. 3.20 corresponds to an eight-bit character code and one marker bit. Most of the character codes represent data being, for instance, arbitrary symbols of the ASCII code. For this searching algorithm it is necessary to introduce some auxiliary codes for special delimiters and values; there are many unused bit combinations in the ASCII code for this purpose. Symbols of special character codes are defined in Table 3.3.

One may make a convention that the first field in the record always indicates its total length, whereas the beginning of this field is completely identifiable by λ. The other fields are then called *information fields*.

168

Table 3.3. Symbols of reserved character codes

Symbol	Meaning
λ	start of the record
ρ	end of the record
σ	demarcates the name (e.g., attribute) and the associated data (e.g., value) that occur in one information field
φ	end of any field
ε	end of an *empty* record
ν	end of a *nonempty* record
β	blank character position

Processing Operations. In the real device the records pass the reading heads; since the beginning of a record along the track is not restricted in any way, the various processing circuits, one per reading head, are normally in different phases. During one revolution, however, any external control such as buffering of one character of the search argument is always identical for all processing circuits, and it is thus assumed that a similar processing step will be carried out for all records during one revolution of the disk.

To describe the various processing operations, it will be equivalent to assume that some processing element makes several recurrent passes over the record from left to right, every time looking only for one particular character and its associated marking, and being able to change the found symbol or its marker or any symbol or marker to the right of it. The operation performed in one pass is defined by a *machine instruction* a set of which is defined, e.g., in Table 3.4.

Table 3.4 lists the instructions which describe the mode of operation of the processing element (processing circuit) during one pass (revolution). A *program* can be formed of these instructions. When studying programs written out of these instructions, it may be helpful to imagine that every instruction takes at least one pass to be executed. So, for instance, the *search* instructions set markers which are then later studied and modified by the *propagate, expand,* and *contract* introduction at the next passes. In reality, when searching for a matching string, only one character at a time can be held in the buffers of the processing circuits. The searching begins by

Table 3.4. The set of machine instructions

Instruction	Definition
search for $s_1 s_2 \ldots s_n$	Find all occurrences of the string $s_1 s_2 \ldots s_n$ and set the marker in the character which *immediately follows* s_n (This instruction applies to a single character, too.)
search for marked $s_1 s_2 \ldots s_n$	Same as before except for a string to match its first character must be marked; this marker is then reset
search for ψs	Find all occurrences of a character which has the relation ψ with s and set the marker of the following character. (E.g., if ψ is >, then for a character to qualify, its numerical value must be greater than that of s. Possible relations are <, $\leq$, >, $\geq$, $\neq$
search for marked ψs	Analogous to "search for marked $s_1 s_2 \ldots s_n$"
search and set s	Find all occurrences of character s and set their markers
propagate to s	Whenever a marker is found, reset it and set the marker in the first character s following it
propagate i	Whenever a marker is found, reset it and set the marker of the i:th symbol to its right
expand to s	Whenever a marker is found, set the markers of all characters following it up to and including the first s
expand i	Whenever a marker is found, set the markers of the first i characters following it
expand i *or to* s	Whenever a marker is found, set the markers of the first i characters following it but only up to and including s if s occurs earlier
contract i	In a row of contiguous markers, reset the first i ones
add s	Add the numerical value of s to the numerical value of all marked characters and replace these by the character representing the sum
replace by s	Replace all marked characters by s

searching for the first characters of the search argument. When a matching character is found, the marker in the *next* position is set. At the next pass, if a marker is found *and* the corresponding character matches, a new marker is set, etc. In this way the matching proceeds recursively, one character at a time, which takes as many passes as there are characters in the search argument.

Comment: The following list of instructions is almost directly from [3.50] but in view of the fact that the programming examples given in the original work seem to need some revision, one might alternatively study a simpler instruction set which uses more special characters (see an example below). To these instructions which refer to the searching phase only, there must be added the operations associated with *reading* and *writing* of records. The purpose of searching operations is to leave markers only in such records which satisfy all searching criteria and which, accordingly, can be located by these markers in the actual reading process. Writing must be preceded by a program which finds an empty and large enough slot for a record and marks this location available. For reading and writing, a host computer may be needed.

There are plenty of standard methods to implement the instructions given in Table 3.4 and the associated stored program control; the logic described in [3.50] gives some outlines for them. No attempt is made here to go into details. Similar logic principles will be discussed in Chapter 6. The block diagram given in Fig. 3.21 shows the most important parts of the processing circuits and the common control.

Example:

This program marks all records which contain the word 'magnet' followed by at least 2 nonblank characters. It is assumed that all markers have been reset before searching, and if the word is found, a marker is set in the 'ρ' character.

Program	*Comment*
search for magnet	if found, the next character is marked
expand to β	a row of markers is first set; (β=blank);
contract 2	if 'magnet' is followed by less than 2
	nonblank letters, all markers are deleted
propagate to ρ	

Comment

The programs could be made simpler if two types of "don't care" symbols were defined:

δ = any character including blank

κ = any character excluding blank .

Then, for instance, to find all words which have 'magnet' as the root followed by 2 to 6 nonblank characters, the following program might be used:

search for magnetκκδδδδβ

propagate to ρ

Reading of matching records from a rotating device looks difficult because the markers which indicate matching are at the trailing end of the record. There are now two possible ways for reading the marked records. In one of them, the descriptors and attributes are located in front of the above-mentioned markers but the information to be read out is written behind these marker positions. If necessary, duplicates of the descriptors could be written in this area. Another way is first to read the length of every record, found behind the λ character. If at the end of the record a marker is found, the program waits until the beginning of this record at the next revolution. The beginning address is obtained from the address of the marker by subtracting the length of the record.

Hardware Organization. The overall organization of the byte-serial CAM is shown in Fig. 3.21. There are external control circuits common to all processing circuits and they are connected to a stored-program control device, for example, a host computer. Each of the processing circuits, one per track, contains a buffer for the character to be compared, as well as a few flip-flops and some logic functions for the sequential operations which are connected with the reading and writing of markers. It may be recalled that the data representing the search argument, and data read from and written into the memory are transferred to and from the processing circuits one character at a time. Since input to and output from the rotating device are always through the processing circuits, these must be selectable by address decoders.

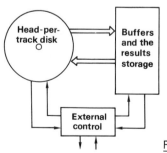

Fig. 3.21. Byte-serial CAM organization

No attempt is made here to describe the detailed control circuitry. Its implementation is straightforward, but a bit lengthy to be explained. The reader may consult the original article [3.50].

There is one particular detail worth mention. In the instruction "search for [marked] ψs" it is necessary to perform a magnitude comparison of two characters which are assumed to have a numerical value. Notice that the ASCII codes of numerals are natural binary numbers to which only a constant 10 110 000 has been added. Thus the equality and magnitude comparisons can be made by arithmetic subtraction. This is most easily implemented in 1's complement notation where the negative of a number is obtained by inverting all bits. If X and Y are integers and $\overline{X}$ is the 1's complement of X, the arithmetic subtraction is equivalent to

$$Y - X = Y + \overline{X} + 1 \ . \tag{3.16}$$

Notice that the addition of 1 is equivalent to setting the carry bit to 1 in the least significant position. The result zero means matching of X and Y, and an overflow (carry) from the most significant bit means that $Y \geq X$.

Related Systems. Rotating memories with various storage organizations and hardware are described by HOLLANDER [3.51], WARREN [3.52], ROTH [3.53], SLOTNICK [3.54], HEALEY et al. [3.55], MINSKY [3.56], HOUSTON and SIMONSEN [3.57], and CHYUAN SHIUN LIN et al. [3.58].

3.7 Functional Memories

In the usual CAMs, those parts of the search argument which have to match with the stored words are defined by setting up a *mask*. The masking pattern is the same for all stored words, and it thus excludes the same bit positions in all locations from comparison. On the other hand, in the byte-serial content-addressable search it was possible to define the "don't care" characters individually for all strings because the coding capacity of the characters allowed the use of special symbols. This naturally leads to a question whether the binary '0' and '1', as well as the "don't care" value $\emptyset$ could be set in the usual CAMs, too, i.e., at bit positions which can individually be defined for every stored word. If this were possible, then those parts of information which are regarded to be uncertain could be masked and skipped upon retrieval.

Another possible area of application for CAMs which can be masked within the memory is in the implementation of logic functions for computation and

control, in the way described below in this section. Content-addressable memories which have provisions for the within-memory masking are named *functional*
memories (FMs); for the implementation of logic functions, they usually include some auxiliary circuits for the collection of responses from the matching words, too. These constructs, as introduced by FLINDERS et al. [3.59],
as well as GARDNER [3.60,61], have the same objective as the microprogram
memories, i.e., to make the detailed machine operations and the control logic
programmable. The FMs in fact allow the representation of truth tables in a
highly compressed form.

It turns out that for the representation of the three symbols 0, 1, and
∅, at least two binary storage elements per bit cell are needed. Pairs of bit
values can be assigned in four different ways to three symbols, e.g., 0 ⇔ (1,0),
1 ⇔ (0,1), and ∅ ⇔ (0,0). In a hardware design, two bistable circuits together
with some comparison logic are needed for every bit cell.

3.7.1 The Logic of the Bit Cell in the FM

One schematic implementation of the FM bit cell, somewhat similar to that of
the CAM bit cell shown in Fig. 3.3 but without addressed readout, is presented
below in Fig. 3.22.

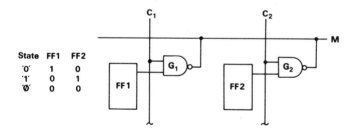

Fig. 3.22. Content-addressable reading function of the FM bit cell

The central idea applied in this circuit is the following. If no masking in
the search argument is used, in searching for '0' one has $C_1 = 0$, $C_2 = 1$,
and in searching for '1' there is $C_1 = 1$, $C_2 = 0$. Assume now that both flip-
flops are in the 0 state corresponding to value '∅'; since there are only
gates G_1 and G_2 connected to the M line, contrary to the double gates G_6 and

G_7 of Fig. 3.3, no mismatching signals can be caused on the M at any value combination of C_1 and C_2 whatsoever.

Only a few electronic implementations of the above principle have been suggested, although all basic circuit constructs of the CAMs are in principle amenable to the FM.

3.7.2 Functional Memory 1

The first of the FM types [3.19] is intended only to implement disjunctive forms of Boolean functions. They all can be expressed in a general form

$$F = \bigvee_P (A_1^{p_1} \wedge A_2^{p_2} \wedge \ldots \wedge A_n^{p_n}) \qquad (p_1,p_2,\ldots,p_n) \in P \qquad (3.17)$$

where the superscript p_i, $i = 1,2,\ldots n$ is assumed to attain one of the values 0, $\emptyset$, and 1, P is a set of combinations of the superscripts, and $A_i^{p_i}$ is an operational notation with the meaning

$$A_i^0 = \overline{A}_i, \ A_i^\emptyset = 1, \text{ and } A_i^1 = A_i \ .$$

The expressions $\overline{A}_i$ and A_i are named *literals* corresponding to an independent variable A_i.

The notations expressed in (3.17) have now a very close relationship to the so-called *compressed truth table* which is obtained, e.g., by the well-known *Quine-McCluskey method* (cf, e.g., [3.62]). The following example, Table 3.5, shows a usual and a compressed truth table, respectively.

Table 3.5. An example of truth tables

Usual				Compressed			
A	B	C	F	A	B	C	F
0	0	0	0	0	1	$\emptyset$	1
0	0	1	0	1	$\emptyset$	0	1
0	1	0	1				
0	1	1	1				
1	0	0	1				
1	0	1	0				
1	1	0	1				
1	1	1	0				

In the combination of rows, the $\emptyset$ signifies a "don't care". It is also possible to combine rows including $\emptyset$'s in the same positions whereby new rows with more $\emptyset$'s are obtained.

When comparing the compressed table with the Boolean expression (3.17) in which the same simplifications have been made as those which have led to the compressed table, then *every row in the truth table corresponds to one product term in (3.17), and the row is identical with the word* $(p_1, p_2, \ldots, p_n)$.

Obviously, if the prevailing value combination of the independent logic variables is compared with all rows of the table, the matching of a row in all specified bit positions (i.e., when no attention is paid to the $\emptyset$'s) can only occur at the rows shown. A comparison of this type differs from the masked comparison operations discussed with the CAMs in that $\emptyset$ corresponds to a mask which is set *within the table* and not in the search argument.

It is now possible to implement the compressed truth table using FM hardware, by storing the left half of the table in a special memory where content-addressable comparisons are performed. The memory cell, however, must then be able to attain one of three possible state values denoted by 0, 1, and $\emptyset$, respectively, and the bit comparison shall be masked out at $\emptyset$.

The FM is also suitable for simultaneous representation of several Boolean functions. The FM then consists of two parts, named the Input Table and the Output Table, respectively. As an illustration, we shall consider the incrementation of 8421-coded numbers mod 16, according to [3.18]. The simplified Boolean functions for the result bits (R_i) together with the carry (C), for all combinations of source bits (S_i) are given as follows:

$$C = S_3 \wedge S_2 \wedge S_1 \wedge S_0$$
$$R_3 = (\overline{S}_3 \wedge S_2 \wedge S_1 \wedge S_0) \vee (S_3 \wedge \overline{S}_2) \vee (S_3 \wedge \overline{S}_1) \vee (S_3 \wedge \overline{S}_0)$$
$$R_2 = (\overline{S}_2 \wedge S_1 \wedge S_0) \vee (S_2 \wedge \overline{S}_1) \vee (S_2 \wedge \overline{S}_0)$$
$$R_1 = (\overline{S}_1 \wedge S_0) \vee (S_1 \wedge \overline{S}_0)$$
$$R_0 = \overline{S}_0 \; .$$

$$(3.18)$$

The combined truth table is shown in Table 3.6. The $\emptyset$'s of the Input Table, and the 0's of the Output Table are indicated by blanks, and this convention shall be followed throughout the rest of this section.

In the hardware implementation of the combined truth table, the Input Table has a similar FM counterpart as that described above, with MATCH signals obtained as outputs at every word line (row). Several logic sums are formed by OR circuits, one for every Boolean function. The inputs to these OR cir-

Table 3.6

Row	Input Table S_3 S_2 S_1 S_0	Output Table C R_3 R_2 R_1 R_0
1	$\quad\quad\quad\quad$ 0	$\quad\quad\quad\quad\quad$ 1
2	$\quad\quad$ 0 $\;$ 1	$\quad\quad\quad\;$ 1
3	$\quad\quad$ 1 $\;$ 0	$\quad\quad\quad\;$ 1
4	$\quad$ 0 $\;$ 1 $\;$ 1	$\quad\quad$ 1
5	$\quad$ 1 $\;$ 0	$\quad\quad$ 1
6	$\quad$ 1 $\quad\;$ 0	$\quad\quad$ 1
7	0 $\;$ 1 $\;$ 1 $\;$ 1	$\quad$ 1
8	1 $\;$ 0	$\quad$ 1
9	1 $\quad\;$ 0	$\quad$ 1
10	1 $\quad\quad\;$ 0	$\quad$ 1
11	1 $\;$ 1 $\;$ 1 $\;$ 1	1

cuits are taken from those word lines which correspond to the bit value 1 in the columns of the Output Table.

It may be recalled that the main objective in the introduction of functional memories was the implementation of logic operations by programming. Accordingly, any specified hard-wired operations such as that described above in which word lines were connected to OR circuits according to the functions to be implemented, should not be allowable. The Output Table can now be made *programmable* by providing every location in it, i.e., every crossing of the word lines and columns by a usual flip-flop which can be read by the word line signal, and connecting the output circuits of all flip-flops of one column by a Wired-OR function. The value 1 is written into all flip-flops in which the Output Table has them, and so the vertical output line at every column will receive a resultant signal which corresponds to the hard-wired operation diescribed above. We shall revert to a similar programmed output operation with Functional Memory 2 below.

Search-Next-Read. Table 3.6 can further be compressed by introducing the Search-Next-Read function, as named by FLINDERS et al. The above example, due to its special properties, may yield a rather optimistic view of the applicability of this method, but similar cases may occur rather often. The central idea is that some input and output rows, not necessarily adjacent ones, may resemble each other. For example, in this example, input row 2 has a 1 in the same position as output row 1 has it; if a read operation were possible in the *Input Table*, output row 1 could be represented by input row 2. If we would now make the convention that after searching for input rows, the *next* input rows are automatically read, output row 2 could be deleted. Similarly, the output rows 2, 3, 4, 5, 7, 8, and 9 could be deleted, because they have 1's in the same position as the input rows 3, 4, 5, 6, 7, 9, and 11, respectively, have them. (Notice that rows can be easily *reordered* in order to represent as many output rows by next input rows as possible.) A separate problem arises with output rows 6, 10, and 11, which cannot be represented by the next input rows. The solution is that these left-over output rows are added behind the corresponding input rows in the table, but now every row is provided with an additional *tag bit*; for words not allowed to occur as search arguments, as for the three ones mentioned last, this tag bit is 1, and it is 0 for the rest of the words. The source word, the search argument, now has an extra bit in this position and its value is 0 during searching; during reading it is 1. The words provided with a tag 1, therefore, cannot match with any search argument during searching, and do not interfere. During reading, the next to the searched word is always read. Table 3.7 shows the more compressed table.

Table 3.7

Row	Tag	C	S_3 R_3	S_2 R_2	S_1 R_1	S_0 R_0
1	0					0
2	0				0	1
3	0				1	0
4	0			0	1	1
5	0			1	0	
6, input	0			1		0
6, output	1			1		
7	0		0	1	1	1
8	0		1	0		
9	0		1		0	
10, input	0		1			0
10, output	1		1			
11, input	0		1	1	1	1
11, output	1	1				

For instance, if the number obtained by incrementing 4, (i.e., 5) had to be generated by the functional memory, the search argument would be 000100. This matches with rows 1, 5, and 6 (input), and, therefore, rows 2, 6 (input), and 6 (output) are read out. The ORing of the respective bits yields 100101. By neglecting the leading 1, the tag, the result is 00101 which represents 5.

The tag concept can be developed further. By using different combinations of tags for different sets of rows, the same table can be used to represent different functions, just by providing the search argument by a function code which must match with particular tags in the searching operation. For further suggestions, see [3.59].

3.7.3 Functional Memory 2

The previously discussed functional memory was intended for a direct implementation of two-level logic, i.e., for disjunctive Boolean forms. With a minor amount of additional circuitry, certain multilevel logic expressions can advantageously be represented by a functional memory which thus can further be compressed. The first-level operation in these expressions is always AND, and the second-level operation is OR. The logic operation on the third level is EXOR (EXclusive OR), and the fourth level implements the ANDNOT function. These particular operations on different levels were selected for this implementation because it will be easy to transform truth tables into such expressions, as will be demonstrated below. In particular, the EXOR and ANDNOT forms can be found easily, e.g., by direct inspection of Karnaugh maps as illustrated by a couple of examples.

Let us first consider the Karnaugh map shown in Fig. 3.23a. The intersectional area of the two terms shown by dotted lines cannot be represented by their Boolean sum, but EXOR is a function which is directly suited for this purpose: the EXOR of these terms has zeros in the intersection.

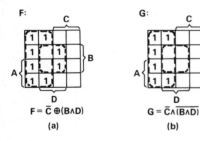

$$F = \bar{C} \oplus (B \wedge D)$$

(a)

$$G = \bar{C} \wedge \overline{(B \wedge D)}$$

(b)

Fig. 3.23a,b. Finding terms for Functional Memory 2: a) for the EXOR function b) for the ANDNOT function (see text)

Another example, the Karnaugh map of Fig. 3.23b represents a case in wich a small fraction of 1's are missing from an otherwise "complete" Boolean product term. By multiplying the "complete" term by the negation of another suitable term, i.e., by forming the ANDNOT of these terms, the 1's are covered.

A heuristic procedure for finding a simple three-level expression (not necessary the absolutely simplest one) is thus to try normal methods for covering the 1's in a Karnaugh map by the simplest terms. If, then, one otherwise had an almost "ideal" solution except that a few 1's were missing from some terms, or that some O's would exist in the intersections of two "complete" terms, the solution can be "patched" by introducing the third level as indicated in these examples.

Whatever procedure for finding a multilevel Boolean expression of the above type is utilized, the result is then readily implementable by Functional Memory 2. It consists of two partial tables, the Input Table and the Output Table. The former is similar to the Input Table of Functional Memory 1, and the searching operation is similar, too. The fundamental difference lies in the Output Table which has a special column, consisting of usual flip-flops and an EXOR circuit for every Boolean function which has to be represented by this memory.

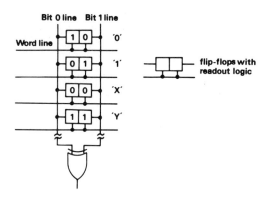

Fig. 3.24. One column of the Output Table of Functional Memory 2

Consider Fig. 3.24 which shows one column of the Output Table. A *pair of flip-flops* is shown schematically at every word line. The operation of these flip-flops is such that whenever the word line is activated (equivalent to reading to the flip-flops), the states of the flip-flops are made to appear on the Bit 0 line and Bit 1 line, respectively. The output circuits of all flip-flops connected to the Bit 0 line and Bit 1 line, respectively, are such

that a Wired-OR function is implemented at the corresponding line. (If the
word line signal is 0, then the corresponding flip-flops are not read, and
the contribution to the lines is zero from this row, irrespective of the bit
values stored in the flip-flops).

For brevity of notation in the Output Table, the state combinations in
the output flip-flops are denoted as indicated in Table 3.8:

Table 3.8. Abbreviated notation of states in the Output Table

Flip-flop states	10	01	00	11
Abbreviation	0	1	X	Y

The Bit 0 and Bit 1 lines are connected to an EXOR circuit. As a whole,
the logic operations performed by this column, as explained below in more
detail, are signified by the name *Read Right-EXOR-Left*.

Implementation of four-level logic expressions by means of Functional
Memory 2 needs a careful study and shall be discussed in the following with
the aid of examples relating to various partial operations on the different
levels.

1) Implementation of AND operations (first level) occurs in the Input Table.
   The results are the Boolean product terms which correspond to word line
   outputs.
2) Implementation of the logic sum of these Boolean products (second level)
   as well as the EXOR of two such partial expressions (third level) is done
   using the Read Right- EXOR-Left operation. Assume that a three-level Boolean
   expression found, e.g., from the Karnaugh map is of the form

$$F = (A_1 \vee A_2 \vee \ldots \vee A_M) \oplus (B_1 \vee B_2 \vee \ldots \vee B_N) , \qquad (3.19)$$

where the A_i and B_j, i = 1...M and j = 1...N are Boolean product terms
already computed by the Input Table, and their signals occur on the word
lines. Information is now written into the output column in the following
way: at all word lines which correspond to terms of the type A_i, the value
combination 10 is written into the corresponding pair of flip-flops, and
at word lines corresponding to the B_j terms, the value combination to be
written into the flip-flops shall be 01. Thus, if there were only value
combinations 10 and 01 in the Output Table, the expression F of (3.19)
would be implemented by the output columns, as can readily be seen.

3) If there were pairs of flip-flops with value combinations 00 in this column, then they would not be able to contribute to the output function F, irrespective of the word line becoming active. This value combination (abbreviated X) can, therefore, be written for all those Boolean product terms which have been computed in the Input Table but do not appear in the Boolean function corresponding to this column.

4) If, on the other hand, the value combination 11 = Y were written in the flip-flops, then, due to the EXOR function at the output, the function F would attain the value 0 whenever the corresponding word line became active. This value combination is then used to implement the fourth-level ANDNOT operations, as exemplified below.

The state Y is used in the Output Table if the corresponding Boolean product term Q_k, k = 1,2, ... occurs in the Boolean function in a position able to inhibit the whole function: that is, with a function of the form

$$G = F \wedge \bar{Q}_1 \wedge \bar{Q}_2 \wedge \ldots \qquad (3.20)$$

where F was implemented as before, the pairs of output flip-flops have Y at every word line corresponding to the inverted product terms $\bar{Q}_k$. If any word line corresponding to state Y is activated, it will yield a 1 at both bit lines of the column, with the result that the EXOR function then attains the value 0. Note that the previous form, by the well-known formulas of DeMorgan, can be represented in the ANDNOT form,

$$G = F \wedge \overline{(Q_1 \vee Q_2 \vee \ldots)} \ . \qquad (3.21)$$

Combination of the results from cases 1 through 3, results in that, with the Read Right-EXOR-Left, the following type of Boolean function can be implemented:

$$G = [(A_1 \vee A_2 \vee \ldots) \oplus (B_1 \vee B_2 \vee \ldots)] \wedge \overline{(Q_1 \vee Q_2 \vee \ldots)} \ . \qquad (3.22)$$

An Example, Derived from the Karnaugh Map. The map shown in Fig. 3.25 is represented by the Boolean function

$$F = (\bar{A} \wedge \bar{C}) \vee (\bar{A} \wedge D) \vee (\bar{B} \wedge \bar{C}) \vee (\bar{B} \wedge D) \ , \qquad (3.23)$$

but it has a striking correlation to a very simple form $F' = \bar{C} \vee D$. Only three squares ought to be "carved out" from F'.

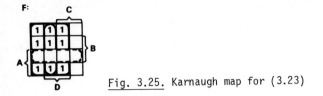

Fig. 3.25. Karnaugh map for (3.23)

The dotted line represents a term that now can be used to inhibit F' for the three mentioned squares to yield F, and the corresponding Boolean function would then be

$$F = (\overline{C} \lor D) \land \overline{(A \land B)} \qquad (3.24)$$

The functional memory for this functional form is represented in Table 3.9.

Table 3.9

Input Table				Output Table
A	B	C	D	F
		0		1
			1	1
1	1			Y

When the word line corresponding to the term $A \land B$ is activated, the Y state in the output bit cell yields a 1 at both output lines of this column, and so the EXOR of these signals yields the result 0.

Increment Table Solved by the Read Right-EXOR-Left. Without derivation, the increment table (Table 3.6) is now shown in a compressed form in Table 3.10 using Functional Memory 2 and this solution can easily be verified. The "don't care" states are shown by blanks in the input table, and the state notations defined in Table 3.8 have been used in the Output Table.

3.7.4 Read-Only Functional Memory

It may not always be realized that simple logic circuits have a close relationship to content-addressable memories. The wiring of a logic circuit is

Table 3.10

Input Table	Output Table				
S_3 S_2 S_1 S_0	C	R_3	R_2	R_1	R_0
0	0	0	0	0	1
0	0	0	0	1	X
0	0	0	1	X	X
0	0	1	X	X	X
	1	X	X	X	X

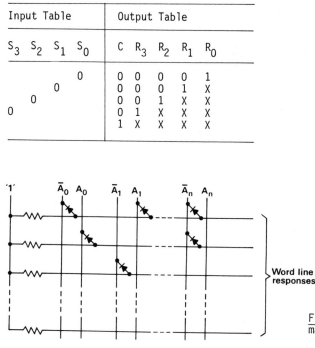

Fig. 3.26. Switching matrix as a read-only FM

representable as a double *switching matrix* (cf Sect. 5.4.4), and here we shall consider its first level only, depicted in Fig. 3.26.

The diodes connected to a horizontal line from vertical lines implement the logical AND operation over respective literals. It is now possible to regard this matrix as a *read-only functional memory*, or more accurately, the left-hand part in either Functional Memory 1 or 2. This is shown in the following way: If a diode connected to an A_j line, j = 0,1,...,n is regarded to represent a *permanently stored value 1*, a diode connected to an $\overline{A}_j$ line a *permanently stored 0*, and a diode *missing* at a vertical line from both A_j and $\overline{A}_j$ the *"don't care"* value $\emptyset$, respectively, then *the matrix is logically equivalent to the left-half array of functional memories*.

If there is a diode connected to either A_j or $\overline{A}_j$ for all j and for all the vertical lines, i.e., no $\emptyset$ values occur, then the array is logically equivalent to a *usual CAM*.

Read-only content-addressable memories have been described in [3.63-72].

3.8 A Formalism for the Description of Micro-Operations in the CAM

All the digital searching algorithms, including those discussed in the previous sections, can be defined using a formal notation. This formalism might
be used in the design of microprogrammed associative memories. It is based on
a subset of the APL language of IVERSON [3.73] which has been used in the
description of CPU functions in large systems, and the present form has been
worked out by FALKOFF [3.74]. The main advantage of the language of IVERSON
in this connection is that micro-operations such as masking, substitution,
as well as compression and expansion of the dimension of logical vectors
(i.e., vectors with logical variables as components), can be expressed in
a very compact way.

Statements. The elementary operations are expressed by statements numbered
by rows. Although several operations sometimes might be performed in parallel,
they are here shown on separate lines for better clarity. The statements are
executed in numerical succession except for when the order of a sequence is
changed, whereupon a jump on a branching condition is made.

Notation. The content of a *bit cell* is a binary scalar. Contents of *operational
registers* are binary vectors, and the *memory matrix* corresponds to the contents of an array of bit cells. *Functional variables* can be implemented by
hardware, the circuit logic, and they are integral parts of the memory. The
constants are logic values defined either by hardware connections, or by
binary values buffered in registers of the control logic. Notice that the
indices in the following ascend to the right and downwards, always starting
with the value 1.

Machine Variables

   M   memory matrix, an r by n array
   x   argument vector, 1 by n
   m   mask vector, 1 by n
   s   result vector, r by 1
   $\left.\begin{matrix} g \\ \ell \\ t \end{matrix}\right\}$ auxiliary result vectors, r by 1

Functional Variables

   X   matrix of r rows, each identical to x
   P   matrix of operators, same dimension as M

Constants

    ε row or column vector of binary ones

    ε^j row or column unit vector: the j:th component is 1, all others are 0

    α^j row or column prefix vector: the i:th components, $i \leq j$ are 1, all others are 0.

Statements

Let a, b, and c denote arbitrary vectoral arrays, and let arbitrary logical vectors be denoted by u and v. Let $\bar{u}$ be the Boolean complement of u. Matrices formed of vectors are denoted by A, B, C, and U, respectively. Then it is defined:

Operation	Definition
C ← a×b	C is formed as the *outer product* of a and b, i.e., $C_{ij} = a_i b_j$ where the subscripts denote array indices.
c ← v/U	c is a column vector that is formed as the *OR-reduction* of each row of U, i.e., $c_i = 1$ if at least one $U_{ij} = 1$, $1 \leq j \leq n$.
c ← ∧/U	c is a column vector that is formed as the *AND-reduction* of each row of U, i.e., $c_i = 1$ if all $U_{ij} = 1$, $1 \leq j \leq n$.
c ← u/A	*compression* of A by u: C is obtained from A by deleting all columns of the array for which u has the value 0.
C ← u\A	*expansion* of A by u: C has the same number of columns as u has components, with the columns consisting of zeroes where u has zeroes, and the other columns picked up from A in the same order. Notice that u/C = A and $\bar{u}$/C = 0, a zero matrix.
C ← \A,u,B\	*"row mesh"* of A and B: the columns of C are obtained from the columns of A and B retaining their order but always taking a column from A when u = 0, and from B when u = 1. Notice that u/C = B, $\bar{u}$/C = A.
c ← +/u	c is the arithmetical sum of the component values of u.
c ← (u≠v)	if c, u, and v have the same dimension, c has ones where the components of u and v disagree and zeroes where u and v agree.
x ← x*	x attains a value x* that is given externally.
a : b	branch on comparison of a and b: transfer to another line c may be indicated, e.g., as in the following: "if =, go to c" which means that if a = b, a jump to the line c is to be made, otherwise the next line is executed.

186

c ← a∗b	the arithmetic or logic operation which is signified by ∗ is applied to all the respective components of a and b and the results are substituted for the components of c in the same order.
c ← aRb	here R is a general relation over the respective components of a and b, e.g., = , and the value of the respective component of c is 1 if the relation holds and is 0 otherwise.

Examples of the Use of This Notation. In the following, the rows are denoted by superscripts and columns by subscripts. So M^k is the k:th row of the memory matrix. A comparison relation of M^k with the unmasked argument vector x is denoted by $M^k = x$ or $\overline{M}^k \neq x$, and the result is a row vector in both cases; in the former case it has ones in the places where the bits of M^k and x agree and 0 otherwise. In the latter relation the bits are 1 if the respective components disagree and 0 otherwise. The comparison for *equality* of all rows in the store with the argument vector, and a subsequent substitution of the matching results into a flip-flop s(k) of the results store, is denoted by

$$s(k) \leftarrow \wedge/(M^k = x) \quad \text{using AND-reduction, or}$$

$$\overline{s}(k) \leftarrow \vee/(M^k \neq x) \quad \text{using OR-reduction.}$$

The masked comparison can be expressed in a variety of ways. For example, it could be described by

$$s(k) \leftarrow \wedge/(m/(M^k = x)) \tag{3.25}$$

which means that a comparison of all the respective bits of M^k and x is made, and in combining the results, only those bits corresponding to components with the value 1 in the mask vector m are taken into account. (In this formalism, 0 means disabling a component.) In practice, however, the comparison circuits at the masked bits are disabled, i.e., these comparison operations are cancelled. For this purpose the *operator vectors* 1 and = are introduced. Because 1 and = are operators, relations are obtained from them by applying these expressions to a second operand vector. So, for an arbitrary vector a, the relation 'a1ε' is defined to have the value 1 for all components (identity comparison), and 'a=ε' means the usual comparison. By *row meshing*, the following *variable* operator vector is defined:

$$q \leftarrow \left| 1\varepsilon, m, = \varepsilon \right| \tag{3.26}$$

in which an identity relation holds for the components of m that are 0, and the comparison relation for components that are 1. The notation

$$s(k) \leftarrow \wedge/(M^k q x) \qquad\qquad (3.27)$$

is the masked comparison operation that has occurred, e.g., in the wired-logic example.

Parallel-by-Bit Equality Search. Let us assume that the following operations can be implemented by the circuit logic so that they need not be computed in separate steps (this is where the following algorithm differs from that presented by FALKOFF):

$$X = \varepsilon \times x \ , \quad P = (\neq \varepsilon \times m) \ . \qquad\qquad (3.28)$$

The matrix P is an operator matrix of the same dimension as M. Then the algorithm for parallel-by-bit equality search, including masking, is expressed as

1) $x \leftarrow x^*$
2) $m \leftarrow m^*$
3) $s \leftarrow \overline{\vee/(\overline{MPX})}$

At the third step, use is made of the results store in which a one is first set in all positions.

Serial-by-Bit Equality Search. A cycle counter is represented by a variable index j which represents the contents of a counting register or storage location. Again it is assumed that there exists circuit logic to implement the functional variable X described above. The j:th columns of M and X are denoted by M_j and X_j, respectively. This time the algorithm is written as

1) $x \leftarrow x^*$
2) $m \leftarrow m^*$
3) $j \leftarrow 0$
4 $s \leftarrow \varepsilon$
5) $j : n$;   if = , stop.
6) $j \leftarrow j + 1$
7) $m_j : 0$;   if = , go to 5.
8) $\overline{s} \leftarrow \overline{s} \vee (M_j \neq X_j)$;   go to 5.

Still Another Algorithm for Equality Search. If the content of any memory element can be inverted by a control signal derived from the argument vector, the following (unmasked) comparison algorithm can be used. If a bit of x is 0,

188

the corresponding column of M is then inverted, and after search, it is converted back to the original value. The two loops are counted by a scalar
functional variable called *alternator*, and denoted by a.

1) $x \leftarrow x^*$
2) $a \leftarrow 0$
3) $M \leftarrow \backslash \overline{\overline{x/M}}, x, x/M \backslash$
4) $a : 1$; if = , stop.
5) $s \leftarrow \wedge/M$
6) $a \leftarrow 1$; go to 3.

Serial-by-Bit Magnitude Search. The comparison algorithm of (3.9) shall be
represented in this notation using two results stores g (greater than) and ℓ
(less than). The argument is not masked. (Note: The array indices increase
to the right.)

1) $x \leftarrow x^*$
2) $j \leftarrow 0$
3) $g \leftarrow 0$
4) $\ell \leftarrow 0$
5) $j : n$;   if = , stop.
6) $j \leftarrow j + 1$
7) $t \leftarrow g$
8) $g \leftarrow g \vee (\overline{\ell} \wedge \overline{X}_j \wedge M_j)$
9) $\ell \leftarrow \ell \vee (\overline{t} \wedge X_j \wedge \overline{M}_j)$; go to 5.

Constant-Argument Search for Maximum. Following the general outlines of maximum search given in Sect. 3.4.5, the following algorithm is presented. The
bits of the argument are not changed by the algorithm. Instead, the *jth* bit
is unmasked. Candidates for further search are those for which it cannot be
determined with certainty on the basis of the higher bits that they are smaller
than the maximum.

1) $x \leftarrow \varepsilon$
2) $j \leftarrow 0$
3) $s \leftarrow \varepsilon$
4) $t \leftarrow s$
5) $j : n$;   if = , stop.
6) $j \leftarrow j + 1$
7) $m_j : 0$;   if = , go to 5.
8) $m \leftarrow \varepsilon^j$
9) $\overline{s} \leftarrow \vee/(MPX) \vee \overline{s}$

10) +/s : 1 ;   if > , go to 4; if = , stop.

11) s   ← t ;   go to 5.

Variable-Argument Search for Maximum. If the argument bits can be changed by the algorithm, an alternative for maximum search is the following:

1) j ← 0

2) j : n ;   if = , stop.

3) j ← j + 1

4) m_j^* : 0 ;   if = , go to 2.

5) x_j ← 1

6) m ← $m^* \wedge \alpha^j$

7) s ← ε

8) $\bar{s}$ ← v/(MPX) v $\bar{s}$

9) +/s : 1 ;   if > , go to 2; if = , stop.

10) x_j ← 0 ;   go to 2.

3.9 Survey of Literature on CAMs

Review Articles. In addition to the survey articles of HANLON [1.3], MINKER [1.4], and PARHAMI [1.5] mentioned earlier, general reviews on CAM have been published by ESTRIN [3.75], AUSLEY [3.76], FLYNN [3.77], FULLER [3.78,79], COMPUTER COMMAND and CONTROL [3.80], McATEER et al. [3.81], SLADE [3.82], SOHARA [3.83], CAMPI et al. [3.84], CHOW [3.85-87], HAAS and BLEVIS [3.88], SEEBER and LINDQVIST [3.89], HAAS et al. [3.90,91], BARNES and HOOTON [3.92], CHONG et al. [3.93], BARTLETT [3.94], BARTLETT et al. [3.95], CASHERA [3.96], ELECTRON [3.97], ELLIS [3.98], MINKER [3.99], LEILICH [3.100,101], WOLF [3.102,103], MOTSCH [3.104,105], and ELECTRONICS [3.106].

Special Constructs. Some solutions intended to introduce some auxiliary features in conventional CAMs are listed below: special array control structures [3.107-112]; CAMs which are hybridized with other memory functions [3.113-117]; special control of conventional memories [3.118]; multiaccess memories [3.119]; storage of extensible tables with variable word-length in usual CAMs [3.120]; a CAM with a dynamic loading feature for efficient usage [3.121], and a feature for many-to-many argument comparison [3.122].

Chapter 4   CAM Hardware

In this chapter, several detailed circuits and physical principles for the implementation of CAM functions are presented. It may be necessary to point out that not all of them have been used in practice; it is quite possible that certain solutions, although patented, may not prove practicable. It may be said definitely that *semiconductor CAM circuits* which are based on approved logic switching principles have already established their status in computer technology. *Magnetic elements,* although sometimes extensively studied, must now be considered obsolete. An important exception are the *magnetic-bubble memories* which are beginning to replace disk memories; it seems possible to add active searching and sorting functions to them. The oldest *cryotronic* principles, although once extensively studied, too, can no longer compete with LSI semiconductor circuits in packing density. On the other hand, the most modern superconducting switches, the *Josephson junction devices*, have the highest packing density of all known electronic switches, and they are seriously considered for ultra-high-speed computers. These devices have already been applied to buffer memories, although they too are still at an experimental stage of development. Finally, the *optical,* especially *holographic* memories ought to be mentioned. The general trend has for some time been away from them since comparable packing densities are achievable by LSI circuits.

4.1   The State-of-the-Art of the Electronic CAM Devices

Hardware implementations for CAMs have existed since 1956 [1.10] and CAM arrays currently exist as off-the-shelf semiconductor components. However, since the leading principles of electronic components have changed at a frantic pace during recent years, one may understand why solutions for the CAMs have not crystallized. One aspect is that the CAM components are needed in much smaller quantities than the circuits used in random logic and standard memory devices, which keeps their prices still high and their technology lagging compared to that of other circuits. Some promising new designs have been suggested and studied by computer simulations but there are only a few commercial CAM components, fabricated by a rather conventional technology, in arrays of 64 bits at maximum which are available at this writing. Since the future development of CAM technology cannot easily be forecast, I have tried to concentrate in this chapter only on the explanation of the basic principles of operation of these devices, and have left aside details of their prevailing specifications as well as structures of auxiliary circuits thereby needed. In any case, this review contains mainly such designs which can be mass-produced at reasonable prices and which, therefore, can most seriously be considered in practice. The main purpose in my presentation has been to render possible the understanding of the functional principles and the state-of-the-art of this technology.

Contrary to what is generally believed, the degree of complexity of a bit cell in a hardware content-addressable memory (CAM) need not be high, at least in principle; cf, e.g., the extremely simple solutions presented in Sect. 4.2.1. It is even noteworthy that a CAM can be built of quite usual, readily available standard memory modules by a new organization as shown in Sects. 3.4.1,2 and 4.3.1,3. Nonetheless CAMs have not yet acquired a position in computer technology which theoretically they could have. When looking for reasons for it, one may find a few: 1) The above-mentioned lack of demand which does not call for mass production. The fact that CAMs have not found their way into the main memories of general-purpose computers except as small auxiliary devices may be because it has not yet become clear what would be the standard way of programming a computer with content-addressable memory, and accordingly, what architecture should be selected. Widespread adoption of new designs cannot be expected until new disciplines for their use are generally accepted. One may recall that the boom of the third-generation computers was to a large extent due to certain identical basic solutions applied by most known manufacturers. 2) The simplest and cheapest principles for CAM bit cells have not yet achieved a stage of technological development which would guarantee a high stability and wide noise margin. 3) The more reliable CAM bit cells already developed involve many transistors because of the mixed modes of access (addressed and content addressable), and are, therefore, rather expensive. 4) Coordinate selection which in usual memories allows a high directly addressed capacity is not readily amenable to implementation in CAM arrays; the linear-select mode of addressing is normally used. 5) The control organizations of CAM systems, because of the rather large word-length normally

used, are more complex and expensive than with usual memories; large CAMs
may, therefore, not become cost effective except in quite special devices
and applications.

4.2 Circuits for All-Parallel CAMs

The logic structures shown in Fig. 3.3 can have many implementations by the
various logic families, first of all by TTL, ECL, and MOS. Since these circuits are usually realized by large-scale integration techniques, it is possible and even necessary to include an appreciable number of active elements
per bit cell, in order to retain ample safety margins against noise, and to
make the interfacing with other units easier. However, because it is also
desirable to tend to higher storage capacities, some suggestions have been
made ultimately to simplify the basic bit cell. It may be stated that the
more complex circuit structures have been commercially available at least ten
years, whereas the simplest constructs are still at an experimental stage of
development.

In addition to active electronic circuits, this section also reviews some
superconductive devices, old and new ones, by which CAM devices might be implementable.

4.2.1 Active Electronic Circuits for CAM Bit Cells

An Example of Content-Addressable Reading Function. Before showing the detailed constructions of CAM circuits, it may be informative to examine the
isolated content-addressable reading function in a very simple form. Fig.
4.1 delineates a TTL (transistor-transistor-logic) flip-flop which has two
control lines with signals C(0) and C(1) for the content-addressable reading
of the state of the bit cell as well as a word line which carries the resulting mismatch signal $\overline{M}$.

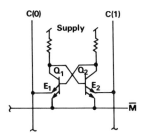

Fig. 4.1. Content-addressable reading in a TTL
flip-flop

If both control lines are held at a potential which is about -0.2 V or more
negative, they will draw all the collector currents and keep the emitter-base
junctions E_1 and E_2 cut off. This cell then cannot contribute to the word
line current. This corresponds to the resting state, or alternatively, the
situation in which the bit position is masked. Assume now that $C(0) = 0$ and
$C(1)$ is made equal to 1 (the 0 value corresponds to a voltage which is equal to
or less than -0.2 V, and the value 1 to, say, voltages over +2 V); if the
flip-flop was in the '0' state (Q_2 on), then the collector current of Q_2 will
switch from the $C(1)$ line to the word line. If, on the other hand, the state
of the flip-flop was '1' (Q_1 on), then the $C(1)$ signal has no effect on the
distribution of currents. Conversely, if it is taken $C(0) = 1$ and $C(1) = 0$,
then the word line will receive current if and only if the state of the flip-
flop was '1'. Let the signals $C(0)$ and $C(1)$ represent the search argument bit
in the same way as in the circuits of Fig. 3.3. It may be obvious from the
foregoing that a current which is switched from this flip-flop to the word
line indicates a *bit mismatch*; only if the search argument matches with the
states of the flip-flops at all bit positions, will the word line be current-
less ($\overline{M} = 0$).

A Complete TTL Bit Cell. A practical solution for a bit cell in a commercial
bipolar all-parallel CAM is shown in Fig. 4.2. It is actually a hybrid of a
TTL flip-flop and of a special diode-logic, emitter-follower comparison circuit [4.1].

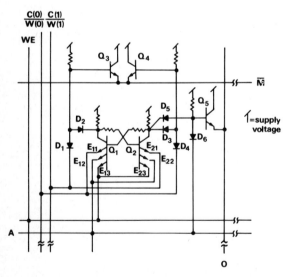

Fig. 4.2. TTL CAM bit cell

The TTL flip-flop of this circuit differs from that of a usual TTL random-access memory cell in that there are extra resistors in series with the base leads. Because of these resistors, a sufficient voltage swing is guaranteed for the diode logic. The latter, together with the emitter-followers, implement an AND-OR function. The OR operation is in fact a Wired-OR; the $\overline{M}$ line is assumed to have a proper impedance towards the ground whereby any single emitter follower (Q_3 or Q_4 in Fig. 4.2) is able to raise the potential of the $\overline{M}$ line high. The same lines are used for writing and reading of information (cf the signals $W(0)$, $W(1)$, $C(0)$, and $C(1)$ in Fig. 3.3, respectively), and so an additional WE (write enable) signal is needed to select either mode of operation.

During *writing*, the WE signal has a high potential. For words with a low address-line potential, this has no effect on the state of the flip-flops. In a word which has a high address-line potential, the new state of the flip-flop now depends on the potentials of the first emitters E_{11} and E_{12}. One of these (the one corresponding to the side to which the current is switched) is kept at a low potential, whereas the opposite emitter is now given a positive potential which is sufficient to cut off the current from that side. This state remains in the flip-flop after the potential of both E_{11} and E_{12} has been returned to its previous low value.

The logic operations for *content-addressable reading*, for which the logic gates G_6 and G_7 were provided in Fig. 3.3, are in the TTL bit cell implemented by the diodes D_1, D_2, D_3, D_4, and the transistors Q_3 and Q_4.

The addressed reading of all the bits of a word, selected by the address signal AD, is performed by a normal selector gate function implemented by the diodes D_5 and D_6, and the transistor Q_5.

Using high-speed components, e.g., transistors and diodes with Schottky barrier junctions, the access delay can be made as low as 10 to 15 ns.

TTL cells have been described by ASPINALL et al. [4.2], KINNIMENT et al. [4.3], HILLIS and HART [4.4], HUGHES [4.5], and SERT JOURNAL [4.6].

ECL Bit Cell. The bipolar content-addressable bit cell can also be implemented by ECL (emitter-coupled logic); a typical circuit consisting of 28 bipolar transistors and six diodes per bit is shown in Fig. 4.3 (without detailed explanation which is left to the reader [4.7]). The speed of ECL is significantly higher than that of TTL, but the supply voltage must be stable in order to guarantee a noise-free nonsaturated operation. Difficulties in the achievement of even quality necessary for the operation of this circuit are perhaps the reason why the price of ECL memories is still rather high. How-

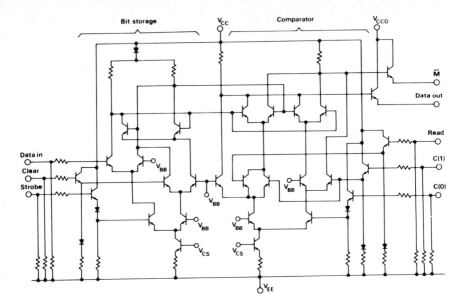

Fig. 4.3. ECL CAM bit cell

ever, it is decaying quite fast with time and, according to some forecasts, will approach the price of TTL memories in the 1980s.

Special Bipolar CAM Bit Cells. It is obvious that plenty of new circuit principles for CAM bit cells can be developed. Since price has been a limiting factor in the acceptance of CAMs, several attempts have been made to simplify the basic circuitry. Two of such designs, as suggested by MATSUE [4.8], MURPHY [4.9], and BERGER et al. [4.10] are mentioned here. In all of them, the state of a flip-flop can be set and read in linear-select as well as content-addressable mode using special output transistor stages: for reading or searching, they behave as amplifying gates, whereas for the writing of information into the flip-flops, the output transistor is used in the backward direction, by using the output current obtained from the base to set the flip-flop. It was suggested by MURPHY that the flip-flops can be provided with simultaneous horizontal and vertical sense logic; this is a feature that might also be used in other memory circuits. We would also like to mention a special two-way memory organization of CHU [3.15] which makes extensive use of this feature.

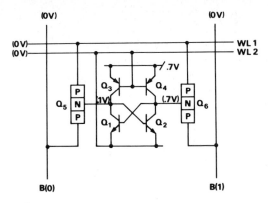
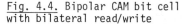

Fig. 4.4. Bipolar CAM bit cell with bilateral read/write

The bit cell of BERGER et al. is shown in Fig. 4.4. The two PNP junction structures represent symmetrical transistors. During searching, their lower P regions act as emitters, whereas during reading, the upper P regions are emitters.

The voltage values shown in parenthesis in Fig. 4.4 are stable state values when Q_1 is conducting. The state of the flip-flop formed of Q_1 and Q_2 can be *read* (in the linear-select mode) by raising the potential of the word line 1(WL 1) by about 0.8 V, i.e., by forcing a current into WL 1. The output transistors Q_5 and Q_6 then act as grounded-base logic gates, and the current is transmitted to either B(0) or B(1), depending on whether Q_1 or Q_2 is conducting. In the condition shown in Fig. 4.4, the current flows into B(0), and it must be noticed that this is equivalent to sensing a 1. During *searching,* the output transistors are used in the reverse direction. To search for 1's, the potential of B(1) is raised by 0.8 V, and to search for 0, the same is done for B(0). The *mismatches* are indicated by a current on WL 1: for example, if the state is 1 as indicated in Fig. 4.4, and B(0) is activated, then Q_5 conducts. If the state were 0, Q_5 would have been cut off.

The *writing* of information is done by controlling the output transistors in the backward direction, i.e., by sinking enough current at the collectors of the flip-flop through the base of either Q_5 or Q_6. The *write enable* operation is implemented by holding the word line WL 2 at about -0.9 V, and impressing current on the selected bit line.

A Bipolar FM Bit Cell. One possible implementation of a cell for FM is a double TTL flip-flop of which the content-addressable reading function, similar to that depicted in Fig. 4.1, is shown in Fig. 4.5 [3.59].

198

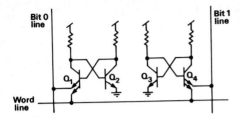

Fig. 4.5. Content-addressable
reading in a TTL FM bit cell

The states of the cell, as shown in Fig. 4.5, are defined in the following
ways: state '0', Q_1 and Q_3 on; state '1', Q_2 and Q_4 on; state 'Ø', Q_2 and
Q_3 on. The fourth state, with Q_1 and Q_4 on, is forbidden in Functional Memory 1, while it is utilized in Functional Memory 2. The bit value 0 is found
from the memory by holding the bit 0 line high and the bit 1 line low. The
bit value 1 is found with the bit 0 line low and the bit 1 line high, respectively. If masking is needed, both bit lines are held low. The matching of
all specified bits of a word is indicated by the absence of mismatch, i.e.,
by the absence of word line current in the searching operation. If the cell
is in the Ø state, it does not contribute to the word line mismatch current
for any bit line control condition whatsoever. The logic MATCH signal is obtained by complementation of the logic value of the word line currents. In
the following it is assumed that true MATCH signals are always obtained from
the complete memory array.

Other Bipolar Circuits. Further designs of CAM circuits based on bipolar
components have been presented by BIDWELL and PRICER [4.11], REPCHICK [4.12],
HILBERG [4.13], PRICER [4.14], and ORLIKOVSKII [4.15].

MOS Bit Cells. The circuits of KOO [3.4], BURN and SCOTT [4.16], WALD [4.17],
LEA [4.18], SHEAD [4.19], as well as PRANGISHVILI et al. [4.20] are very
much related in principle. In each of them, the bilateral nature of MOS transistors is utilized in order to implement an AND function during reading and
writing. All of the MOS bit cells mentioned have been implemented by integrated circuits, but it is difficult to guess which of them will be preferred in mass production. Making an arbitrary choice, the circuit of WALD
shall be described here. It is depicted in Fig. 4.6.

The operation of this circuit can be studied in two parts, first dealing
with the writing and reading of information in the addressed mode. The second
discussion is then related to the content-addressable search. *Writing* is performed in a similar way as in random-access MOS memories. The word control
line W is activated whereby the switching transistors Q_3 and Q_4 are made to

conduct. One of the bit lines B(0) and B(1) is kept low (grounded) whereby
the flip-flop formed of Q_1 and Q_2 is set to a corresponding state. If B(0)
is grounded, the new state will be 1. During *addressed reading*, the word
line W is also activated, but the bit lines B(0) and B(1) are now connected
to high-impedance sense amplifiers. Depending on the present state of the
flip-flop, current is now impressed on one of the bit lines through Q_3 or Q_4,
thus indicating the state of the flip-flop. Only one of the word lines is
activated at a time, so a Wired-OR operation is implemented on the bit lines.

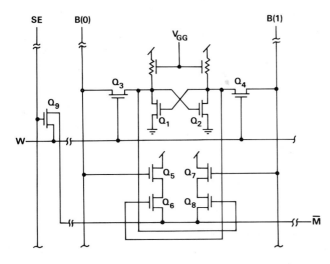

Fig. 4.6. MOS CAM
bit cell

The *content-addressable reading* follows the general principles already
discussed with the logic of Fig. 3.3. Now the W line is connected to a high-
impedance reading amplifier. The series connection of Q_5 and Q_6 forms an in-
verter-logic AND circuit which is used to detect the mismatch between B(0)
and the flip-flop output (collector of Q_2). Similarly, Q_7 and Q_8 detect the
mismatch between B(1) and the opposite output of the flip-flop. If neither
of these operations detects a mismatch, the $\overline{M}$ line potential remains low;
otherwise it is switched to the supply voltage. All bit positions of a word
are connected to the common $\overline{M}$ line whereby a Wired-OR operation is implemented.
The $\overline{M}$ line could be used as such for the indication of the searching results:
however, in order to keep the number of pads for external connections on a
memory chip low, it is customary to use the W line for address selection as
well as for match output. This implies that the reading and writing circuits
must be dynamically separated from each other using an extra switch Q_9 which

conducts only during the content-addressable search (whereby the search-enable signal SE is kept high).

In order to reduce power dissipation in the circuit, the transistors of Q_1 and Q_2 may be kept in a low-current "standby" state during storage, and a high current during writing and reading operations can be switched on by short strobing pulses superimposed on the supply voltage V_{GG}.

Other Designs for MOS Bit Cells. The previous circuit has in practice been proven to have good electrical characteristics for large-scale integration. However, a bit simpler versions of the circuit exist of which those of IGARASHI et al. [4.21], IGARASHI and YAITA [4.22], LEA [4.23,24], BANKOVSKI [4.25], and BANKOVSKI and SIMANAVICIUS [4.26] are very much related.

The bit cell of LEA, together with its control circuit [4.23] is shown in Fig. 4.7. This circuit is in many respects similar to the bipolar designs of MATSUE [4.8], MURPHY [4.9], and BERGER et al. [4.10] described earlier, and has the same logic of operation. The symmetric bipolar switching transistors have been replaced by the MOSFET transistors Q_5 and Q_6 which are inherently able to act as bidirectional current gates. A writing time of 25 ns and reading times (addressed as well as content-addressable) of about 10 ns are achievable.

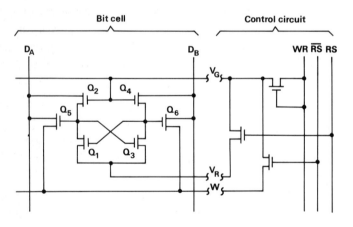

Fig. 4.7. MOS CAM bit cell with bilateral read/write. (The control signals correspond to those explained with Fig. 4.8)

An ultimately simplified circuit, with only *four* MOS transistors per bit cell has been invented by MUNDY [4.27-29]. In order to make its understanding easier, one half-cell of it, containing only two transistors, is first ex-

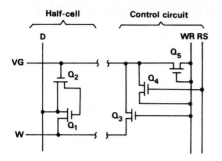

Fig. 4.8. Simple MOS CAM bit cell

plained with the aid of Fig. 4.8. This analysis follows closely that of LEA
[4.30,31]. The half-cell is able to act as a simple addressable memory element in addressed writing and reading modes, as will be explained below. The
MOS transistors are of the P-channel type, and the binary information in the
cell is stored in the form of electrical charge at the gate capacitance of
Q_1. Presence of a charge corresponds to value 1, and absence of it to the
value 0. There are neither constant supply voltages nor reference potentials
in this cell, and its state transition, as well as transfer of information
from it, are completely mediated by three lines with signals W, D, and VG,
respectively. The operation of the cell is "dynamic" in the sense that the
stored data will be lost by charge leakage unless special "refreshing" operations are performed intermittently with the aid of external circuits (the
VG control cell). These operations are equivalent to read/write cycles. The
D (data) as well as the W (write) signal assume one of the states logical 0
(+12 V) or logical 1 (0 V). (The negative logic convention is thereby applied.) The WR, RS, and VG signals assume different voltage combinations
depending on the operations to be performed. For their description, three
operational states, termed State 1, State 2, and State 3, respectively, have
been defined in Table 4.1.

Table 4.1. Operational states of the VG control cell

State	W [V]	WR [V]	RS [V]	VG [V]
1	0	- 8	+12	- 2
2	+12	- 8	+12	+11
3	X	+12	- 8	+12

X = either 0 or +12

The cell is in the *standby* condition when the VG control is in State 3.

Writing: The VG control circuit is in State 1, whereby transistor Q_2 is on. The zero voltage (logical 1) on the W line corresponds to the selection of a particular word location (address), and data corresponding to the D signal is written into the gate capacitance of Q_1 via Q_2.

Masked Writing: The selection of a word by the W signal can be inhibited by State 2 in the VG circuit.

Addressed Reading: A word is selected by setting the W line at +12 V (logical 0). The VG control circuit is otherwise in State 3. *All* D lines (bit lines) thereby assume the logical value 1. If the state of the storage cell is 0, then Q_1 will be on, and a current will flow between the D and W lines. No current flow is detectable if the state of the cell is 1. Presence of current on a particular D line onto which the respective bits from all words are connected is an indication of State 0 at this bit position in the selected word.

Content-Addressable Search: In order to make the above MOS memory design content-addressable, two half-cells are needed to form a bit cell, depicted in Fig. 4.9. There are two data lines denoted D_A and D_B.

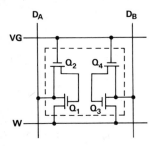

Fig. 4.9. Complete bit cell corresponding to Fig. 4.8

By convention, the state of the bit cell is said to be 1 if the left half-cell is in State 0 and the right half-cell in State 1, respectively. Conversely, the state of the bit cell is 0 with the left half-cell in State 1 and the right half-cell in State 0. The two other value combinations (0,0) and (1,1) of the half-cells are forbidden in this design but at least the first of them may be utilized in the functional memories described in Sect. 3.7. This kind of complicated arrangement is necessary because it will be easy to indicate a match between the data line signal value 1 and the corresponding state of the half-cell, whereas the half-cell states 0 are not directly detectable by matching. For this reason, the matching bit-cell states must be detected by searching for half-cell states 1 from the left and the right, respectively. Assume that the D_A line is held at the logical 0 value (+12 V), and the logical value of D_B is 1. The logical value of W shall be 1 (0 V). If the left half-cell is in the 0 state, then Q_1 will be on, and current on the *W line* indicates mismatch. Accordingly, if $D_A = 1$, $D_B = 0$, and the right half-cell is in the 0 state, then Q_3 will be on, and again, mismatch of the bits is indicated by current flowing on the W line. Only if there is a *matching at all bit positions,* the W line will be current-less.

Masking of Bits in Content-Addressable Reading: In this mode of operation, $D_A = D_B = 1$ (0 V), and since W = 0, too, no current can flow to the W line in this bit position irrespective of the stored state of the bit cell.

Other MOS Circuits. Miscellaneous solutions for MOS elements for CAM arrays have been presented by BURNS et al. [4.32], McKNIGHT [4.33], and CARLSTEDT et al. [4.34].

Special Active Circuits. A few other electronic circuits for CAMs have been suggested. Tunnel diode logic circuits have been presented by CORNERETTO [4.35], WANG and RUEHLI [4.36], as well as KOERNER and NISSIM [4.37]. Electro-luminescent elements have been suggested by FRENCH [4.38,39].

4.2.2 Cryotron-Element CAMs

The first experimental content-addressable memory was built and reported in 1956 by SLADE and McMAHON [1.10]; that construction was based on the super-conducting switching element named *Cryotron*. It seems that great promises were initially held by Cryotrons in computer technology; they are very simple in construction and reliable. The main disadvantage is a need for a constant supply of liquid helium for cooling these circuits to low temperatures. The first content-addressable memory elements were made of tantalum and niobium wires; later on, in 1960, SLADE and SMALLMAN [4.40], as well as SMALLMAN et al. [4.41] demonstrated thin-film versions of these elements which are ame-nable to automated production.

The Cryotron memories have been superseded by magnetic and semiconductor devices. However, in view of the experimental work recently done on other superconducting devices for digital electronics, for instance, the Josephson junctions that operate on low-magnetism quantum-mechanical principles, it seems that superconducting devices, at least partly, will be revived. We shall review some data published on these devices in Sect. 4.2.3. This review is not only included for historical interest; it exemplifies principles which are transferable to the more modern constructs, too.

The Cryotron memory can be made either bit-serial or bit-parallel. In this context the latter is more interesting, and the all-parallel thin-film version shown in Fig. 4.10 will be reviewed.

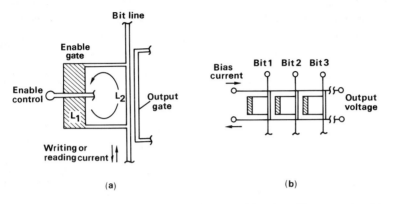

Fig. 4.10a,b. Cryotron memory: a) bit cell, b) cell array (enable control not shown)

The basic bit storage mechanism in the thin-film Cryotron memory is a persistent current flowing in a closed superconducting loop, the memory cell, either in the clockwise or counterclockwise direction (Fig. 4.10a). There is a leg L_1 in this loop, named "enable gate" which is shown shaded in Fig. 4.10a. This leg can be switched into the resistive state through an external magnetic field created by a control current in the enable electrode (named "enable control" in Cryotrons). Without control current, this leg will be superconducting. Writing of information into the bit cell occurs by controlling the "enable gate" into the resistive state whereby a *writing current* + I or - I, corresponding to either of the bit states, is conducted through the leg L_2 (named "output control") using peripheral control circuits (bit lines). When the "enable control" current is switched off and the "enable gate" becomes superconducting again, the writing current flowing in leg L_2 due to the self-inductance of the latter will be diverted through the "enable gate" and continue flowing in the loop.

Content-addressable reading of information from the bit cell is effected by switching a *reading current*, either positive or negative, through the bit lines. It will be superimposed on the L_2 current. The leg L_2, also named "output control", is simultaneously a control electrode to the "output gate", and a sufficiently large resultant current flowing in it will be able to switch the "output gate" into the resistive state. This occurs if the directions of the persistent and reading currents are the same in L_2, i.e., if the logic value of the reading current matches with the bit state.

In order to facilitate the all-parallel content-addressable searching, it is necessary only to connect all "output gates" corresponding to a word in parallel, as shown in Fig. 4.10b. As long as at least one of them is superconducting, the resistance of this parallel connection will be zero. If and only if all stored bit values match with the logic values of the reading currents, will all "output gates" of the word become simultaneously resistive, and a bias current through this parallel connection will create an output voltage over it.

Although the "output control" and the "output gate" are shown in parallel in Fig. 4.10a, in practice they overlap. The "enable gate" L_1 is 20 to 30 times wider than the "output control" L_2.

It has been estimated that the readout of a 4000-word by 25-bit memory is theoretically possible in approximately 20 ns. The size of the Cryotron is currently no longer comparable with other solutions; 10^6 bit cells in this design need a volume of one cubic foot. The new low-magnetism devices, however, may change this situation.

There exists voluminous literature on Cryotron CAM networks, suggested as well as implemented. Some of them have given rise to contemporary semiconductor embodiments, and many circuits are even more complex than the latter; they combine in themselves complex processing functions such as multiple-match resolution, sorting, ordered retrieval, etc. Within the scope of this text it is impossible to review all of these divergent approaches. Their details can be found in [4.42-103].

4.2.3 Josephson Junctions and SQUIDs for Memories

The packing density of conventional Cryotron elements cannot be made particularly high because of the appreciable magnetic fields needed to control the gates, and even more so because of the relatively high power consumption per gate in the resistive state; in a densely packed structure, extraction of heat presents a problem. For these reasons, the expectations initially held for Cryotrons have not materialized in commercial designs.

Quite different promises are held by the other superconducting elements which are known as the *Josephson junction device,* and the *Josephson interferometer* or *SQUID* (Superconducting Quantum Interference Device). The oper-

ation of the Josephson junction device is based on the nonlinear voltage-
current characteristics of the Josephson junction which are due to the quantum-
mechanical tunnel effect thereby taking place. The operation of the SQUID
(Josephson interferometer), on the other hand, is based on the distribution
of supply current between parallel junctions, whereby the electron waves
interfere with each other.

Josephson junction circuits and SQUIDs are switching devices which can be
used as logic gates as well as memory elements. It seems possible to inte-
grate logic with memories within a very small space. The Josephson junction
device can switch from the superconducting to the resistive state at very
nearly the theoretical speed of some ps (10^{-12}s). As the operating voltages
of the junction are in the millivolt region, the average power consumption
is very small, of the order of microwatts per junction. The power-delay product, which is a figure of merit, is several orders of magnitude better than
that of semiconductor circuits. The SQUID can be switched between two (or
more) different superconducting states whereby the power dissipation, and
the figure of merit, become better yet.

Digital Circuits Based on Josephson Junctions. There is plenty of research
in progress on the development of superconducting digital circuits for super-
fast computers. The Josephson junction thereby applied most often operates
as a magnetically controlled switch, very much in the same way as the Cryotron
[4.104-110]. A single junction is presented schematically in Fig. 4.11a, and
its static voltage-current characteristic curve is delineated in Fig. 4.11b,
with the load line corresponding to a resistive load R and supply voltage V_g.
The steep, practically vertical branch of the curve is due to the quantum-
mechanical tunneling current which causes no voltage drop. The peak value of
it is called the *Josephson threshold current* and denoted I_m. The other branch
is due to current in the resistive state of the junction. Notice that this
branch has a valley (V_v, I_v). Actually there should be a negative-resistance
part of the curve visible, as in tunnel diodes.

An external magnetic field, for instance that caused by one or several
control electrodes near the junction, is able to lower the curve as shown in
Fig. 4.11b. Assume now that a load resistance R is connected in parallel with
the junction, and a supply current I_s is fed through this combination. The
operating point is obtained as known by drawing a *load line* with slope 1/R
through the point (0, I_s). The intersections of this line with the curve
define the operating points. The biasing of the device is usually such that
$I_s < I_m$, or that in the absence of any control field at the junction there

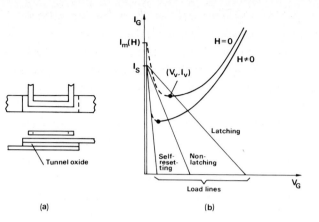

Fig. 4.11a,b. Josephson junction: a) electrode structure, b) characteristic curves and load lines at zero and nonzero external magnetic field strength H, respectively

will be at least one stable operating point at $(0, I_s)$. In order to make the device switch from this point, it is possible to lower the peak value I_m by the application of control current to a value $I_m(H) < I_s$ whereby a switching cycle, similar to that encountered in tunnel diode circuits, is triggered. This cycle ends in different ways depending on the value of R. Three principally different cases with different values of R can now be distinguished:

1) The load line always makes a second intersection with the right-hand branch.
2) There is a second intersection only in the absence of control current.
3) There is never a second intersection.

With a typical junction, these cases may correspond to values of R of, say, 0.3 Ω, 0.2 Ω, and 0.1 Ω, respectively.

In the first case, termed the *latching operation*, the circuit makes a transition which converges to the second stable point. The exact form of the trajectory which is usually oscillatory depends on the circuit reactances. In order to reset the circuit back to the operating point $(0, I_s)$ it is necessary to lower the supply current for a moment in order to shift the load line below the valley point (V_v, I_v).

In the second case, named *nonlatching operation*, there will be a trajectory which also converges in the second stable point as long as the control current is on; after the control current is removed, there will be an automatic transition to the only remaining stable point $(0, I_s)$.

In the third case, named *self-resetting operation*, there will never be a second stable operating point, and a monostable multivibrator action takes place whereby, after a delay determined by the circuit reactances, the trajectory returns back to $(0, I_s)$ even though the control current is held on.

It is obvious that the Josephson junction device can operate as a threshold-logic gate since the net control field is the sum of the fields caused by several adjacent electrodes. If one electrode alone were able to cause a field H such that $I_m(H) < I_s$ the device would then be equivalent to an OR gate. If it needs the simultaneous activation of all electrodes, then the logic operation is AND. Other Boolean functions are implementable by weighting and reversing some of the control currents.

A two-input logic circuit, with the load resistor connected via a transmission line, is shown in Fig. 4.12. The conductors of the line can act as control electrodes to further circuits.

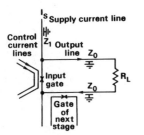

Fig. 4.12. Josephson junction logic circuit

SQUID Logic Elements. The Josephson interferometer or SQUID is in effect a system of two or more parallel Josephson junctions between which the supply current can be made to distribute in discrete steps. Switching to a new current mode is then equivalent to a threshold-logic operation the result of which can be detected as a change of current in one of the legs. The physical principle of operation is explained below in the context of SQUID memory cells. These devices were earlier used mainly to detect extremely weak magnetic fields, but in principle, they can replace the single-junction Josephson device in the circuit of Fig. 4.12. The use of SQUIDs as digital switches might be motivated for the following reasons: 1) The operating currents, and accordingly, the power consumption can be made much smaller (almost by a factor of ten) than those of the single-junction devices. This is due to the mode of control by magnetic field which affects the effective circuit inductances (Fig. 4.13b), and thus the distribution of current between the junctions.

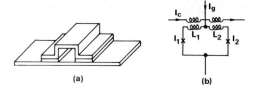

Fig. 4.13a,b. Two-junction
SQUID: a) electrode structure,
b) approximative equivalent
circuit

2) The different physical effects utilized allow more flexible geometric
design, and the minimum widths of the control lines are solely limited by
the resolution of photolithography applied in fabrication. A typical two-
junction SQUID is shown in Fig. 4.13a.

Memory Circuits. In logic circuits, as mentioned above, it is possible to replace the single junction by a special SQUID (preferably with three parallel
junctions). A bistable circuit is then implementable as a logic flip-flop.
A much simpler method, however, is to use *parallel junctions as a loop* which
carries a persistent current. This current creates a magnetic flux which must
be an integral multiple of a flux quantum $\phi_0 = h/2e = 2.07 \cdot 10^{-15}$ Vs where
h is Planck's constant and e the elementary charge. It is also in practice
possible to have a persistent current corresponding to only one flux quantum
which then determines the theoretical lower limit of storable energy
[4.111-116].

The persistent current which is superimposed on the supply current is a
special closed-loop superconducting state named *vortex mode* of superconduction,
and it is identified by the number of flux quanta n accompanied by it. It
can be set, or information can be *written* into the cell by including a transition from one vortex mode to another. Assume that a two-junction SQUID is
used, and n = 0 and 1 correspond to binary values '0' and '1', respectively.
Fig. 4.14 shows the loop current I vs the control current I_c indicating that
the current modes overlap, i.e., for a particular I_c, both modes are possible.
There is also a region without overlapping and in which the device is uniquely
switched into the corresponding mode. Assume that the control current is biased to the value I_{cb} which is symmetrical with respect to the two modes. If the
device was in the '0' mode, deviation of I_c to a value $I_{cb} + \Delta$ shown in Fig.
4.14 sets the '1' mode, and again, when the device is in the '1' mode, resetting to '0' is effected by application of control current $I_{cb} - \Delta$.

Reading of the vortex state can be made in two different ways. In the
first of them, change in the flux by a quantum causes an induction voltage

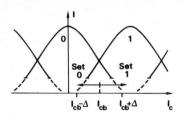

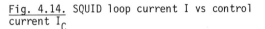

Fig. 4.14. SQUID loop current I vs control current I_C

the integral of which is $\int V\,dt = \pm\,\phi_0$. The voltage can be detected by another junction biased close to its threshold.

Notice that switching between the vortex states is only a change of super-conducting mode. It is also possible to make the SQUID switch into the resistive state whereby a higher voltage and power dissipation are exhibited. The high voltage swing due to the transition can also be used for the *reading* of the cell state: for instance, if the device is initially in the '1' state, it can be made to switch into the resistive state by simultaneously pulsing I_c as well as I_g high enough. This transition does not occur if the device was in the '0' state.

Reading of the memorized state by either of the above methods is destructive since the state must be changed for its detection, and it must be returned to the original value. After the superconducting-to-resistive transition, the superconducting state must be reset by momentarily switching off I_g.

Content-Addressable Cell. A CAM cell which has been built and organized according to the above-described principles has been suggested by BANKOWSKI and HAMEL [4.117].

4.3 Circuits for Bit-Serial and Word-Serial CAMs

The main motive for the introduction of bit-serial and word-serial CAM architectures has always been ultimately to lower the cost of the bit cell. One of the word-parallel, bit-serial memory technologies which was accepted for the first large-scale CAM installations is the *plated-wire* memory principle discussed below in Sect. 4.3.2. It was later replaced by LSI random-access memory modules not only because of higher speed but also because of the skew-addressing mode of operation used.

It is to be mentioned that plenty of magnetic memory principles based on annular or multiaperture cores have been suggested and implemented, too (cf, e.g., [4.118].) The main drawback of magnetic devices is the rather long access time, which in these applications typically ranges from 100 ns to 1 μs. Nonetheless, magnetic-core memories have survived even in conventional computer technology until these days because they have one particular merit: information stored in them is nonvolatile even after power blackout.

After discussion of semiconductor RAM circuits, the following presentation will review briefly only some of the simplest magnetic-core CAM implementations. After that, the dynamic memories based on shift registers, CCDs, and magnetic-bubble devices are discussed because they might be applied for bit-serial as well as word-serial operation. The clock frequency of these circuits, however, is not comparable to that achieved with delay-line dynamic memories. To the knowledge of this author no commercial architectures of CAMs have been based on these principles, but it may nonetheless be useful to mention them here since the technologies may in near future develop quickly enough to allow very simple and cheap implementation of content-addressable search by these devices, at least in special applications.

4.3.1 Semiconductor RAM Modules for the CAM

The memory component needed for reading and writing by bit slices, as explained in Sect. 3.4.2, shall be a linear-select random-access memory (RAM) module. Since with bit-serial operation it would be desirable to have a small access time, maybe the Schottky barrier diode clamped bipolar (TTL-type) memory modules would be most suitable. An example of commercial 256-word by 1-bit memory module with built-in decoder and Wired OR output is the Intel 3106 shown in Fig. 4.15; it is to be noted that no control is necessary for reading, whereas for writing, the $\overline{WE}$ pin must be activated. If these modules are to be connected to a larger array, up to 2 K words, of storage locations, the *chip select* (CS) inputs can be used.

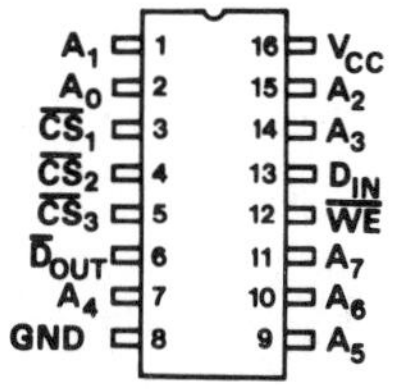

Fig. 4.15. RAM module for word-parallel, bit serial CAM. Legend: A_0 through A_7 = address input; CS_1, CS_2, CS_3 = chip select; D_{IN} = data input; D_{OUT} = data output; WE = write enable; V_{CC} = supply voltage; GND = ground

212

As explained in Sect. 3.4.2, this same module is directly applicable to skew addressing.

4.3.2 Magnetic Memory Implementations of the CAM

Annular-Core Implementation. Consider the usual or annular ferrite core used in memories. We might rest content with the bit-slice operation, in which case a usual linear-select magnetic-core memory could be used as a CAM. However, since the manufacturing technologies of ferrite core memories easily allow some modifications, we shall here consider the special solution depicted in Fig. 4.16 in which four lines are threaded through the core: the *write line* which carries the current I_W, the S_W (*word sense*) and the S_B (*bit sense*) lines, as well as *bit line* with current I_B on it. The two directions of magnetization are clockwise and anticlockwise, henceforth defined as the binary states '1' and '0', respectively. Along with the principles applied in the linear-select magnetic-core memories (cf, e.g., [3.21]) these states are written into the memory elements in two phases: first, all cores on a word line are reset to '0' by the application of current $I_W = -I_C$, where I_C is the current value corresponding to the coercive field strength. After that, I_W is set equal to $+I_C$, and the bit line current to 0 or $-I_C/2$, depending on whether '1' or '0', respectively, has to be written in this bit position. In the latter case, namely, the resultant current through the core is $+I_C/2$ which is not able to change the direction of magnetization.

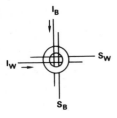

Fig. 4.16. Magnetic core with signal wires for CAM operation

Assume now that all words with the '0' state in this bit position have to be found. When the I_B current equal to $+I_C$ is applied on the bit line, all cores in this bit column with the state '0' will switch to '1', thereby inducing a voltage signal in the S_W line. If the state was '1', no induction voltage due to state change will be detected. Similarly, if all words with the bit state '1' have to be searched, the bit current is set to $-I_C$. The

induction voltages on the S_W lines will be amplified, rectified, and detected by reading amplifiers familiar in usual ferrite memory technology.

Another possibility for content-addressable reading is to take $I_B = -I_C$
for all bit positions which are read in succession. The "bit slice" is thereby read in a normal way. If the search argument bit was 1, then the word
line signals directly indicate matching rows. If the search argument bit was
0, the word line signals have to be inverted in the output circuit.

Since the reading of magnetic elements is *destructive*, i.e., the stored
state value is lost in reading, the information read out must immediately be
written back in the same place. After this kind of content-addressable reading, if the current $I_B = +I_C$ was used for searching the bit '0' values, then
the rewriting must be done by holding I_B equal to $-I_C$ and setting the word
current I_W equal to 0 for matching words and equal to $+I_C/2$ for mismatching
ones. If the search for bit values 1 was performed using current $I_B = -I_C$,
then I_B is set to $+I_C$ and I_W equal to 0 for matching words and $-I_C/2$ for
mismatching ones, respectively.

Addressed reading is carried out by setting the I_W current on the selected
word line to $-I_C$ whereby an induction voltage will be detectable on all S_B
lines corresponding to the '1' state in this bit position. The read information is written back as in usual addressed writing.

Special annular-core CAM solutions have been suggested by HUNT et al.
[4.119], BEISNER [4.120], and TUTTLE [4.121].

Multiaperture Cores. Some improvements to the electrical characteristics of
magnetic-core memories such as an improved signal-to-noise ratio are brought
about by the *multiaperture* switching cores, which also facilitate some extra
logic operations per core. Those structures which have mostly been used in
practice are the *Biax element*, the *Transfluxor*, and the *LEM core*. Reviews of
these are found in the books of CHU [4.122], as well as of IL'YASHENKO and
RUDAKOV [4.118]. Some samples of their applications in CAMs can be found in
[4.123-136]. With these and similar devices, the readout time can further
be compressed by careful timing and partial overlapping of the control currents. While multiaperture devices are readily fabricatable in the laboratory,
their automated mass-production presents some difficulties. Multiaperture
cores have, by and large, been replaced by other technologies.

The Plated-Wire Memory. In the basic design of this device, there is usually
no provision for addressed reading since this memory is normally used for
location of variables in parallel computations. Its physical principle is the
following. When a nonmagnetic substrate wire is plated with a thin nickel-iron

214

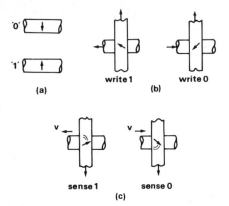

'0' ☐ ↓
'1' ☐ ↑

write 1   write 0

(a)   (b)

sense 1   sense 0

(c)

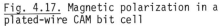

Fig. 4.17. Magnetic polarization in a
plated-wire CAM bit cell

coating under the influence of a static magnetic field, caused by a current
flowing through the wire, the coating becomes magnetically anisotropic: it
will have two easy directions of magnetization, the clockwise and the anticlockwise one with respect to the axis, as indicated in Fig. 4.17a with small
arrows. The resulting magnetic polarization can be made to switch between
these two states by the application of an external magnetic field. Let us
call the stable states the '0' state and the '1' state of the bit cell, respectively. A current-carrying conductor strap passing the surface will create an external magnetic field that is orthogonal to the external field caused by the current flowing through the wire. The resultant of these two fields
turns the magnetic dipoles in either way shown by the small arrows in Fig.
4.17b, depending on the direction of the current in the wire. After the currents are turned off, the dipoles will relax to the '1' or '0' state, correspondingly. This is the operation in which information is written into the
magnetic bit cell. Notice that in absence of the strap current, the current
on the wire alone is able only to turn the magnetization by an angle that is
much smaller than 90 degrees, and cannot cause a flipping of the state. The
strap current during reading of information is opposite to that during writing,
and the wires are then connected to the read amplifiers. Because of the strap
current, the magnetization vector is rotated from the stable state as shown
in Fig. 4.17c. The direction of rotation depends on whether a '1' or a '0'
was stored in the cell. But a change in the magnetic flux through the wire,
due to the rotation of the magnetic polarization, will now cause an induction
voltage the direction of which depends on the flux change as indicated in
Fig. 4.17b. The polarity of the induction voltage is then sensed by the read
amplifier.

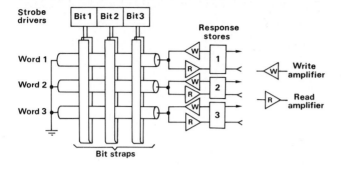

Fig. 4.18. Memory array of a plated-wire CAM

Fig. 4.18. shows the organization of a content-addressable memory array, discussed earlier in a general form in Sect. 3.4.1 [4.137-139].

Since the reading operation is destructive, the retrieved information must immediately be written back using the writing circuits provided at every word line, and strap current corresponding to the prevailing bit value.

The access time in a plated-wire memory is usually smaller than in ferrite core memories, typically of the order of 100 ns.

Plated-wire CAM structures are further described in [4.140-145].

Further Magnetic-Element CAMs. Thin-film magnetic-element memories can in general easily be designed to perform content-addressable searching functions. Examples of these are described in [4.146-153].

Special magnetic-element structures have been presented in [4.154-162].

4.3.3 Shift Registers for Content-Addressable Memory

The two primary objectives in the design of electronic circuits for active semiconductor memories are a high packaging density, and a small power/speed ratio. The MOS (metal-oxide-semiconductor) circuits are advantageous in both of these respects. The linear dimension of a single transistor on the silicon chip is typically of the order of 20 μm, and one bit storage is typically implementable by six transistors.

There are two basic types of MOS shift registers which may be taken into consideration in a word-parallel, bit-serial content-addressable memory, namely, the *static* and the *dynamic* one. One stage (bit position) of a static MOS shift register, together with the waveforms of timing signals needed to effect one shifting operation is shown in Fig. 4.19. Information in the static

216

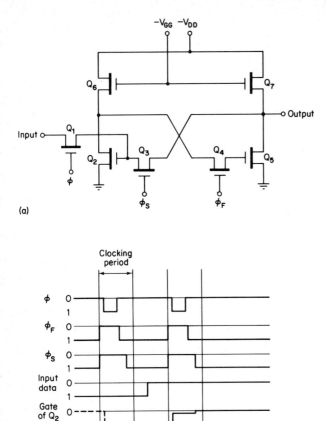

(a)

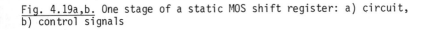

(b)   Time ——→

Fig. 4.19a,b. One stage of a static MOS shift register: a) circuit,
b) control signals

shift register will be saved for indefinite periods of time, and the shifting can be started asynchronously.

Contrary to the above circuit, information in a dynamic MOS shift register
is volatile and will be lost unless the memory is "refreshed", i.e., the
contents are continually shifted end-around at a certain minimum clock frequency. Accordingly, the operation of a dynamic shift register is usually
synchronous. One practical implementation, the *four-phase shift register* together with the waveforms of its four clocking signals, is shown in Fig. 4.20.

a)

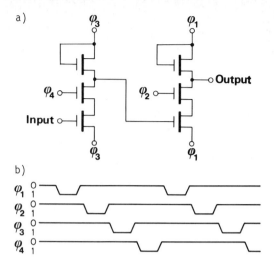

b)

Fig. 4.20a,b. One stage of a dynamic (four-phase) MOS shift register: a) circuit, b) control signals

In this solution, the four clock phases must be divided *cyclically* between the different stages so that the gates open in succession and the bit values "ride on the waves". For more details of these and related circuits, see, e.g., [3.62].

4.3.4 The Charge-Coupled Device (CCD)

Another elementary component for dynamic memory which has the same technology of fabrication as the MOS circuits but does not operate on standard logic signals is the *charge-coupled device (CCD)*. It, too, holds great promises in future computer technology. The construction of a CCD is still simpler than that of the MOS shift registers, and the packaging density is correspondingly higher. In return, its speed of operation is somewhat inferior to that of the shift registers, a feature which in large-scale word-parallel operation is not particularly bad, however. A review of the present state of CCDs can be found in [4.163].

The basic design of a CCD is a silicon substrate, coated with a thin (0.12 μm) insulating SiO_2 layer over which a series of metal electrodes are sputtered. (Fig. 4.21a). Contrary to the MOS components, no metal electrodes are evaporated directly onto the semiconductor in this device. When a sufficiently high voltage is applied on one of the electrodes (positive with P-type silicon) then the induced field will attract minority carriers, with a result that a local conductive channel (N-type in this example), as well

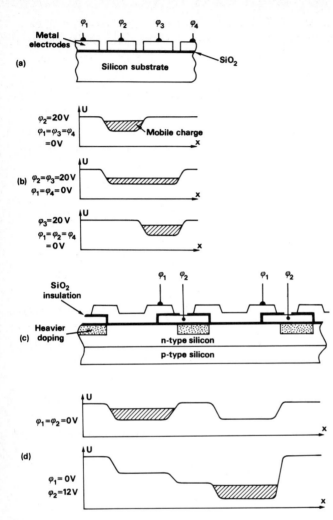

Fig. 4.21a-d. Charge-coupled device (CCD): a) principal structure, b) control voltages and propagation of charge, c) practical electrode structure for two-phase operation, d) two-phase control voltages and potential wells

as a potential well are formed. The channel and the well are capable of holding mobile charges (electrons in this case). By a suitable timing of control voltages applied on consecutive electrodes, the channel and the well can be made to move along the surface (Fig. 4.21b). Any amount of mobile charge contained with them will then be moved, too.

The first task is to load the channel at one end of the structure with a proper amount of mobile charge, whereafter, since the surrounding substrate in neutral state is a good insulator, the charge (and binary information associated with it) will be propagated along the surface. As there is a minor amount of leakage of charge from the moving channel, however, the electrode structure cannot be made indefinitely long, and the memory must be "refreshed" by amplification of the signals received at the other end, whereafter they are let recirculate in the same way as with delay-line memories and the dynamic shift registers.

As the memory capacity of a CCD, or the maximum useful length of the electrode structure directly depends on leakage effects, especially those which occur near the surface, some improvements to this structure have recently been made. One of them is the so-called *buried channel* which means that if, for instance, on a P-type silicon substrate, a thin layer of N-silicon is formed by applied doping, then the potential well, as well as the channel will be shifted in the vertical direction below the p-n junction, away from the charge traps and other causes of leakage at the surface. The efficiency of charge transfer from one electrode location to another of about 99.99 per cent is thereby attainable, with the result that the recirculating bit string can be made several thousands of bit positions long. This is already more than needed in a word-parallel, bit-serial content-addressable memory, and the new types of CCD might be considered as a memory medium in the word-serial designs, too.

One practical electrode structure, together with *two-phase* controlling waveforms, and the shape of potential wells which results from auxiliary implanted electrodes (due to extra P-type doping) are shown in Fig. 4.21c and d [4.163].

Some typical specifications of CCD devices, those already achieved, and projections to the 1980s, respectively, are shown in Table 4.2 [4.163].

Table 4.2. Specifications of CCD devices

Capacity		Linear dimension of a bit cell		Data rate	
Achieved [Kbits]	Projected [Mbits]	Achieved [μm]	Projected [μm]	Achieved [Mbit/s]	Projected [Mbit/s]
64	4	5	1	1-5	5-20

4.3.5 The Magnetic-Bubble Memory (MBM)

The dynamic memory operation, essential to the word-serial, bit-parallel as well as word-parallel, bit-serial, content-addressable memory, will also be implementable by a new technology which holds great promises in computer engineering. This is the technique of the *magnetic-bubble memory (MBM)* [4.164-167]. The MBM is often regarded as the primary rival of disk memory, possibly being able to replace it in years to come. However, the horizons of MBM applications may be much wider than this: there are many possibilities of performing operations internal to the memory medium which will facilitate sorting, list processing, and, as already mentioned, content-addressable searching.

For the time being, magnetic-bubble memories are already at a commercial stage of development and they have been fabricated as packaged units. The MBMs have been applied in special equipments, but have not yet attained the role of a standard option in computer systems. This situation may change in near future.

Physics of the MBM. The memory medium of MBM consists of a thin ferromagnetic layer or film. This must be magnetically uniaxially anisotropic, and the direction of easy magnetization must be perpendicular to the layer (this direction being named vertical in the following). Single-crystal films of synthetic garnets, grown epitaxially on a nonmagnetic garnet substrate, are most widely used. Other possible materials are orthoferrite, hexagonal ferrite, and various amorphous films of metal alloys, e.g., Mn-Bi and Gd-Co.

Information in the MBM is represented by ferromagnetic domains which can be made mobile. In an equilibrium state, ferromagnetic domains are vertical cylinders with varying, usually serpentine cross-section, with magnetization either upward or downward. (Fig. 4.22a). An external magnetic field which has a vertical orientation affects the size and shape of these domains (Fig. 4.22b): with increasing field strength, the cross-sectional area of those cylinders which are magnetized in parallel with the field will increase, whereas the area of the domains with antiparallel magnetization will decrease. At a critical field strength H_{c1} the latter domains suddenly become circular in their cross-section (Fig. 4.22b), or "bubbles" are formed. If the field strength is yet increased from this value, it is possible to shrink the diameter of the bubbles to a third of the original size before they suddenly disappear. This happens at another critical field strength H_{c2}. The range $[H_{c1}, H_{c2}]$ which is favorable for the existence of bubbles depends on the material properties, on the thickness of the layer, as well as on ambient

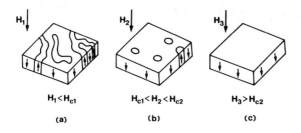

Fig. 4.22. Ferromagnetic domains of the MBM at different magnetic field strengths H

temperature. For garnets and orthoferrites, the necessary magnetic flux density is about a hundred Gs; for hexagonal ferrites, it is roughly 10^3 Gs, and for the Mn-Bi alloy, about 10^4 Gs. The typical bubble diameters for the above materials are 500 μm, 5 μm, and 0.5 μm, respectively.

An effect exists which aids in the regularization of bubble patterns: as the bubbles are equivalent to small bar magnets, they repel each other, and thus keep a minimum mutual distance.

Magnetic bubbles can be made visible if the surface is illuminated by polarized light. (Cf also Sect. 4.4.1). Up to the present time, optic effects have not been used in commercial devices to read the information, however.

Assume now that the magnetic field has a nonzero gradient in the horizontal direction. As the bubbles tend to grow at the side where the field is weaker and shrink at the opposite side, the result is that they start moving in a direction which is opposite to that of the gradient. Apparently the size of the bubbles thereby must change, too, but in practical constructions bubbles move freely only short distances. The horizontal velocity of the bubbles is appreciable: they may move a distance equal to their diameter in about hundred ns.

The initial configuration of ferromagnetic domains is rather random and hence the number and location of the bubbles would be random, too. Nonetheless it is possible to control the production and organization of bubbles in various ways. Annihilation of bubbles can be done by the application of a strong local magnetic field. For the generation of new bubbles, a bubble is first stretched in a locally weaker field of a particular form, whereafter application of a stronger field concentrated in the middle of the bubble will split it in two. For the application of such local magnetic fields one may fabricate tiny circuit loops for control current directly on the magnetic layer, using metal-film evaporization techniques.

Storage Mechanisms of the MBM. It may have become clear from the foregoing that the bubbles are made to carry binary information: if the spacing of the bubbles can be controlled and maintained during their movement, then an existing bubble may represent the value '1' and an empty location in a row of bubbles is equivalent to '0'. In fact, bit densities of $10^6/in^2$ (155 kbits/cm^2) are readily achievable. But although bubbles can be made to move distances which are long when compared with the diameter, nonetheless special provisions are necessary to preserve the stability and configuration of bubble trains.

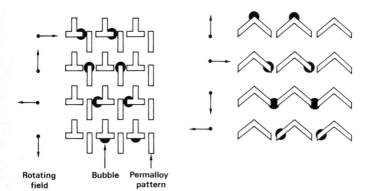

Rotating field   Bubble   Permalloy pattern

Fig. 4.23. Magnetic control structures and propagation of bubbles in them at different phases of the rotating magnetic field

The propagation of bubbles can be controlled and synchronized in two basic ways. In the so-called *field-access method,* the field gradients are defined locally and dynamically using auxiliary patterned magnetic structures deposited onto the surface. These structures, examples of which are shown in Fig. 4.23, usually consist of Ni-Fe (Permalloy) films, and using a homogenous, rotating external magnetic field, usually generated by two-phase coils and superimposed on the static bias field, bubbles can be "pumped" from one control structure to another at a particular phase of the field. It is possible to make rather long paths of control structures, limited only by the dimensions of the device and resolution achievable in the fabrication process, along which the bubbles can be guided in synchronism with the external field.

There is another possibility to control and guide the movement of bubbles, too: this is done with the aid of magnetic fields which are generated by

suitable structured conductors of electrical current. This is called the
control function method.

After the above provisions it is only necessary to construct the transmitting and receiving circuits in a proper way in order to facilitate a dynamic memory operation. At the transmitting end, bubbles are generated by
control current loops which reduce the magnetic fields of particular size
and duration, and information can be read at the receiving end by magnetic
induction caused by the bubbles when they traverse a pickup loop. Magneto-
resistive detectors are often used, too.

Although the operation of the MBM is dynamic, nonetheless the storage is
not volatile. This is made possible by maintaining a bias field, equal to the
field strength necessary to keep the bubbles circular, using permanent magnets. In the case of blackout of control power, the bubbles just stop their
movement and are stabilized by the remanence fields at the magnetic control
structures. During the dynamic operation, the power consumption is rather
small, anyway: the power dissipation in a complete 100-kilobit memory pack-
age is less than 1 W. (The standby power is zero.) It has been estimated
that the control power needed by a 10^6-bit memory medium only would be 40 mW.

As the transmitting and reading circuits are relatively expensive in comparison with the memory medium, magnetic bubble memories are advantageous
when the number of parallel channels is rather small, e.g., they would be
suitable for peripheral devices with eight-bit channels.

Architectures of Addressed MBMs. For dynamic memory operation, the bubble
trains must be made to recirculate in closed loops which is equivalent to a
unidirectional shift register operation. Most typical for addressed MBMs is
the major-minor-loop organization that is primarily intended to reduce the
access time. Consider Fig. 4.24 which schematically presents the main functions of this architecture.

For example, in the 92 Kbit memory chip TBM 0103 of Texas Instruments
Inc. there are 157 minor loops, each with 641 bubble positions, connected
to one major loop. Only 144 minor loops are used for operation; the remaining
13 are spare circuits, intended to replace faulty loops.

Recirculation of bubbles in a loop does not require their regeneration.
For the coupling between the major loop and every minor loop a special electrode structure, capable of transfer (switching) of bubble trains, is needed.
Moreover, for input and output purposes, the major loop must have electrodes
for generation, conditional replication, and detection at the major loop.
Propagation and transfer of bubbles within and between loops occurs in synchronism with the control fields and a system clock.

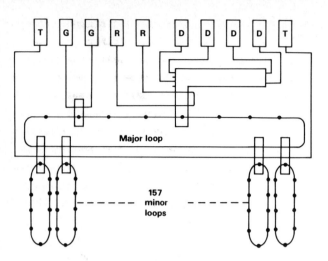

Fig. 4.24. Major-minor-loop organization of a MBM. T = timing, G = bubble generation, R = bubble replication, D = bubble detection

New bubbles are created in the major loop by the externally controlled *generate-current electrode*; this, however, only implements the writing of bit value '1' which takes about 10 µs. To write a '0', a bubble must be removed from the train. This is done at the *replicate electrode* which either passes the bubble or annihilates it. *Detection* of bubbles for readout from the major loop is also made at this electrode structure. If the magnetoresistive effect is used for detection, bubbles are elongated under a chevron-type Permalloy structure to maximize coupling. Output signals have a level of a few mV and they must be amplified.

Transfer of information between the major and minor loops is made at *transfer positions* at the tops of the minor loops. A record is usually stored so that it occupies the same bit position in the minor loops. If a record is suitably positioned in the major loop, writing of it into the minor loops or reading of it can occur in parallel over all bits.

In order for blocks of data to be correctly transferred between the major and the minor loops, they must enter the major loop correctly timed such that, e.g., the segment in the minor loop into which a block is written will be rotated into the right position in synchronism with the contents of the major loop, whereafter the transfer can begin.

The major-minor-loop organization, as mentioned, was intended to speed up addressed reading and writing. For the selection of a particular loop and the segment in it, an external counter system is necessary, very much analogous to the selector system used to define a track and sector in a magnetic-disk memory.

Several alternative architectures, however, can be devised, too. It seems that independent storage loops have some value in practice, too, especially in the content-addressable organization discussed below. The interactive loop organization, on the other hand, can be developed in many ways. The following are only a few ideas: 1) Loops differing in length by a fixed amount can be used to shift or shuffle bubble trains relative to each other in mutual read-write operations. 2) Propagation of bubbles in the loops can be made bi-directional, which further aids in shuffling, sorting, and many kinds of editing, indexing, and stack operations. 3) Architectures with several major loops allow, e.g., interleaving operations. 4) The loops may have junctions, and the flow of bubbles can be steered into alternative branches depending on control signals acting at the junctions (*"bubble ladder"*). This is an especially powerful method for sorting. 5) Clock rates can be made different for different loops, and the trains can be stopped ("frozen") and restarted. This operation preferably ought to be program-controlled.

An Architecture for Content-Addressable MBM. The first suggestion for bubble CAM was based on logic operations performed by the electrode structure [4.167]. It seems, however, that the word-parallel, bit serial CAM architecture, in which MBMs are used for delay lines in the word locations, is particularly suitable for this technology. The most central idea in making a content-addressable MBM cost-effective is to implement the results storage by magnetic-bubble technology, too.

MURAKAMI [4.168] was first to propose a word-parallel, bit-serial CAM using shift register bubble memories. A somewhat simpler design, devised by LEE and CHANG ([4.169], cf also [4.166]) will be reviewed here. This structure implements the *parallel equality-match operation* and also opens the passages at the outputs of all matching words for their readout.

The mechanism used to register matching conditions is based on the so-called *loadable-latch* control of bubble propagation. A bubble, because of magnetic polarization caused by it, can be latched at a small Permalloy disk which belongs to the control structure. Latching occurs only at the coincidence of the bubble and an electrical current simultaneously flowing in a control conductor (Fig. 4.25); without this current the train of bubbles will pass the disk. After a bubble has been latched at a disk, however, the magnetic dipole moment of the bubble will be enough to repel other bubbles and to prevent their flow along their intended path.

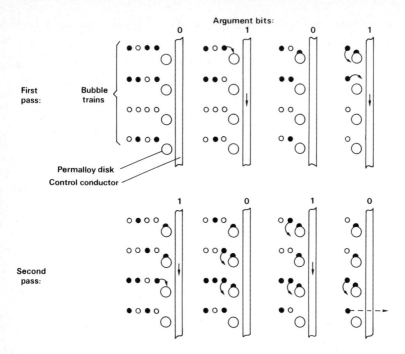

Fig. 4.25. Loadable-latch organization for a content-addressable MBM, shown at eight different phases of operation (according to [4.169])

Consider the series of pictures in Fig. 4.25 which schematically shows the bubble paths. Bubble trains representing the stored words, with '1' indicated by a solid circle and '0' by an open one, approach the Permalloy disks, one per word location. It is assumed that the disks were "empty" in the beginning. (Latched bubbles can be removed by using a control current in the opposite direction.) The external search argument is represented as a sequence of control current impulses in synchronism with the incidence of bubbles at the disks. Assume that the first argument bit is 0 whereby the conductor is currentless. The passage of a bubble (or a vacancy) is yet unhampered. Assume now that the next argument bit is 1 corresponding to control current in the conductor. Two possibilities then exist: either there is a bubble in the train or there is not. An existing bubble will be drawn to the disk where it is latched; if there is no bubble, nothing happens during this bit cycle. In other words, the latch will detect the matching of bit value 1 in the train with that of the search argument but nothing else. It will be easy to see that after the complete train of bubbles representing a stored

word has been received, the latch has become loaded if there is at least one
coincidence of bit value 1 between the search argument and the bubble train.

Matching by logical equivalence means, however, that also the bit values
0 have to be compared. This is done in a second pass during which the bits
of the search argument as well as those of the bubble train are inverted. In
the MBM, inverting of bits occurs at the I/O position of the loops. During
this second pass of operation, again, a bubble is latched at the disk if
there is at least one coincidence of the current pulse with an existing
bubble in the train. After both passes, *the latch will remain unloaded if
and only if there was a mismatch at every bit position of the original search
argument and the original bubble train.* In other words, this mechanism detects the logical equivalence of the stored information with the *logical
complements* of the search argument applied. Naturally, the search argument
then must be the complement of the keyword information.

Masking of bits is simple in this method: it is necessary only to keep
the control conductor currentless in both passes when handling the corresponding bit position.

The mechanism presented above does not yet involve any provision for multiple-match resolution; this feature must be implemented by additional bubble
logic or by an external electronic circuit. A 256 Kbit MBM array complete
with signal connections but without external magnets is shown in Fig. 4.26.

Further Works on Bubble CAMs. A number of articles have been published on
MBMs applied to content-addressable search. KLUGE [4.170] describes a simultaneous readout logic implementable by the MBM technology. Various features
of MBMs have been described by AVAEVA et al. [4.171,172], ALLAN [4.173],
LEE and NADEN [4.174], IL'YASHENKO et al. [4.175-179], and NADEN [4.180a].
An up-to-date review of magnetic-bubble devices has recently been worked
out by ESCHENFELDER [4.180b].

4.4 Optical Content-Addressable Memories

In search of a cheap and high-density medium for CAMs, and in an attempt to
implement computing operations in a highly parallel way, optical methods have
also been considered. This study belongs to a broader field of research in
which methods for the implementation of logic functions by optic, electro-
optic, or magneto-optic principles are being developed.

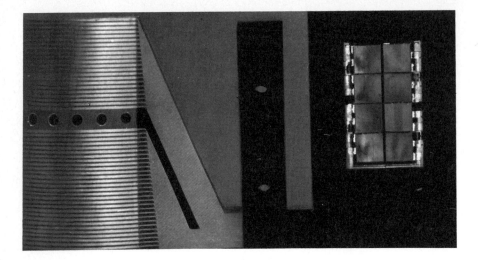

<u>Fig. 4.26.</u> A 256 Kbit MBM array with connectors (by the courtesy of Hitachi, Ltd.)

   The electro-optic components (optical fibers, optically active media, and tiny semiconductor lasers) are finding their way into telecommunication and into computing circuits. On the other hand, certain integral transforms such as the two-dimensional convolutions are readily computable in parallel by Fourier holography. There is, however, a fundamental obstacle in the way of optical memories, and this is the memory medium itself. Rapid and reversible changes in optical transmission or refractive index, comparable to those occurring in electrical components, have not been achieved. On the other hand, the magneto-optical effects, or changes in the reflection or transmission of magnetized materials might be used in memories read optically; the writing of information is then most advantageously done by usual electromagnetic means. The all-optical memories ought to be regarded as *read-only memories*.

   There is one further aspect which explains the difference in readout especially between the electronic and optic memories. Consider, for instance, the combination of bit matches by Wired-AND, or the parallel connection of output gates in the Cryotron memory. It is obvious that the implementation of a logic function by parallel switches has excellent electrical characteristics in regard to stability and noise margins. Contrary to this, most of the optical implementations of matching operations are based on discrimination of a signal level which results from many superimposed wave intensities; it is finally the stability of the discrimination circuits which determines the maximum capacity of these memories. It may be mentioned that discrimination methods for the resolution of parallel matches have been tried in electronic memories, too, but they have not been shown practicable. Parallel matching on the basis of superimposed signal levels cannot be used in magnetic memories at all since the variations in the signal '0' and '1' levels, due to partial switching and reversible polarization, are too large to allow this.

In this section a straightforward suggestion for a parallel-readout (content-addressable) memory by magneto-optical means is presented first. After that suggestions for content-addressable memory based on holography are discussed.

It has been claimed that optical memories would possess the storage capacity necessary for *extensive files*. For the time being it is too early to verify this, because even the usual (addressed) optical memories exist mainly as laboratory versions. As for *content-addressable holographic memories*, it will be neccessary to point out that they have not yet been implemented as mass memory versions even in laboratory. Nonetheless the presentation of their principles may be useful to show the feasibility of this principle.

4.4.1 Magneto-Optical Memories

The memory elements in this solution are thin-film dots of magnetizable material. The writing of information in a magneto-optical memory is made as in a conventional magnetic thin-film memory, in a way analogous to that described with the plated-wire memory. For the content-addressable reading of information from the magnetized dots which is more interesting in this connection, the longitudinal magneto-optical effect as suggested by SMITH and HARTE [4.181,182] can be used.

In Fig. 4.27, a collimated, plane-polarized beam of light falling on the magnetic dots is passed through electro-optical films, one for each bit line. Light is then reflected from the magnetic dots to an analyzer which consists of a set of photodetectors, one per word. The light intensities reflected from all bits in a word are summed up. In front of the photodetector there is another polarizing sheet, the analyzer. Its polarizing direction is such that if there were no extra change in the polarization plane by the electro-optical films associated with the polarizer or by the dots, all light would be absorbed by the analyzers. The reading of bits can now be performed using

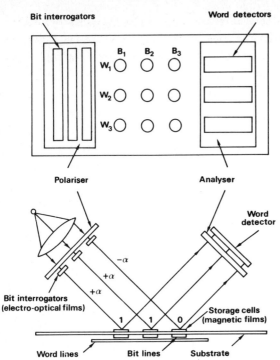

Fig. 4.27. A magneto-optical CAM

the electro-optical effect, by rotating the polarization of the incident light by an amount $+\alpha$ if the bit value 1 is searched, and by $-\alpha$ if a search for the bit value 0 is made. The value of α is determined so that if the magnetization of the dot corresponds to bit value 1, a rotation $-\alpha$ is further caused by the longitudinal magneto-optical effect, and if the stored bit value is 0, the rotation is $+\alpha$. Now, if all bits in a word agree with those of the search argument, no light is received by the word detector. If at least one bit disagrees, light transmitted by the corresponding bit or bits passes through the analyzer.

Another magneto-optical principle for all-parallel CAM has been suggested by AHN and LIN [4.183]. In their idea the magnetizable film carries information in the form of magnetic bubbles (cf Sect. 4.3.5). The film is transparent, and another film, covering the storage plane, is used to represent the search argument. Reading of information is based on the Faraday effect, the change of light polarization in longitudinal magnetic field. Opposite polarization at respective positions in the two sheets is supposed to guarantee unhampered passage of light. In order that this principle operate, it has to be demonstrated that the magnitude of the Faraday effect is large enough. Also another addressed writing and reading principle than that suggested by the inventors has to be devised.

4.4.2 Holographic Content-Addressable Memories

On the Principle of Holography. Consider Fig. 4.28a which depicts a photographic plate P and two beams of coherent light, one falling upon the plate from direction A, and the other from direction B, respectively. Let the sources of light be modulated spatially, for instance, by their transmission through templates or diapositives. It is also possible to produce the beams A and B by groups of sources of coherent light, e.g., by semiconductor lasers which can be controlled by electrical signals to form binary patterns.

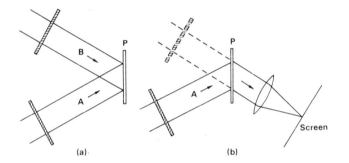

(a)   (b)

Fig. 4.28a,b. Principle of Fresnel holography: a) recording, b) recall

By diffraction, the wavefronts from A and B form a spatial interference pattern at P. The transparency of the plate, by exposure to the light, will then be changed. After development and fixing, the photographic plate will store information from the pair of patterns which were simultaneously displaced at A and B, respectively, in this *hologram* [4.184-186].

When the photographic plate P is later illuminated by the pattern from A or B alone, the wavefronts behind P, due to diffraction which occurs during their passage through the hologram, will assume information from the missing pattern, too. In particular, if P were illuminated by the A pattern only, then the diffracted waves could be collected by a lens system to form a projection image of B on the screen placed behind the photographic plate, as in Fig. 4.28b. A theoretical explanation of this phenomenon can be made, for instance, in the following way. If the complex amplitudes of the electromagnetic fields at plate P, corresponding to the light sources at A and B are denoted by $F_A(r)$ and $F_B(r)$, respectively, where r is the spatial coordinate vector at P, then the transmittance $T(r)$ of the plate is assumed to change directly proportionally to the resultant light intensity, or

$$\Delta T(r) = -\lambda [F_A(r) + F_B(r)][F_A^*(r) + F_B^*(r)] \qquad (4.1)$$

where λ is a proportionality factor depending on the sensitivity of the film as well as exposure time, and the star as superscript denotes the complex conjugate. If the plate P is assumed to be very thin, then, upon illumination of the hologram by waves from A alone, the complex field immediately behind P is

$$F(r) = [T(r) + \Delta T(r)] F_A(r)$$

$$= T(r) F_A(r) - \lambda [F_A(r) F_A^*(r) + F_B(r) F_B^*(r)] F_A(r)$$

$$- \lambda [F_A(r) F_B^*(r) F_A(r) + F_B(r) F_A(r) F_A^*(r)] \quad . \qquad (4.2)$$

Now $F_A(r) F_A^*(r)$ and $F_B(r) F_B^*(r)$ are the wave intensities which are assumed constant with r over plate P, and for simplicity, they both can be normalized to unity. On the other hand, $- \lambda [F_A(r) F_B^*(r)] F_A(r)$ is a term in which the expression $F_A(r) F_B^*(r)$ represents a so-called destructive interference and is not able to produce any image. The noise caused by this term may be neglected. It is further possible to assume that the initial transmittance of the plate is $T(r) = T$, i.e., constant with r. For the field behind the hologram it is, therefore, possible to write

$$F(r) \approx (T - 2\lambda) F_A(r) - \lambda F_B(r) \quad . \qquad (4.3)$$

In other words, the wavefronts behind the hologram are as if an attenuated image of the A pattern, as well as a reconstruction of the B pattern multiplied by $-\lambda$ were seen.

Assume now that exposures from several pairs of different patterns (A,B) are recorded (superimposed) on the same photographic plate. One might expect that the different holograms were confused and information would be lost. This is in general true. It can be shown, however, that under certain con-

ditions, an information pattern associated with a particular key pattern can
be reconstructed from this mixture with reasonable selectivity. Assume that
several pairs of patterns with field strengths $F_{Ak}(r)$ and $F_{Bk}(r)$, respectively,
with k = 1, 2, ..., N, have been recorded on plate P in successive exposures.
Later illumination of the hologram by a pattern with field $F_A(r)$ then results
in field intensity behind P which, on similar grounds as before, can approximately be written as

$$F(r) \approx (T - 2N\lambda) \, F_A(r) - \lambda \sum_{k=1}^{N} \, [F_{Ak}^{*}(r) \, F_A(r)] \, F_{Bk}(r) \quad . \tag{4.4}$$

The *recollection*, represented by the sum over N terms, is then a *linear mixture* of images of the B patterns, with relative intensities that depend on
the degree of matching of the field patterns $F_{Ak}(r)$ with the field pattern
$F_A(r)$ that occurs during reading. So if the A pattern used during reading
as the *search argument* would be identical with one of the earlier A patterns
used as *"keyword"*, then for one value of k, $F_{Ak}^{*}(r) \, F_A(r)$ would be equal to
unity, and the corresponding term in the sum, representing the associated
field $F_{Bk}(r)$, would dominate. The other terms in the mixture in general have
variable phases and represent superimposed noise due to "crosstalk" from the
other patterns. If the various A patterns were independent random images or
randomized by modulation with irregular "speckles", or if they had their nonzero picture elements at different places, the noise content would probably
remain small. When the information patterns must represent arbitrary information, however, this assumption is in general not valid.

On the Concept of Addressed Storage Location in Holographic Memories. As
mentioned above, when a content-addressable memory is used to store arbitrary
(normally binary) patterns, a large storage capacity cannot be achieved by
the previous superposition method on account of the crosstalk noise between
the patterns. It will then become necessary to store the partial holograms
that represent different entries on locally separate areas on the film, corresponding to *addressed storage locations*. The addressed writing and reading
of the small holograms can be made by the same technique that has been applied
with conventional holographic memories; this will briefly be reviewed below.
When it comes to content-addressable reading, however, special provisions
are necessary.

A holographic memory plane may comprise, for example, an array of 100 by
100 small holograms, each with an area of 1 mm by 1 mm. Every small hologram

stores typically an amount of information equivalent to 10^4 bits; thus the total capacity is 10^8 bits. In order to write a small hologram in one of these 10^4 storage locations, the holographic memory needs a delicate light switching system to make the beams A and B point at the location. This is equivalent to the *address decoding system* existing in all normal memories. The switching of the light beams, controlled by electric signals which define the address, is made by so-called *deflectors* the operation of which can be based, for instance, on polarization switching in electro-optic crystals, and subsequent passage of the beam through birefringent prisms. For a more detailed discussion of the principles of beam deflection, see, e.g., the article of KULCKE et al. [4.187].

The storage density, referring to surface area of the photographic plate, can be increased at least by a factor of 10 to 20 by making the hologram three-dimensional; it is then named *volume hologram*. The fields of the waves from beams A and B, of course, interfere anywhere where the waves meet, and in particular within a *thick* photosensitive emulsion used for recording. It can be shown [4.188-189] that if several holograms are recorded in the same volume using A and B beams with different tilts, then, provided that the differences between the incident angles used in different exposures are greater than a certain limit depending on the thickness of the hologram, the stored information can be reconstructed without any crosstalk from the other patterns.

Thus, using multiple sources of light with different incident angles, every small hologram can be made to correspond to several storage locations. The control circuits used to select a particular angle naturally belong to the address decoding system.

In the following, the CAM principle will be exemplified with thin holograms only.

Addressed Writing and Content-Addressable Reading. The small holograms, corresponding to the various stored entries, had to be written using narrow coherent beams confined to them only. Content-addressable reading of the memory, on the other hand, has to be made in an all-parallel fashion, by interrogating all entries simultaneously and in parallel. Consider first the *addressed writing* of information. Since with the normal photosensitive materials the holographic memory operates in the read-only mode, it is possible to record the small holograms using a mask with a small aperture in front of the plate during writing, and the position of the mask can then be adjusted mechanically. If it were possible to make the memory traces erasable, then the aperture, too, ought to be electrically controllable. Two different principles to implement a movable aperture have been suggested [4.190,191]. In both of them the fields of beams A and B are made broad enough to cover the whole memory plate, but during writing, only a portion of them corresponding to a small hologram is activated. In the first principle, the photosensitive plate is covered by another plate which in ambient light and in that due to the A and B beams only is opaque, but becomes transparent at light intensities exceeding a certain threshold. Such materials exist [4.192] although their technology has not been developed for this application. It is then possible to hold the broad A and B beams steady and to use a third, narrow beam with constant intensity which is deflected towards the location in which a small hologram has to be written. If the intensity of the third beam is sufficient, it will exceed the threshold and make the cover plate transparent to the A and B beams. In the second principle a magnetizable film is used as the storage medium. The interference pattern of the light waves is only used as a heat source which warms up the film locally. In the presence of a constant magnetic field the film will be magnetized at cool places, whereas if an area is heated above the Curie temperature, the magnetic polarization will be destroyed. A third beam of light is then used to bias the temperature over the small hologram to the vicinity of the Curie temperature. Reading of magnetic holograms is effected by polarized light the reflectivity for which depends on magnetic polarization of the surface.

The *content-addressable reading* of the holographic memory is done by exposing all small holograms in parallel to a broad A beam which corresponds to the prevailing *search argument*. The matching of this beam with information in the small holograms can be detected as explained below. A particular problem arises, however, with how to deal with multiple matches which are commonplace in all content-addressable memories. In the following it shall be explained in detail how the matching results are recollected, and how the associated data in the matching locations are again read out in entirety. This discussion contains in it ideas from the works of SAKAGUCHI et al. [4.190,193] and KNIGHT [4.191,194], but the particular solutions presented here have been selected mainly for clarity of explanation.

To begin with, it will be necessary to discuss the imaging in Fig. 4.29, in order to clarify the basic optical matching operation. Assume that the field strength corresponding to the recollection of the B pattern (in the output plane) is integrated using a detector, one device per small hologram. Assume first that the B beam had no information written in it, having a field strength which is constant (with amplitude 1) over its cross section. It is then known from the theory of holography that, at least when the tilts of the A and B

beams are small, the distribution of the electromagnetic field over the hologram corresponds to the two-dimensional Fourier transform of the spatial distribution of the field amplitude in the A pattern. Assume, without loss of
generality, that the field amplitude of the A pattern is a real function $A(x)$
where x is the spatial coordinate vector of the plane in which the A pattern
is defined. In other words, it is assumed that (for a unit area)

$$F_A(r) = \int_{S_x} e^{-jrx} A(x)\ dS_x \tag{4.5}$$

where S_x denotes the plane corresponding to x. Assume now that the field of
an A pattern recorded on the hologram is $A_k(x)$, with corresponding field at
plate P equal to $F_{Ak}(r)$. When the field on the output plane, corresponding
to the recollection of the B beam, is integrated spatially and its intensity
is detected as shown in Fig. 4.29 (notice that during reading the real B beam
has been switched off), for the output it is obtained

$$I_B = \lambda^2 \left| \int_{S_r} F^*_{Ak}(r)\ F_A(r)\ dS_r \right|^2 = \lambda^2 \left| \int_{S_x} A_k(x)\ A(x)\ dS_x \right|^2 \tag{4.6}$$

where the last expression results from general properties of Fourier transforms. When it was assumed that $A_k(x)$ and $A(x)$ are real amplitude distributions, (4.6) then states that the output is directly proportional to the
square of the *inner product* of the patterns $A_k(x)$ and $A(x)$.

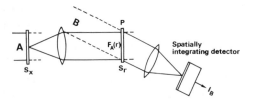

Fig. 4.29. Response from a hologram with spatial integration of the output
beam intensity

In practice, the patterns $A_k(x)$ and $A(x)$ are formed of binary picture
elements corresponding to one of two light intensities; one of the values is
usually zero. The output I_B is then zero if and only if all light spots of
$A(x)$ match with dark spots in $A_k(x)$.

It was assumed in the above analysis that the B pattern had a constant intensity. It can be shown that the same analysis basically applies to a system in which the B beam carries an information pattern.

There was also another feature which on purpose was neglected above, namely, that the small holograms were formed by interference from three beams. The narrow control beam, however, has no effect in the readout since its "recollection" points at a direction where detection of light intensity is not made in this example.

A *masked equality search* can now be implemented in the following way. Every bit position in $A_k(x)$ and $A(x)$ must be provided by two squares, one which is light and one which is dark. Let us call these values 0 and 1, respectively. The *logic value* of the bit positions is '0' if the value combination in the squares is, say, (1,0), whereas the bit value is '1' if (0,1) is written into the squares. If the value combination is (0,0), then it corresponds to the masked bit position; in the previous content-addressable memories, masking was usually made in the search argument, whereas in the functional memories discussed in Sect. 3.7, masking can also be due in the stored information. The value combination (1,1) shall be forbidden.

Consider now the integral in I_B; the integrand is zero at all places in which either $A_k(x)$ or $A(x)$ has a masked value. If and only if the bit values of $A_k(x)$ and $A(x)$ at all unmasked positions are logic complements of each other, will the output I_B be zero. This suggests a straightforward implementation of equality match operation: the bit values of the pattern to be used as search argument are *inverted* logically, and these values, together with the masking markings, then constitute the pattern $A(x)$. A match in unmasked portions is indicated by absence of light in the output detector corresponding to this small hologram.

In holography, the coherent light beam corresponding to $A(x)$ is during reading spread simultaneously over all small holograms. If in the output plane there is an intensity detector for each of the small holograms, their responses then correspond to *multiple response detection*, and may be buffered by flip-flop states in a *results store*. The rest of a holographic content-addressable memory system is then analogous to that described earlier, e.g., in Sect. 3.3.3. The responses can in turn be handled by a multiple response resolver, and the address code thereby obtainable is used to control the final readout process.

Addressed Reading. During writing, a constructive interference between *three* beams, namely, the A beam, the B beam, and the narrow control beam was formed. The combination of the A and control beams, or either beam alone, is then

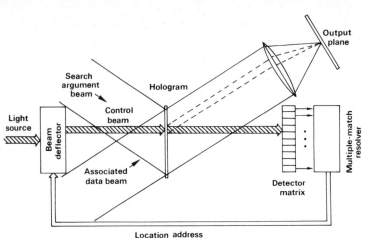

Fig. 4.30. Complete holographic CAM system organization

effective to be used for holographic readout. In fact, it was the A beam
alone which was used to implement the content-addressable readout. For the
addressed reading, the control beam can be used, respectively. Since it was
assumed to be provided with a controllable deflector mechanism, it is only
necessary to control it by the address code obtained from the multiple-match
resolver. The complete holographic CAM system is delineated in Fig. 4.30.

Holography may be used for addressed reading if the amount of information
stored and retrieved is large, and since the provision for this is automatically included in the system. It may be obvious, however, that the outputs from
the multiple-response resolver may be used to address a conventional memory
system from which the associated data are read out. In this case the holographic memory is only used for the location of matches.

Finally it may be mentioned that *proximity search* is directly amenable to
computation by the above system. If the bit values in the search argument
and the $A_k(x)$ disagree in few positions, the output I_B will not be zero but
attain a small value. It is then possible to adjust the detection threshold
of the devices in the detector matrix to give a response when the disagreement
is below a specified limit.

The characteristic property of holographic content-addressable memories
which distinguishes them from other CAMs is that one entry may contain plenty
of information; this feature, together with the possibility of performing a
proximity search by simple means are suggestive for the use of holographic
CAMs in file searching.

Studies on Holographic CAMs. One of the first experiments to superimpose a great number (about 1000) of holograms on the same film was performed by LaMACCHIA and WHITE [4.195]. The array organization of holograms, however, has been found necessary for high capacity. Photodiode detector arrays for this organization have been studied by THIRE [4.196] and MALINA and CHOMAT [4.197]. An ingenious principle which implements the readout by MBM shift registers has recently been invented by MINNAJA and NOBILE [4.198-200]; cf also a review of GARELLI [4.201]. Further works on holographic associative memories have been presented by NISHIDA et al. [4.202], ALEKSAKOV et al. [4.203], FATEHI and COLLINS [4.204], BYKHOVSKY et al. [4.205], MÜHLENFELD [4.206,207], MÜHLENFELD et al. [4.208], AKAHORI [4.209], PRANGISHVILI et al. [4.210], MOROZOV [4.211], VASILIEV et al. [4.212], and VASILIEV [4.213].

Chapter 5   The CAM as a System Part

Many systems and organizations exist in computer technology and data management in which the CAM is included as an auxiliary memory unit or an otherwise essential system part. This chapter reviews a few such examples: advanced
memory architectures and control structures implementable by CAM circuits.
The main emphasis is in pointing out the functional role of the CAM in the
system, and its interaction with the other units. Some of the most important
applications are mentioned in the proper context, and some related hardware
principles can be found in Chap. 6. The particular applications discussed in
this chapter are:

1) Virtual memory.
2) Dynamic memory allocation.
3) Content-addressable buffer.
4) Programmable logic.

5.1 The CAM in Virtual Memory Systems

5.1.1 The Memory Hierarchy

If a single technology for memories existed by which a high storage capacity
and a short access time would simultaneously be achievable at low cost, the
organization of a memory system in a general-purpose computer would be very
simple: all information needed for computation would be stored in a large
one-level memory. However, although digital computers have existed for over
30 years, and from the beginning their architecture was designed around memories, no such solution has been shown practicable. It is true that the
present semiconductor technology already allows the construction of mass memories with a reasonably short access time, say, below 100 ns, but another
technological trend also exists, namely, development of extremely fast memories compatible with the fastest arithmetic and logic circuits. The shortest
access times achieved in small memories are of the order of 1 ns and there

are physical limitations such as the finite velocity of electrical signals, because of which large memories cannot be made that fast. So, even in principle, there will exist memory devices with different optimal speed/capacity configurations.

Nonetheless it was demonstrated long ago that it is possible to combine memory devices of different kinds in a way which allows a very high average processing speed, and yet there can be an access to masses of data; this is because the usual digital computers (the so-called von Neumann computers) process pieces of information in a serial manner. Even if there existed memories in the same system with as high a difference in access times as 1 to 10^{10}, by a clever scheduling the fastest unit could continuously be kept busy without significant waiting times. Such an operation can be made possible by the employment of a *memory hierarchy* which has been known as long as slow archival storages such as magnetic tape units have been combined with fast electronic central processing units.

The memory system of a large computer is organized in several levels depicted in Fig. 5.1. Small computers may involve only a subset of this organization. Starting from the bottom, at the lowest level there will be an *archival storage*, usually a set of magnetic-tape units. In special data base computers, archival memories with as high as 10^{12} bit capacity can be constructed by special optoelectronic or magnetic-film technology. In the magnetic-tape units, data records are not identified by addresses but for retrieval they have to be scanned sequentially until a proper identifier is found. The archival memory is also frequently called *tertiary storage*. The next level upwards, the *secondary storage* usually consists of magnetic disk or drum memories or slow magnetic-core memory units. Often the secondary storage is named the *backing storage*. The memories of the uppermost level, comprising the *primary storage* are usually housed in and functionally belong to the central processing unit (CPU). Although there will also be other types of storage devices in the CPU such as flip-flops, logic latches, registers (even indexed ones), and microprogram control memories, the primary memory is operationally distinguished from the other storage devices by the fact that it is the part in which the memory-reference instructions find their operands on the basis of addresses.

Communication and scheduling of information between the various levels of memory hierarchy is taken care of by the *operating system* which is part of computer system programs. Physically the mode of transfer of data is different between the different levels. The archival memory and the disk memories usually communicate with the CPU and each other via *channels* which

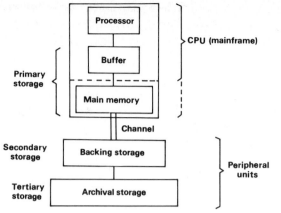

<u>Fig. 5.1.</u> Memory hierarchy

are transmission systems using eight or nine parallel lines, i.e., one byte
wide. One stored byte is usually equivalent to a character code, most often
expressed in the ASCII. The data transfer paths between the slow ferrite-
core memories and the CPU are normally wider, one to four word lengths, and
the transfer may be *interleaved* in the sense that the reading command is
simultaneously given to several memory units or memory banks, but the data
words are read in escalated order, being transmitted through common lines.
Interleaving is advantageous in speeding up data transfer if the access time
of a memory unit is significantly longer than the transmission time.

Interleaving. There exist many organizational solutions in computer engineering for which the principal motivation has been to circumvent handicaps
characteristic to a particular technology. One of them is *interleaving* which
was introduced to speed up retrieving of large but slow ferrite-core memories
used in the big third-generation computer installations. These memories, used
as backing storages, were physically separate from the CPU (mainframe), whereas the primary storages (mainframe memories) were parts of the CPU. Communication between the backing storages and the CPU was through data paths, the
interconnection cables, including as many as 128 parallel lines.

By virtue of the electronic technology used, the transmission delays,
and especially the intervals at which the CPU was able to receive parallel
words from the data path were much shorter than the access times of the ferrite-
core memories, although the latter were of the order of 1 μs. In order to
fully utilize the transmission band width of the data path, the latter was
used to multiplex information from several sources. A typical large third-

generation computer simultaneously utilized four memory banks, often housed
in four separate cabinets. The idea was to simultaneously issue the address
codes and the reading command to all four memory banks, whereby the addressing
used in the system was such that the four words fetched from the different
banks had consecutive addresses. In other words, the address codes sent to
the banks had all but the two last bits identical. In the above example,
4-way interleaving was applied. Solutions with 2-way to 16-way interleaving
have existed.

As stated above, the reading operation was performed in parallel on all
memory banks whereby the data read out were simultaneously obtained and buffered in the respective memory registers. From these they were transferred
via a common data path in an escalated order (cf Fig. 5.2).

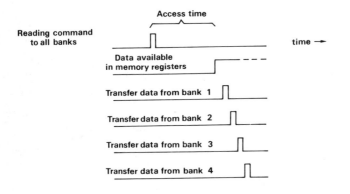

Fig. 5.2. Interleaving

Although interleaving is normally associated with virtual memory, it must
be pointed out that before the principle of virtual memory was generally
adopted, some large computers had a "lookahead" feature with, say, four registers for the buffering of the current instruction or data and three forth-
coming instructions or data, respectively, likely to be fetched next.

In contemporary computer systems, information (data and program codes) is
usually organized in *files* which may be created in any part of the memory
system. One application program may use several files. If files are not used
for longer times, they will be pushed downwards in the memory hierarchy. In
the scheduling of computing jobs, the operating system keeps those files to
be used next ready for transfer into the primary storage.

5.1.2 The Concept of Virtual Memory and the Cache

In order to relieve programming work, and to make programs transferable between computers with different architectures, it is desirable to be able to
use similar machine instructions at least for computers within the same series of models. The individual units may differ in the number and capacity
of their memory devices; nevertheless it is possible to use an assembly code
with similar memory-reference instructions, and moreover, to have only a
single range of memory addresses in spite of many different memory devices
used. This becomes possible by a special organizational principle named
virtual memory. This principle is commonplace in contemporary computers,
and as already stated, it has been found advantageous mainly from the programming point of view.

When designing a computer architecture, one of the basic features is the
address capacity of machine instructions, i.e., the range of absolute addresses in the memory system which can be directly indicated in a machine-
code instruction (direct addressing). The idea applied in virtual-memory
computers is to make this range fairly large, *corresponding to that of the
backing memory*. For instance, if as the backing memory a slow ferrite-core
memory is used, as the case is with the IBM System/360 computers, this range
could be 4 M (1 M = 2^{20}), corresponding to a 22-bit address. The majority of
contemporary computers use disk memories in the backing storage whereby the
address range might preferably be 16 M, corresponding to a 24-bit direct
address. Now it is to be noted that the operands in memory-reference instructions are thought to be fetched from the smaller but faster primary
storage during real computations; how does this comply with references to the
backing storage? The answer is that the fast memory is made "transparent" to
the programmer. This becomes possible by a *buffering* principle explained below with the aid of Fig. 5.3.

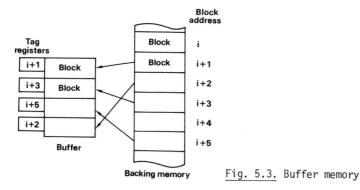

Fig. 5.3. Buffer memory

Paged Virtual Memory. Assume that the backing storage is divided into *blocks* *(pages)* with, say, 256 words each. Assume that there were 4 M words in the backing storage; then each word could be identified by giving a 14-bit *block-address,* and an 8-bit *word-address* or within block address. The block address is hereupon named *tag.* Now assume that the primary storage has a capacity which is a multiple of 256. Actually the primary storage could be fairly small, in our fictive example 1 K, corresponding to four sections, one block each. The primary storage is hereupon named the *buffer.* A control logic will be assumed in the computer system which is able automatically to transfer or copy the contents of any block of the backing storage to any block section in the buffer. The CPU, during computation, then could directly make use of the information stored in the fast buffer provided that it were known from which block in the backing storage it was taken. For this purpose every section of the buffer is provided with a *tag register* into which the corresponding block address is automatically written every time when a block is transferred. Assume now that the CPU issues an instruction referring to a word in the backing memory, and the corresponding block has already been transferred to the buffer. When the 14 most significant bits of the address part of the instruction are compared with the contents of the tag registers, one of the latter is found to agree indicating that the needed word has been buffered in the corresponding section. Thus, using the eight least significant bits of the address, the relative position of the word within the block will be found, and the word can be read out of the buffer.

Apart from considerations associated with the automatic transfer of information, there will arise some principal problems. For instance, there are only four sections in the buffer of this example. When, and in which order shall they be filled, and what is to be done when all are full? Obviously the sections can be filled in any order since a tag is sufficient to identify a block. As to the remaining parts of the question, obviously if the buffer was initially empty the transfer of the first block could be initiated at the first memory-reference instruction; the next references are then likely to be made to the same block. We shall discuss this argument, associated with the philosophy of buffering, below in Sect. 5.1.3 in more detail. As the first rule-of-thumb, it may be stated that the transfer of a block must be initiated every time when the sought word does not already exist in the buffer, and this can be easily deduced from the fact that the contents of none of the tag registers agrees with the 14 most significant bits of the address given in the instruction. Now assume that all sections are already full. It seems very likely that the block in the backing storage to which the latest instruction

referred is more important than some of the old blocks in the buffer, since
the next references are also likely to be made to the same new block. It
seems wise to replace one of the old blocks in the buffer by the new one.
Which one, however, will be abandoned? There are actually two algorithms
applied for this decision. By one of them, the block to be deleted is selected at random. A more sensible strategy might be to delete that block which
was referred to the longest time ago. These *replacement algorithms* will be
discussed in Sect. 5.1.4 in more detail.

It has been customary to restrict the usage of the word "virtual memory"
to a memory organization of the above type in which data are buffered by blocks
or pages, e.g., 256 words in size, and in which the backing storage is a
relatively slow device such as disk memory. In principle, however, the same
strategy could be applied between any two levels of addressable memories
with greatly different speeds. In particular, another type of virtual memory
which seems to become accepted even in small computers, primarily due to developments in semiconductor component technology, is the *cache*. This is a
buffer which works according to the same principle as the virtual memory but
which, however, is mainly intended to speed up the access time of the primary
storage up to the ultimate limit. In a cache, usually an extremely fast bipolar buffer memory is associated with a larger conventional semiconductor
memory. If a cache buffer is used, then the range of direct addresses is that
of the largest configuration of the primary memory. As for swapping of pages
between the primary and secondary storages, different criteria are then used
by the operating system.

It may be stated that *by virtue of the buffer, the computer may seem to
have a one-level memory with capacity that of the secondary storage and speed
which is essentially the same as that of the primary storage.*

5.1.3 Memory Mappings for the Cache

When compared with the virtual memory organization implemented between the
primary and secondary storages, there are certain characteristic features of
the cache because of which certain address mappings will prove more effective
than those used in the usual virtual memory. First, the capacity of the primary storage is usually a few decades smaller than that of the secondary
memory. Since the primary memory now takes the role of the backing memory,
the cache buffer, being the fastest part of the memory system, then must be
yet smaller. This means that only fragments of a program being run can usually
be stored in it at a time, while in a paged memory, a complete procedure may

be stored on one 256-word page. The relative difference in speed between the buffer and the primary storage is also much smaller, say, 20 to 1, as compared with that of the primary-to-secondary memory difference which may range 10^3 to 1 or higher. For this reason, the transfer of data between the buffer and the primary storage must be made in rather small chunks, using wider pathways (with more parallel lines), whereby the block size must be selected smaller.

It is difficult to give any universal rules for the dimensioning of the cache because different programs may make widely different patterns of memory references. The optimization must be performed by benchmarking runs on a mixture of typical programs, say, operating systems, FORTRAN, ALGOL, COBOL, etc. procedures, and possibly assembly-language programs.

Freely Loadable Cache. The simplest buffering principle similar to that discussed in the preliminary example of virtual memory is hereupon called *freely loadable* since any block of the primary memory can be buffered in any section of the cache. (Originally this mapping was called "fully associative" [5.1] but it may be advisable to avoid an inflatory usage of the attribute "associative".) If there are plenty of sections in the cache, the set of tag registers can be replaced by a single all-parallel content-addressable memory (CAM); agreement of the most significant digits of the address issued by the CPU, the *virtual address*, with the tags stored in the CAM is then made by a content-addressable reading operation. (This is just the place in which the CAMs have their most important application in general-purpose computers.) For comparison with the other memory mappings, the freely loadable cache is once again depicted in Fig. 5.4. The primary storage which supplies the cache is in this discussion called *backing memory*.

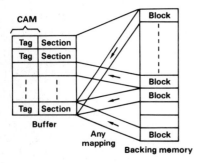

Fig. 5.4. Memory mapping in the freely loadable cache

It may be recalled that the central motive of buffering is that in usual computing programs there will be a high probability for subsequent memory references to be made into the same block. Especially in the case of cache it may further be noticed that subsequent memory references are often clustered in rather small chunks of consecutive addresses. This is primarily due to the following facts: 1) Most high-level and assembly language programs make extensive use of instruction loops. To a large extent the loops result from representation of data in indexed arrays, whereby in order to process them, small loops of program code are run iteratively with the consequence that the same set of machine instructions, usually stored in consecutive addresses, is fetched many times. 2) The data from indexed arrays can be read starting with the innermost index, i.e., from sets of consecutive locations which can be easily buffered. (It may be useful to notice that pieces of data as well as stored program code are equally buffered in various parts of the cache.)

The most central design parameters of the buffer are the block size and the number of blocks in the buffer, and they depend on the patterns of memory references. The *efficiency* of the cache may be expressed in terms of the *average miss rate*, briefly called *miss rate* which tells the percentage of memory references for which an access to the backing storage has to be made if the sought information does not reside in the cache. The miss rate directly relates to the speed and cost of computation; for instance, if the miss rate were one percent, and an access to the backing storage including the block transfer time would be 100 times slower than an access to the primary storage, then the average speed of computation would be roughly half of that attainable if all information could be held in a one-level storage with access time equal to that of the cache.

As stated earlier, an accurate evaluation of performance may be a complicated and very much case-dependent task; however, at least to get an order-of-magnitude idea of the quantitative figures, the dependence of the miss rate on the buffer and block size is delineated in Fig. 5.5. This result [5.2,3] roughly deflects those simulation results obtained in the design of the IBM System/360 Model 85 computer which was the first commercial computer with cache organization.

When for a given buffer capacity the block size is increased, the miss rate first decreases since the block size starts to approach the size of the memory reference clusters. After that, an impairment of performance is due to the fact that with fewer blocks in the buffer the many clusters in a program can no longer be simultaneously buffered, and jumps cause more frequent misses.

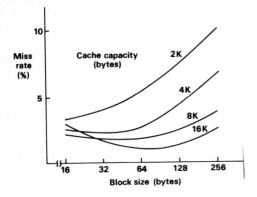

Fig. 5.5. Dependence of miss rate on buffer and block size

The first large computers with buffer memory were IBM System/360 Models 85 and 95, as well as System/370 Models 155, 165, and 195. All of these are byte-oriented computers. A typical backing memory in them was a large and slow ferrite-core memory, divided in several (typically four) memory banks housed in separate cabinets. The address capacity selected for the machine instruction was 16 M bytes, and justified by simulations such as those shown in Fig. 5.5, a block size of 64 bytes was selected. There were 262 144 (256 K) blocks in the backing storage, and a typical size of a freely loadable cache was 128 blocks. The tag word capable of identifying 256 K blocks thus had to be 18 bits long.

Direct-Mapping Cache. From the organizational point of view, the freely loadable cache is simplest. At the time when the first cache memories were designed (around 1965-67) it was thought that the access time of a 18 bit by 128 word CAM would be significantly higher than that of smaller arrays. Nowadays this is no longer the case. Nonetheless, a significantly smaller tag memory than the previous method can be used if the blocks of the backing storage are allowed to be mapped only into particular sections of the cache such that, e.g. the number of the section in the cache is equal to the block address modulo 128; see the exemplification in Fig. 5.6. This is named the *direct-mapping* cache principle.

The format of an address word in the direct-mapping cache is shown in Fig. 5.6. The 11 most significant bits are now sufficient to identify the transferred block since the section into which a block can be buffered is now predetermined, and is given by the next 7 bits. The restriction imposed upon the locations of blocks in the buffer effects the replacement algorithm and then an optimal set of blocks can no longer be maintained in the buffer;

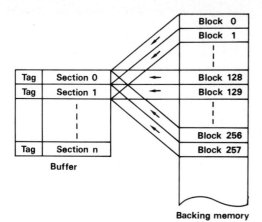

Fig. 5.6. Direct-mapping cache

notice that transferring a block into the cache determines which one of the old blocks must go.

"Set-Associative" Cache. This principle, like the previous one, was also intended to make the tag memory smaller, although in a different way. In Fig. 5.7, the principle applied is demonstrated.

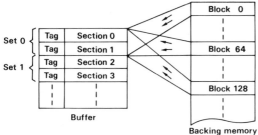

Fig. 5.7. "Set-associative" cache

The cache is divided into *sets*, with a small number (e.g., two) of blocks in each. Analogously to the direct mapping principle, a block from the backing storage can be mapped only into a set the number of which is the block address modulo 64; the order of the block within the set can be arbitrary, however.

In this solution, too, the tag can be shorter (12 bits) since it only needs to identify a group of 64 blocks in the backing storage. The six following bits identify the within-block address of a byte. The tag is only one bit longer than that of the direct-mapping cache, and the freedom in location of the block within a set makes it possible to maintain a better set of buffered blocks in the cache since one is now free to decide which one of the blocks within a set will be replaced.

The "set-associative" as well as the direct-mapping cache were mainly reviewed above since they are in use in computers existing at the present time. Nonetheless, with the present CAM technology they would hardly be adopted for new designs.

Sector Cache. The purpose of this design was to radically decrease the number of tag words in the cache so that instead of a CAM, a small set (in the example below, 16) of faster special registers could be used. The idea is to divide the backing storage as well as the cache into *sectors*, each one capable of holding, say, 16 blocks. This was in fact the detailed design adopted to the IBM System/360 Model 85 (see Fig. 5.8).

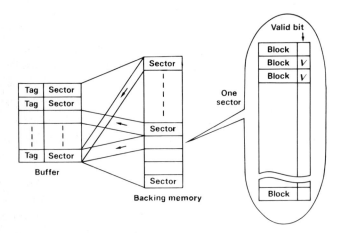

Fig. 5.8. Sector cache

A sector of the backing storage can map into any sector of the cache, and each sector of the cache needs only one tag register. Now, in contrast to the direct-mapping approach, blocks must be mapped congruently *within the sector*, i.e., the relative address of the block within the sector remains the same.

In the first commercial design, as mentioned above, the cache was able to hold 16 sectors with 16 blocks each, which is the same number as in the other alternatives exemplified above. It must be noted, however, that there is no need to transfer an entire sector to the cache in buffering, and this would indeed take too much time. It will be enough to transfer only the block needed.

This will leave some unused capacity in the buffer, and furthermore it is necessary to have a "usage flag", or a "valid bit" for every block in the cache to show its occupancy.

Blocks from different sectors cannot be mixed in one sector of the cache since the tag is common to a sector; in replacement, a whole old sector must be emptied. This is an extra restriction when compared to the other designs. The control of the sector cache is obviously more complicated than with the other methods.

5.1.4 Replacement Algorithms

It may have become clear that one of the central problems in buffering is to predict what set of addresses are likely to be referred to next since this will affect the size of the buffer needed. Another related problem is to predict which sets of addresses already buffered are going to be needed in the most distant future because it would then be possible to determine the block that is next in the order to be replaced by a new one. One may assume that address references in large are ergodic processes whereby predictions can be based on past events; on the average, those addresses which were used a longer time ago are also likely to be used later. A strategy in which the block into which any memory references were made the longest time ago is replaced by a new one is named the *LRU algorithm* (*least recently used* block is to go) [5.4,5].

Bookkeeping of the Usage of Blocks. Assume for simplicity that the freely loadable cache organization is used. The sections for blocks in the cache are numbered, and every time when an address reference is made into a section, the corresponding number is recorded in a special list. In order to comply with the speed of computing, this list must be implemented by hardware. It is not quite obvious, however, which list principle should be used. The first one which comes into mind might be the list named *queue* (FIFO, first-in-first-out) that must be modified, however, because of possible occurrence of the same number in a sequence which then would be placed in several positions of the list. Consider the storage structure shown in Fig. 5.9a into which a number enters at the left end, and all contents of the list are thereby shifted one step to the right.

The number at the right overflows and it would represent the oldest item in the list, provided that all numbers were different. Assume now that a number may occur several times in the sequence. The problem thereby caused is that when the new number enters, the same number already occurring in the

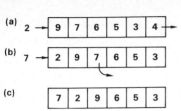

Fig. 5.9. Ordering of block references in a queue

list must be withdrawn (Fig. 5.9b) and the gap thereby formed must be closed by shifting only those items in front of it (Fig. 5.9c).

Instead of the FIFO, another list organization named *stack*, *pushdown list*, or *LIFO* (last-in-first-out) can be used. The stack may be visualized as a vertically ordered array in which every new item is placed at the top; the oldest item resides at the bottom. If a number to be entered already exists in the stack, it must be withdrawn and all numbers above it pushed down by one step.

Identification of a number in the list (by content addressable search or by a slower process of serial scan), and restriction of the shift operation to the front end are possible but a bit complex solutions. The hardware implementation of the list must combine in itself the CAM and shift register operations. For this reason there would be considerable interest in alternative solutions. One of them is a tricky method which uses a simple binary matrix without any shifting operations. This principle was applied for the System/360 Model 85 cache control. Consider Fig. 5.10 which exemplifies this method for a four-block buffer. The circuit implementation will be shown later in Fig. 5.12.

	1	2	3	4
1	0	0	0	0
2	1	0	1	1
3	1	0	0	1
4	1	0	0	0

Initial state

Order: 2,3,4,1

	1	2	3	4
1	0	0	0	0
2	1	0	0	1
3	1	1	0	1
4	1	0	0	0

Reference to 3

3,2,4,1

	1	2	3	4
1	0	1	1	1
2	0	0	0	1
3	0	1	0	1
4	0	0	0	0

Reference to 1

1,3,2,4

Fig. 5.10. Implementation of the LRU algorithm by a binary matrix

Records of the block numbers are represented by certain binary numbers automatically generated in the following way. In the matrix, the row representing the number of block is set full of ones except for the column with the same block number; this column is written full of zeroes. It is very easy to see that the binary numbers so formed on the respective rows are always ordered in the same way as the four latest block references (although these numbers are not identical with block numbers), and moreover, after the cache has been filled up, the least recently used block is represented by a row of zeroes. If there is a zero decoder for each row, that one giving a response can then be used to control writing into the corresponding section of the buffer.

Yet another solution is the *numbering counter method* [5.5]. A sequential number is formed in a special numbering counter at every memory reference made to the cache. Every section is provided with a register into which the sequential number is written upon reference to this block. The smallest number stored then indicates the least recently used block. The most difficult problem in this design is to compare the sequential numbers fast enough, in order to determine their minimum. This operation may be done by a special logic circuit which in effect implements the content-addressable minimum-search algorithm earlier described in Sect. 3.4.5. In other words, a small word-parallel, bit-serial CAM with special search argument bit control logic is needed for this purpose.

One problem with the numbering counter method arises when the counter overflows. In this case the contents of the registers associated with the blocks must be renumbered. It has turned out that in order to avoid extra complications of control, the LRU order may at this point be forgotten and all blocks renumbered in their physical order. Of course, this causes a disturbance in the LRU order but if overflows of the numbering counter are relatively rare events (e.g., with 12-bit numbering counters once in every 4096 memory references), this disturbance does not last long and its effect on the *average* performance of the cache is negligible.

Random Replacement. In view of the complexity of control circuits needed to implement the LRU replacement algorithm, sometimes a much simpler rule can be used, especially if there are many blocks in the buffer and if the ultimate speed is not of paramount importance. Especially in minicomputer environment, the block to be replaced can be decided quite *randomly*. For this purpose, an elapsed-time clock with a binary display register can be used: a number formed of the least significant bits, capable of numbering all blocks of the buffer, may be regarded random enough for this purpose.

The LFU (Least-Frequently-Used) Rule. Although the LRU algorithm is the most commonly used in cache control, and random replacement has the simplest implementation, it may be expedient to mention one further algorithm suggested for this purpose, namely, the LFU (least-frequently-used) rule. The idea in this method is to count references to every block during a certain elapsed period, and to choose the block with the smallest number of counts for replacement. If the address references were describable by an ergodic stochastic process, there would not be much theoretical difference between the LRU and LFU criteria. In view of the possibility of simple implementation of LRU control logic as shown in Fig. 5.10, the LFU, however, has seldom been considered in practice.

5.1.5 Updating of Multilevel Memories

The computing processes normally modify stored data and possibly the program code, too. When these modifications are made in the contents of the buffer, a problem arises since a buffered block is then no longer identical with its primary source in the backing storage. It might seem that this problem does not manifest itself until the block is to be replaced; then the contents of the modified block must be written into the backing storage. This updating principle is named *post-storing*. It must be mentioned, however, that virtual memories are frequently used in multiprocessing environments (cf Sects. 5.1.7, 6.6.1) whereby the same block of the backing storage may be buffered in several places. For this reason, all information in the backing storage ought to be immediately updated. This means revision of the backing storage every time changes are made in the buffer, and this principle is named *through-storing*. Immediate updating increases data traffic to the backing storage but the system control, especially in multiprocessing is simplified; the interlocking considerations caused by different buffers are less problematic. On the other hand, with post-storing, the system performance is better. A choice between these two updating methods is very much case-dependent.

The computations in through-storing are generally slower than in post-storing since every intermediate result word must immediately and unconditionally be written into the buffer as well as into the backing storage. Especially in recursive computing operations it would be significantly faster to iterate a program loop with references only to the buffer memory. In general, writing into the memory occurs much more seldom than reading of data from it, especially in view of the fact that the buffer may also be used to store program code from which the instructions are fetched; thus the difference in the average speed between through-storing and post-storing is not remarkably great. Further one has also to take into account the fact that in post-storing, when a new block is transferred into the buffer, the old one must first be saved by copying it into the backing storage, which will cause an additional delay in the execution of a reading instruction under the miss condition.

5.1.6 Automatic Control of the Cache Operations

This subsection discusses some details of the cache and the automatic control necessary for buffering operations. The example given is fictive and combines ideas from several sources.

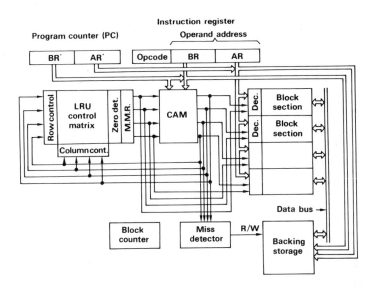

Fig. 5.11. Automatic control of cache operations. Dec. = word address decoder; M.M.R. = multiple-match resolver; R/W = read-write control of the backing storage

Consider Fig. 5.11. The cache operation starts when the CPU issues a memory-reference instruction in its *instruction register*. A similar sequence of operations will be carried out during instruction-fetching; in this case the memory address of the stored program word is given by the *program counter*. The *reading* sequence in both cases shall end with the appearance of the searched data on the *data bus*; it shall be fetched from the buffer of the backing storage. Additionally, if the data is read from the backing storage, the whole contents of the corresponding block shall be copied into the buffer, into a section indicated by the *LRU control*. Transfer of data shall be made via the data bus, one word at a time. As it is most important to have the

sought word immediately available for the CPU, the block transfer commences
with reading of the needed word, whereafter the rest of the words in the block
are read and transferred to the buffer in cyclical order. In the through-
storing principle discussed in this example, there will be no need to write
the displaced block into the backing storage; it can simply be deleted.

In the case in which the memory-reference instruction was that of *writing*
data into a memory location, the cache control in the through-storing method
is rather simple: the data presented on the data bus must unconditionally be
written into the buffer as well as into the backing memory.

In this example only reading of operands from the memory system and writing
the results into it are discussed. The cache is assumed freely loadable and
the LRU algorithm control is assumed to be implemented by the binary matrix
principle shown in Fig. 5.10.

The following notation shall be used in the explanation of micro-operations:
AR = number represented by the set of least significant bits in the address
given in the instruction register of the CPU, equivalent to the "within-
block address" in the buffer as well as in the backing storage; BR = number
represented by the set of most significant bits in the address given in the
instruction register, corresponding to block address in the backing storage;
(BR,AR) = the complete operand address; BUS = set of data signals on the data
bus, one word wide; M(N) = contents of the backing storage location (virtual
memory location) with address N; B(I,J) = contents of the storage location J
in buffer section I; CAM(I) = contents of the storage location I in the CAM;
Z(I) = output of zero decoder associated with row I of the LRU control matrix;
MR(I) = output of the multiple-response resolver with the Z(J),J = 0,1,...,m-1
as its input where m is the number of sections in the buffer; MISS = output of
zero decoder associated with the outputs of the CAM, BC = contents of the
cycle counter which counts the number of words transferred to a block.

In addition to the above definition of machine variables, the following
notation conventions shall be made: the condition for a micro-operation to
be performed on value x of function f is denoted (f = x:). Assignment of
(vectoral) value X to a (vectoral) variable Y is denoted X → Y. Further,
assignment operations are initiated with escalation in time roughly given by
the order of numbering of the rows. (This description of micro-operations
does not actually comply with the formalism known as *register transfer language* [5.6]: actually the numbered rows only show what happens at a particular
time.)

The operation of the LRU control matrix needs a special discussion. The
writing of new information must be made in two steps. First, the row I selected

by a response from the CAM is written full of ones. After that, the column I similarly selected by a response from the CAM is written full of zeroes. Both of these writing operations are conditional on that MISS = 0, i.e., one of the responses is nonzero. One possible solution for a bit cell in the matrix, showing the row and column writing control, is shown in Fig. 5.12.

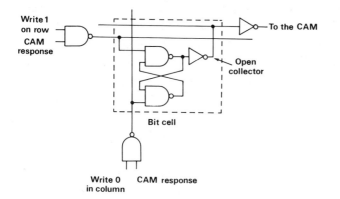

Fig. 5.12. A possible bit cell structure for the LRU control matrix (cf Figs. 5.10,11)

Reading Operations. Assume that the following machine instruction has to be executed: "Read the operand from the virtual memory location (BR,AR) and send it to the data bus", whereby the address is given in the instruction register. This operation includes the following sequential phases:

1) Reading command is given to the CAM.

   <u>MISS = 0:</u>

2) One of the sections of the buffer, say I_0, receives a selection control from the CAM. The location in it is further selected by AR through the decoder.
   Reading command is given to the buffer.
   Row writing command is given to the LRU control matrix.

3) $B(I_0,AR) \rightarrow BUS$
   Column writing command is given to the LRU control matrix.

   <u>MISS = 1:</u>

2) The MISS signal is used as a reading command to the backing storage. The (BR,AR) address has automatically been mediated to the backing storage.

$MR(I_0)$, one of the multiple-response resolver outputs, indicates the section to be replaced.

Writing command is given to the CAM.

3) $m \rightarrow BC$; $BR \rightarrow CAM(I_0)$ (m: number of blocks)

4) $M(BR,AR) \rightarrow BUS$

5) Writing command is given to the buffer.

6) $BUS \rightarrow B(I_0,AR)$

$BC - 1 \rightarrow BC$

7) $\underline{BC = 0}$: Stop.

$\underline{BC \neq 0}$: $(AR + 1) \bmod b \rightarrow AR$ (b: block size = AR capacity)

8) Reading command is given to the backing storage. Return to step 4.

Notice that operand needed by the CPU was fetched from the backing storage as early as possible, namely, at step 4 of the first cycle.

Writing Operations. Assume that the following machine instruction has to be executed: "Write the result given on the data bus into the virtual memory location (BR,AR) using through-storing principle", whereby the address is given in the instruction register. The following sequential phases are included in the execution:

1) Reading command is given to the CAM.
   Writing command is given to the backing storage.

2) One of the sections of the buffer, say I_0, receives a selection control from the CAM. The location is further selected by AR.
   Writing command is given to the buffer.

3) $BUS \rightarrow B(I_0,AR)$

4) $BUS \rightarrow M(BR,AR)$

Notice the timing of commands and transfer operations which is due to different delays.

5.1.7 Buffering in a Multiprocessor System

A multiprocessor system - a set of independent CPUs working cooperatively -
may operate on one or several common primary, secondary, etc. storages; for
speedup of computation, each one of the CPUs may have its own cache (Fig. 5.13).

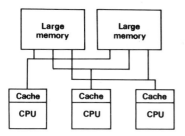

Fig. 5.13. Application of cache buffers in a
multiprocessor system

The operating system of the multiprocessor network shall take care of accesses
to the common memories which have to be made in an escalated order; upon conflict of reading or writing requests, the latter have to be ordered according
to their priority. Some of the CPUs in the multiprocessor network may be dedicated to management of files, e.g., taking care of data traffic between the
primary and secondary storages. It is usually necessary to have some sort of
interlocking control so that the CPUs cannot change intermediate results or
status words of each other, if these have to reside in the common memories.
As mentioned earlier, in a multiprocessor system it is necessary to apply
the through-storing principle of updating between the cache and the backing
storage.

If the CPUs have no working memory other than the cache, the latter should
have a sufficient capacity, say, a few K words. With the present semiconductor
technology this is still an inexpensive solution. On the other hand, such a
cache effectively decouples the CPU from backing memory operations.

Cache memory systems in multiprocessor architectures have been described
by NESSETT [5.7], AGRAWAL [5.8], AGRAWAL et al. [5.9], as well as JONES and
JUNOD [5.10,11].

5.1.8 Additional Literature on Memory Organizations and Their Evaluation

Although by now a lot of experience from virtual memories and operating cache
units in real computer systems already exists, nonetheless it is difficult to

evaluate their performance in universal and simple terms. Let it suffice here to review some recent original works with reference to particular applications.

Principles of memory hierarchies including virtual storage have been discussed from various points of view by KILBURN et al. [5.12], CAMPBELL [5.13], WOLFF [5.14], SCARROTT [5.15], HOBBS [5.16], LINDQVIST et al. [5.17], ANACKER and WANG [5.18], JOSEPH [5.19], MATTSON et al. [5.20], KATZAN [5.21], WILLIAMS [5.22], BENSOUSSAN et al. [5.23], PARMELEE et al. [5.24], BRUNDAGE and BATSON [5.25,26], BERKOVICH et al. [5.27], as well as SCHUENEMANN [5.28]. Implementation of virtual memory in minicomputers has been described by CHASE and GLORIOSO [5.29].

General considerations of cache have been presented by WILKES [5.30,31], GIBSON and SHEVEL [5.32], LEE [5.33], GUNDERSON [5.34], STONE [5.35], KROEGER and MEADE [5.36], BARSAMIAN and DeCEGAMA [5.37], KAPLAN and WINDER [5.38], and NIEDERREICHHOLZ [5.39].

ACKLAND and PUCKNELL [5.40,41] have evaluated the performance of the direct-mapping cache in minicomputer environment, based on simulations; set-associative mappings have been studied by SMITH [5.42,43]; further cache simulations have been made by IIZUKA and TERUI [5.44]. Cache efficiency has also been studied by BENNETT et al. [5.45]. The problem of optimum capacity of the cache has been investigated by CHOW [5.46-48].

Implementation of cache in minicomputers has been studied by BELL and CASASENT [5.49], BELL et al. [5.50], MONROE [5.51], and STRECKER [5.52].

Replacement algorithms have been discussed by WEINBERGER [5.53]. Special hardware for the LRU algorithm is described by CHIA [5.54] and WEINER [5.55].

Description of the IBM System /360 Model 85 which was one of the first large computers with cache can be found in CONTI et al. [5.56] and LIPTAY [5.57]; description of virtual memory in System /370 has been presented by McLAUGHLIN [5.58].

The speedup of execution of instructions by a cache in the control store has been discussed in [5.59]. Particular cache control organizations have been suggested by MASTRANADI [5.60] and HAIMS [5.61].

File organizations and data bases have been described, e.g., in the books of KNUTH [2.1], MARTIN [2.2], FLORES [2.3], DESMONDE [2.4], and LEFKOVITZ [2.5]. Articles on data bases and their hardware architectures from the point of view of content-addressable memories have been written by GELERNTER [5.62], PETERSEN [5.63], HELLERMAN [5.64], HILBING [5.65], ROTHNIE and LOZANO [5.66], FURXHI [5.67] and BERRA [5.68]. An auxiliary storage organization for the PL/I programming language is described by SYMONDS [5.69].

Allocation of memory in large multiprocessor systems presents many problems. These have been discussed by FOTHERINGHAM [5.70], ASPINALL et al. [5.71], HOLT et al. [5.72], and BAYLIS et al. [5.73].

5.2 Utilization of the CAM in Dynamic Memory Allocation

It is customary to write computer programs in small pieces of procedures. For a computational task, a program is usually compiled from many such procedures whereby it is not possible to know at the time of programming in which locations of the hardware memory the program code and its associated data are actually going to reside. For this reason it is simplest to write all procedures relative to origin, i.e., when the instructions refer to absolute addresses, to use an address range starting from zero. The operating system of the computer usually has a special program, the *loader* which automatically takes care of location of the procedures in the main memory in whatever spaces happen to be empty. The loaded program may thereby be split in several parts which are located in different places in the memory space. The address fields of all memory-reference instructions must be changed accordingly.

At a high computing load, especially with the more efficient computers which have the provisions for multiprogramming (several programs loaded at a time and executed in an intermingled order), it is often more advantageous and faster to leave the address fields unaltered. Instead, conversion of the addresses occurs at the time of execution of the instructions with the aid of a special hardware organization. The auxiliary storages thereby needed, the control circuitry, and the overall organization which take care of this automatic address conversion are together named *dynamic loader*, and it is one important example of advanced solutions in computer technology which require a content-addressable memory.

The following example, given in Sects. 5.2.1,2, has been worked out along with the ideas given in [5.74].

5.2.1 Memory Map and Address Conversion

Assume that the main memory is divided into "frames", each one capable of holding one page, e.g., 256 consecutive words of a program. All programs, hereupon named *segments* (of program code) are similarly assumed to be divided into pages; there may be one or more pages in a segment, and the last page may not be full. The same page, however, shall not contain parts from different programs. The data needed by a program are assumed to be included in the segment, too. Upon programming the operand addresses and the line numbers of machine instructions which are referenced in jump instructions are always written relative to the beginning of the segment, the latter having the line number zero.

When a program is loaded into the main memory, the pages belonging to a segment may be allocated in arbitrary frames. With the use of the dynamic loader it is not necessary to make any changes in the address fields of machine instructions upon loading. Instead, in order to find the actual word location in the memory, the true address of operand or jump instruction is computed at the time of execution of the instruction. The conversion into true or absolute addresses is made with the aid of a *memory map* which is stored in a special content-addressable memory. The procedure by which the loading of segments into the main memory is decided, and creation and maintenance of the memory map are explained later. Let us first concentrate in address conversion on the basis of ready memory maps.

The memory-reference address which is given in a machine instruction consists of two parts: the *page number* relative to the origin of a segment, starting with 0, and the *line number* or *within-page address*. The *segments* or programs are identified by a particular identification number. In order to find the actual address, it is necessary only to find the converted page frame number, since the line number is not changed in loading. The CAM used for the page number conversion is shown in Fig. 5.14. Its consecutive locations have a one-to-one correspondence with the consecutive page frames in the main memory.

Only part of the CAM, namely, the fields RPN and SN corresponding to the relative page number within the segment and the segment number, respectively, is needed for this discussion. A content-addressable search with (RPN,SN) as the search argument and the other field NPF (next page frame, discussed later) masked off is then performed. Only one response is assumed to result since segments must have different identification numbers. This response is converted into an absolute *page frame address* by an encoder. The encoded number and the line number together define the complete main memory address. If the machine instruction defined an operand fetch, then the memory address must be transferred to the *address register* AR of the main memory at the execution phase of the operation. If, on the other hand, the instruction defined a jump in the program, then the memory address must first be stored in the *program counter*

265

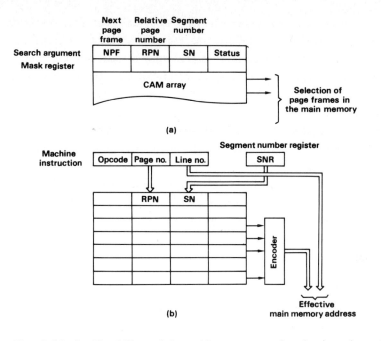

(a)

(b)

Fig. 5.14a,b. The CAM used for address conversion in dynamic memory allocation a) word fields, b) address conversion

PC from which the contents are transferred to the address register of the main memory at the next instruction-fetching phase.

Chaining of Pages in the Memory Map. Dynamic maintenance of the memory map requires that all empty page frames be chained in some order into a single linked list from which frames are then assigned to the program segment that is loaded next. After loading, the list of remaining empty frames must be updated which is a very simple operation as shown below. Whenever a loaded program becomes useless, its page frames can be released by appending them to the list of empty pages. Initially the whole main memory is empty and the pages are linked in consecutive order. After a few loading operations the structure of the list starts looking more random.

The linked list organization can be implemented within the CAM using the *next page frame* field NPF, shown in Fig. 5.14 which is equivalent to a *pointer* discussed earlier in Chap. 2. This field is simply set to indicate the next page frame in the list of empty pages. Similarly all stored program segments are chained whereby the NPF indicates the next page belonging to a segment. It should be noticed that this chaining is in no way necessary from the point

of view of operand or instruction fetching, because the converted address is
always completely defined by RPN and SN. The reason why chaining is maintained for the loaded program segments is that when they become useless, they
can simply be appended to the list of empty pages by changing only one pointer,
namely, the one at the end of the segment, as discussed later.

It should be noticed that due to the chaining convention made above, all
locations of the CAM belong to some chain. The last location in a chain is
denoted by a special value in the NPF field which cannot be a legal page
frame address.

In order to simplify operational notation, the empty pages can simply be
understood as a special program segment with its own identification number.
This segment may be called the *empty segment*.

5.2.2 Loading of a Program Segment

A segment is loaded in two phases. In the first of them, a memory map corresponding to the new segment is constructed in the CAM. In the second phase,
the actual transfer of pages, say, from the backing memory storage into the
main memory is carried out. The following discussion refers to Fig. 5.15
which illustrates the main units of the control hardware.

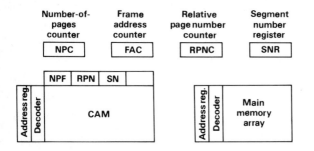

Fig. 5.15. Control hard-
ware for dynamic memory
allocation

Before loading, the operating system checks that the size of the program
segment does not exceed that of the empty segment, and determines the number
of pages. This value is set into the *number-of-pages counter* NPC shown in
Fig. 5.15. An arbitrary identification number, different from that reserved
for the empty segment and from the other numbers already used, must also be
assigned to this segment. The operating system may take care of listing the
segment numbers and their beginning addresses. It is further assumed that

the first frame address in the list of empty pages is available in a special *frame address counter* FAC which will be updated during loading.

The loading starts with copying the address of the first frame of the empty segment from the FAC register into the address register of the CAM, and simultaneously recording it in the list of beginning addresses of programs. This address selects the corresponding location in the CAM into which a new relative page number from the *relative page number counter* RPNC, this time 0, and the new segment identification number from the *segment number register* SNR are stored, using an appropriate mask in writing. [One may recall that in the circuits of the all-parallel CAM shown in Fig. 3.3, masking is implemented by $W(0) = W(1) = 0$.] The contents of FAC and RPNC are incremented by one, and those of NPC decremented by one. After that, an addressed reading of the same CAM location makes the next page frame address from the NPF field available at the output, and this is copied into the address register of the CAM; a new location in the chain is thereby selected. The contents of RPNC and SNR are written into the RPN and SN fields of the new location, respectively, and the above reading and writing phases are alternated until the contents of the NPC become zero. By that time the loading of the segment has been concluded. In order to update the linked lists, it is necessary only to perform an additional writing operation to insert an end marker into the NPF field of the last location.

Since the frame address counter FAC was always updated when pages were loaded, its last value after loading is valid for the storage of the beginning of the empty segment and can be utilized when the next segment is loaded.

The actual loading of a program segment commences with copying the address of its first page frame from the above-mentioned list into the main memory address register, in its frame address part. The same address indicates the beginning of the memory map of this segment in the CAM and is thus copied into the address register of the CAM. The line address begins at 0. It is assumed that the line-address part of the main memory address register is simultaneously a *line counter*, capable of counting upwards modulo 256, whereby the writing of one page into the given frame is easily implemented. No attempt is made here to describe the control of the backing memory which is simply assumed to deliver the whole program segment, one word at a time, on the memory bus upon demand. When one page has been loaded, which is indicated by an overflow from the line counter, the next page address is read from the selected location in the CAM, and this address is copied into the address register of the CAM as well as that of the main memory. Loading continues with the next page, etc., until an end marker is encountered in the NPF field of the CAM.

268

It may be simplest to continue loading, although the last page is not full, until the next overflow from the line counter is obtained.

Releasing of a Program Segment. When a program segment becomes useless, it is not necessary to perform any unloading operations. It will be enough to indicate in the memory map that the corresponding space is available for other programs, and as mentioned earlier, this is effected by appending its linked list into the linked list of "empty" frames, in fact in front of it. The appending is done very simply, e.g., by setting the NPF field (pointer) of the last location in the program segment equal to the contents of the FAC counter, and thereafter updating the FAC counter corresponding to the first frame of the released segment.

Illustrative Example. An example of the contents of a memory map before and after loading of a segment are shown in Fig. 5.16a,b, respectively.

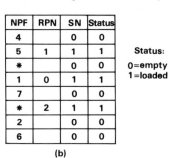

Address	NPF	RPN	SN	Status		NPF	RPN	SN	Status	
0	4		0	0		4		0	0	
1	5		0	0		5	1	1	1	Status:
2	*		0	0		*		0	0	0=empty
3	1		0	0		1	0	1	1	1=loaded
4	7		0	0		7		0	0	
5	0		0	0		*	2	1	1	
6	2		0	0		2		0	0	
7	6		0	0		6		0	0	
	(a)					(b)				

First page address = 3   First page address:
                                      segment 0: 0
End marker: *   segment 1: 3

Fig. 5.16a,b. Exemplification of a memory map a) before loading of a new segment, b) after loading

Comment: The example given in this section has been presented in a simple form, mainly to illustrate the role of CAM in a memory organization.

5.3 Content-Addressable Buffer

In the previous types of virtual memory, the identification of data was still based on an explicitly given address. Another interesting type of buffer which is intended for extensive search operation is completely content-addressable, i.e., it is able to locate the searched information on the basis of its specified parts.

Following the argumentation presented in connection with the addressable memories in Sect. 5.1.1, it may not be wise, for economical, technological, and physical reasons to implement a content-addressable memory using a one-level organization when striving for a high memory capacity; sooner or later there will be a limit beyond which it is more advantageous to adopt a memory hierarchy which uses only a small but extremely fast buffer CAM, provided with a possibility for magnitude-search operations, in combination with a relatively cheap but slower conventional mass memory. Of course, retrieval of data can then not be parallel over all memory locations but only sections of it; it may be recalled, however, that in many practical information retrieval tasks, even completely word-serial scan in small files (say, with a few thousand items) was considered possible. The basic idea applied in a *hierarchical* CAM is now to perform the search by *blocks* such that the reading is word-parallel over every block, and to run through the blocks in a sequential order. Even in rather extensive files, complex content-addressable search operations such as combined magnitude search can be performed.

A Two-Level Hierarchical CAM. One solution for a hierarchical CAM which uses a fast disk memory as the backing storage has been implemented by Goodyear Aerospace Corporation [5.75].

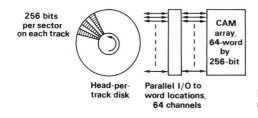

256 bits per sector on each track

Head-per-track disk

Parallel I/O to word locations, 64 channels

CAM array, 64-word by 256-bit

Fig. 5.17. Two-level hierarchical CAM

Fig. 5.17 presents its major parts. The buffer CAM is a bit-serial, word-parallel array, 64 words by 256 bits, capable of performing all the usual modes of search described in Chap. 3. The search time in the CAM is 100 μs at maximum and this may well comply with the speed of buffering. The disk memory has 72 fixed heads, one for every track of which 64 are used to hold data. The purpose is to carry out an exhaustive search of the contents of the disk, and since the disk is rotating continuously but the CAM is loaded and read in turn, in alternating phases, it is found most effective to divide

the tracks into *sectors* 256 bits each, and to read only every second sector
at each rotation. To search the complete disk, two rotations are thus needed.
It takes 100 μs to read one sector and to simultaneously write the data, as
64-bit parallel words, into the consecutive locations of the CAM. Thus a
continuous operation at maximum speed can be guaranteed for the CAM. As there
were 384 sectors in the original design, the whole memory capacity was 6.29
million bits or 24 K words, 256 bits each; the search time was about 76.8 ms.

On the Use of Hierarchical CAM in Large Data Bases. The 24 K words, or 6.29
million bits of the previous design may not yet be regarded as a mass storage
of data. Some special data bases [5.76] may consist of archives with memory
units capable of storing as many as 10^{12} bits of data. Now it is to be noted
that the information stored in such archives is organized in categories of
which only a small subset will be selected for a particular searching task.
Nonetheless, for instance with personal files, it is not uncommon to have a
search be performed over 100 000 to one million records, each one possibly
equivalent to a maximum of 100 character codes. In view of the present CAM
technology it must be stated that: 1) the word length in a content-addressable
array (especially in a word-parallel, bit-serial CAM) can be rather long, say,
512 bits so that a complete record can be stored in one word, 2) present technology facilitates significantly shorter search times than 100 μs, 3) the
loading of the CAM array can be made from several sources, e.g., by *interleaving* techniques described in Sect. 5.1.1.

In view of the high inherent cost of special computer systems dedicated to
file management, the relative costs of even large content-addressable arrays
may remain modest; the maximum capacity is mainly dictated by technical reasons. For instance, if the addressed writing were done in the linear-select
addressing mode, the decoding costs may limit the capacity to, say, 8 K words.

Speedup of Buffering in the Hierarchical CAM. If an electronic, say, bipolar,
all-parallel CAM array were used for buffering, its access time, being usually
less than 10 ns is much shorter than the intervals at which subsequent words
can be transferred from the backing storage. By interleaving, the loading time
of the array can be reduced to a fraction. Notice that the order or location
of the buffered words in the CAM can be arbitrary from the point of view of
content-addressable search.

Another possibility to reduce the buffering time is to divide the CAM
array in several parts which are then loaded simultaneously. For this solution
the addressing system applied in addressed writing must be changed, e.g., in
the way illustrated in Fig. 5.18 which uses independent address registers
and selection circuits.

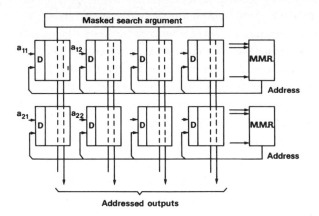

Fig. 5.18. Partitioning of the buffer CAM for fast parallel loading.
D = decoder, M.M.R. = multiple-match resolver, a_{ij} = addresses issued by
parallel sources of data during loading of the array

It may be emphasized that the content-addressable search may be all-parallel
over all partitions. A further fact to notice is that when readout of the
complete locations is made by coupling a response back to the addressable
reading control, the encoded response must be sent in parallel to all decoders
in the bit direction (cf Fig. 5.18).

With the partitioned array, the blocks in the word direction can have an
arbitrary order.

Publications on Hierarchical CAMs. A memory hierarchy between a shift-register
type CAM and drum has been developed by GERTZ [5.77]. Simultaneous access to
memory blocks is described by KOCH [5.78]. Paging in content-addressable memory is reported by CARLBERG [5.79]. An application to a data base system is
described by TAKESUE [5.80].

5.4 Programmable Logic

During the history of development of digital computers there has always been
a tendency to replace wired logic by programming. To a large extent this has
materialized in the arithmetic units, many of which are nowadays controlled
by flexible *microprograms*. The definition of micro-operations is usually made
using random-access memories (RAMs) with rather long words.

Once in a while a suggestion is made that the central control logic, too, might be implementable by programming, especially using content-addressable memories. Although this has been tried in some special systems (cf, e.g., [5.81]), nonetheless the CAMs have not become widely accepted in control circuits. In principle, programmed logic is an elegant method of making the circuitry universal and its design as well as documentation easy. Since the programs can be stored in read-only memories, there is not any loss of information when the supply power is switched off. Anyway, loading of a control memory is a simple task, similar to that of loading a high-level program, provided that the system has a simple bootstrap loader. Maybe the most serious objection against the CAMs in control circuits is that there also exist other, maybe even more amenable alternatives for programmed functions as discussed below in Sect. 5.4.4. Nonetheless, the programming of logic by content-addressable memories deserves a discussion, and we shall attempt to approach this problem systematically.

5.4.1 RAM Implementation

The logic functions are primarily represented by *truth tables*. Assume N independent logic Boolean variables whereby a completely defined truth table has 2^N rows. It is possible to represent several logic functions of the same variables in a common table. Consider the example given in Table 5.1, where x, y, and z are independent Boolean variables and f_1, f_2, and f_3 their Boolean functions.

Table 5.1. Truth table for f_1, f_2, and f_3

x	y	z	f_1	f_2	f_3
0	0	0	0	0	1
0	0	1	0	1	1
0	1	0	0	0	1
0	1	1	1	0	1
1	0	0	0	0	1
1	0	1	0	1	0
1	1	0	0	0	1
1	1	1	0	0	1

The binary number (x,y,z) can now be regarded as an *address* in a *random-access memory* (RAM), and (f_1,f_2,f_3) as a *binary word* stored at (x,y,z). If the contents given in the *right half* of the Table 5.1 are stored in a 8-word by 3-bit memory, the logic values for f_1, f_2, and f_3 can always be obtained by application of address (x,y,z) at the decoder input.

5.4.2 CAM Implementation

Assume that the output functions attain the value 1 on a few rows only, and many of the input value combinations are forbidden or undefined, as is the case frequently in practice. It then seems that a standard memory module with 2^N locations would have plenty of waste capacity, and for this reason there might be some interest in an alternative solution, based on a CAM array or its read-only equivalent. Consider the example given in Table 5.2.

Table 5.2. Defined rows in a truth table with value 1 for f_1 or f_2

A	B	C	D	E	f_1	f_2
0	0	1	0	1	0	1
0	1	0	0	1	1	1
1	0	1	1	1	1	0

The 1's for f_1 or f_2 correspond to the *normal product terms* (cf, e.g., [3.62]) which are assumed familiar to the reader. In this example we have

$$f_1 = (\overline{A} \wedge B \wedge \overline{C} \wedge \overline{D} \wedge E) \vee (A \wedge \overline{B} \wedge C \wedge D \wedge E) ,$$

$$f_2 = (\overline{A} \wedge \overline{B} \wedge C \wedge \overline{D} \wedge E) \vee (\overline{A} \wedge B \wedge \overline{C} \wedge \overline{D} \wedge E) . \tag{5.1}$$

This time we would store the binary values of the *left half* of Table 5.2 in a 3-word by 5-bit CAM with (A,B,C,D,E) its search argument. If now every *row* is thought to correspond to *word-line output*, then, in order to form f_1 it is necessary only to connect the second and third output to an OR gate. In order to form f_2, another OR gate should be connected between the first and second output, respectively. (These OR gates can also be *programmed* as indicated in Sect. 3.7.2). The operation of the circuit is easily proven by stating that

f_1 attains the value 1 if and only if (A,B,C,D,E) matches with the second or the third row, and the converse is true for f_2 for matches on the first and second row, respectively.

5.4.3 FM Implementation

Obviously there is no reason for using memories for logic functions unless an appreciable number of logic gates can thereby be replaced. Now we shall consider the case in which the whole central control logic of a computer or other automatically controlled information processor is to be replaced by programming. A typical minicomputer involves of the order of 50 independent logic variables, and of the order of 50 control functions in which these variables enter as arguments. Normally very different subsets of variables, with only a few variables in each, enter the Boolean expressions.

The complete truth table would have about 10^{15} rows which excludes the RAM solution. With such a high number of rows there is no hope for ending up with a reasonable implementation with the CAM, either. This leaves only the *functional memory* (FM) implementations as discussed in Sect. 3.7.

Below we shall exemplify only the use of Functional Memory 1 in a way which does not require any sophistication in the logic design.

Assume that the control conditions of a computing device have been derived into a set of simplified Boolean expressions which are in the *disjunctive normal form*, or as a logical sum of logical products. (Every product term is a *prime implicant*.) In view of the fact that the storage capacity is relatively cheap, it may not be necessary to strive for absolutely simplest expressions. All of the *product terms* that occur in the expressions are now directly converted back into corresponding rows of the combined truth table in the Quine-McCluskey method (cf Table 3.5 in Sect. 3.7.2, as well as [3.62]). These rows which contain "don't care" values are then stored in an FM array. The word line outputs of the FM array which correspond to ones in a function are connected by a set of OR circuits, one for every Boolean function, in the same way as done in Sect. 4.4.2. Alternatively, the OR functions may be programmed.

Example 5.1:
Assume that the following expressions have to be implemented by a CAM:

$$f_1 = (A \wedge \overline{B} \wedge C) \vee (\overline{D} \wedge E) \vee (F \wedge \overline{G})$$
$$f_2 = B \wedge D \wedge F$$
$$f_3 = (A \wedge \overline{B} \wedge C) \vee (K \wedge L) \tag{5.2}$$

The contents of the FM, using the explicit notation Ø for "don't care", is shown in Table 5.3.

Table 5.3. The FM implementation of logic expressions (5.2)

A	B	C	D	E	F	G	K	L	f_1	f_2	f_3
1	0	1	Ø	Ø	Ø	Ø	Ø	Ø	1	0	1
Ø	Ø	Ø	0	1	Ø	Ø	Ø	Ø	1	0	0
Ø	Ø	Ø	Ø	Ø	1	0	Ø	Ø	1	0	0
Ø	1	Ø	1	Ø	1	Ø	Ø	Ø	0	1	0
Ø	Ø	Ø	Ø	Ø	Ø	Ø	1	1	0	0	1

Notice that the order of rows in Table 5.3 could be arbitrary.

5.4.4 Other Implementations of Programmable Logic

It will be necessary to introduce briefly the most important rivals of content-addressable memories which have been used for the implementation of programmable logic, whereafter a comparison is made.

Programmable Logic Array (PLA). One of the simplest PLA types consists of two switching matrices, implemented by semiconductor LSI technology (Fig. 5.19).

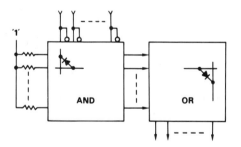

Fig. 5.19. Programmable logic array

In the first matrix, the vertical signal lines correspond to independent logic variables or their negations, and each of the horizontal output lines is equivalent to a logical product term. The product expressions are defined by switching elements (in the example, diodes) connected between the line crossings in a proper way. The horizontal lines are inputs to the second matrix which forms the logic sum of product terms whereby its vertical lines correspond to the logic function outputs. The switching elements in the first matrix must be such that they generate a set of AND functions, and in the second matrix they must generate OR functions, respectively. During the fabrication process a switching element has been provided at every line crossing; proper subsets of elements are activated or passivated, e.g., by a suitable photolithographic process using special masks, or the switching elements can be set by high fields obtainable from automatically controlled programming devices.

The PLA as a Read-Only CAM and FM. Consider once again the first switching matrix of the PLA. As stated in Sect. 3.7.4, it is possible to regard this array as a *read-only content-addressable memory*, either CAM or FM. Consequently, it will be useful to notice that all design procedures developed to Functional Memories 1 and 2 in Sect. 3.7 are as such transferable to PLA devices; the output switching matrix can be designed to correspond to Functional Memory 2.

Universal Logic Modules. While the programming of PLA devices is made by changing their internal hardware structure, the idea applied in the universal logic modules (ULMs) is to define the logic function of a standard unit by auxiliary control signals. It is easy to see that every combinational logic function of n variables can be expanded as a sum of normal products,

$$f(x_1, x_2, \ldots, x_n) = \bigvee_{i=0}^{2^n-1} x_1^{i_1} \wedge x_2^{i_2} \wedge \ldots \wedge x_n^{i_n} \wedge f(i_1, i_2, \ldots, i_n)$$

(5.3)

where each of the superscripts $i_1, i_2, \ldots, i_n$ attains a constant binary value $\in \{0,1\}$. These values together form the binary-number representation of index i. Further $x_j^{i_j}$ is an operational notation with the meaning

$$x_j^0 = \bar{x}_j \quad \text{and} \quad x_j^1 = x_j \text{ for } j = 1, 2 \ldots, n \quad .$$

(5.4)

Notice that $f(i_1, i_2, \ldots, i_n) \in \{0,1\}$ is a Boolean constant. If there existed a standard universal modular logic circuit which would implement the right-hand side of (5.4), then any Boolean function of n variables would be implementable by it by selection of a proper binary value combination for the $f(i_1,i_2,\ldots,i_n)$.

Alternatively, the Shannon expression [5.82]

$$f(x_1,x_2,\ldots,x_n) = \bigvee_{i=0}^{2^{n-1}-1} x_1^{i_1} \wedge x_2^{i_2} \wedge \ldots \wedge x_{n-1}^{i_{n-1}} \wedge f(i_1,i_2,\ldots,i_{n-1},x_n)$$

(5.5)

can be applied which contains fewer terms but in which $f(i_1,i_2,\ldots,i_{n-1},x_n)$ is not constant; it may take on any of the values 0, 1, x_n, or $\overline{x}_n$.

An example illustrating both solutions for the implementation of a universal logic module (ULM) is given below. The circuit which directly implements (5.3) or (5.5) is a *multiplexer* which is available as a standard LSI circuit component [5.83].

Example 5.2:

The two expressions of $f(x,y,z)$ are:

$$f(x,y,z) = [\overline{x} \wedge \overline{y} \wedge \overline{z} \wedge f(0,0,0)] \vee [\overline{x} \wedge \overline{y} \wedge z \wedge f(0,0,1)]$$
$$\vee [\overline{x} \wedge y \wedge \overline{z} \wedge f(0,1,0)] \vee [\overline{x} \wedge y \wedge z \wedge f(0,1,1)]$$
$$\vee [x \wedge \overline{y} \wedge \overline{z} \wedge f(1,0,0)] \vee [x \wedge \overline{y} \wedge z \wedge f(1,0,1)]$$
$$\vee [x \wedge y \wedge \overline{z} \wedge f(1,1,0)] \vee [x \wedge y \wedge z \wedge f(1,1,1)]$$

(5.6)

or

$$f(x,y,z) = [\overline{x} \wedge \overline{y} \wedge f(0,0,z)] \vee [\overline{x} \wedge y \wedge f(0,1,z)]$$
$$\vee [x \wedge \overline{y} \wedge f(1,0,z)] \vee [x \wedge y \wedge f(1,1,z)] \quad .$$

(5.7)

Assume that the function $f(x,y,z) = (\overline{x} \wedge y) \vee (x \wedge \overline{y}) \vee (x \wedge y \wedge z)$ is to be implemented. The two realizations are shown in Fig. 5.20.

Notice that the first implementation is completely programmable: the f values could, if necessary, be taken from a memory circuit. The second implementation is simpler, but it requires yet the definition of its inputs by wiring. Since multiplexers with more than 16 inputs are not readily available, logic networks with a higher number of independent logic variables can be implemented by multiplexer trees.

278

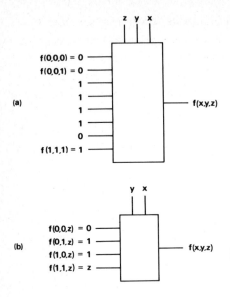

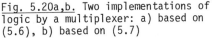

Fig. 5.20a,b. Two implementations of logic by a multiplexer: a) based on (5.6), b) based on (5.7)

Comparison of Programmable Logic Implementations. One of the most important arguments in selecting an implementation for control logic is speed; since the duty cycle of control circuits is high, it is often necessary to trade elegant solutions for performance. The switching times and logic delays of Schottky TTL circuits are of the order of 3 to 5 ns, and Schottky TTL multiplexers have a propagation delay of 5 to 10 ns. A multiplexer tree has, of course, a larger logic delay. Programmable read-only memories and PLAs are usually much slower, with access times of the order of 50 ns. The access times of the fastest CAMs are of the order of 10 ns. It seems that except ROMs and PLAs all of the other methods are preferable for central processors where speed is most important. It has to be emphasized, however, that reliable and cheap FMs have not been fabricated for the time being.

Another question concerns the memory capacity which is needed to replace a certain logic circuit. The number of bit cells in an FM is equal to the product of the numbers of independent variables and logic functions; in random-access memories the capacity grows exponentially with the number of variables. Quantitative comparisons of total costs are extremely difficult to make and they depend on the state-of-art of technology; some attempts have been made with up to about 10 variables, whereby no order-of-magnitude differences were found betweeen the various implementations [5.84]. With a

much higher number of variables, say 50, or more, which is almost the lower
limit in practice, the usual random-access memory falls outside consideration.

Maybe the most important argument in favor of the functional-memory implementation is flexibility which manifests itself in logic design, testing,
correction of errors, and documentation which even with large systems can be
managed; the difference is immense when compared e.g., with random logic
circuits with which any updatings imply new drawings, wiring diagrams, and
much tracing of wiring. For all these reasons the designers of LSI circuits
ought to reconsider the importance of the FM.

5.4.5 Applications of the CAM in Various Control Operations

The use of CAM in control operations of computers has been described by
HOLUM [5.85], BAKER et al. [5.86], GUNDERSON et al. [5.87,88], GONZALES et
al. [5.89], BAIR [5.90], WALD and ANDERSON [5.91], as well as TSUI [5.92].
Implementation of the CAM function by control storage microprogramming is
suggested by HELLERMAN and HOERNES [5.93]. Applications in data acquisition
and processing have been developed by HOOTON [5.94,95], MEYER and STUBER
[5.96], BLANCA and CARRIERE [5.97], as well as JOHNSON and GUNDERSON [5.98].
Further applications have been found in telecommunication exchange systems
[5.99], frame synchronization [5.100], symbol manipulation [5.101], code
translation [5.102], and path finding algorithms [5.103].

Programming of CAM arrays is described by SCHEFF [5.104] and PETERS
[5.105]. Multiwrite algorithms are presented by PORTER [5.106].

Implementation of logic by cellular arrays has been studied by MINNICK
[5.107], YANG and YAU [5.108], YAU and ORSIC [5.109], and YAU and TANG
[5.110]. Field-programmable logic arrays for logic are described by BARRE
[5.111].

Chapter 6   Content-Addressable Processors

This last chapter is intended to show in what ways the structures of content-addressable hardware can further be developed and how CAM functions can be applied to the implementation of other than pure searching functions. In order to increase parallelism and flexibility in searching and computing operations, a few fundamental lines exist of which the following three are believed to represent the basic directions: 1) More logic functions can be added to each cell in an all-parallel CAM, and the cells can be made to intercommunicate locally, in order to distribute the processing algorithms over the memory hardware in a highly parallel manner. 2) An array processor can be built of many higher-level processing elements (e.g., microprocessors) whereby the CAM or an equivalent circuitry may be applied to define the individual control conditions and intercommunication between the processing elements. 3) The results storage of a conventional CAM array, in particular that of a word-parallel, bit-serial CAM can be made more versatile, and for the control and manipulation of the memory array as well as of the results storage, a powerful host computer can be used.

6.1   Some Trends in Content-Addressable Memory Functions

It is somewhat striking that all the new possibilities opened by the large-scale integration technology have not led to novel and more complex internal structures of memories. Obviously the distributed-logic devices, because of the high degree of complexity needed in their cells, have to date been too expensive to make these suggestions practicable for high-capacity memories. A preliminary counterexample given in Sect. 3.6.3 has clearly demonstrated that if *information retrieval* were the only task, and this is for what the distributed logic devices were originally meant, it would seem more profitable to trade off speed and parallelism for a big increase in memory capacity and greatly lowered costs. The situation, however, may change in the near future when certain new technologies such as the CCD and the magnetic-bubble memories become generally accepted. It then seems possible to transfer a greater amount of the active searching and sorting functions to memory units at relatively low cost, and this will certainly have an impact on the new computer architectures.

Although *parallel computations* are implementable by distributed-logic memories, too, it seems that the most effective embodiment of a parallel content-addressable processor is the one based on the third principle, i.e., on a word-parallel, bit-serial CAM provided with flexible word and bit-slice addressing and a results storage with efficient sequential processing logic. We shall later give a few examples of architectures which have been imple- mented with this principle as a basis. When this solution is applied, rather long storage locations (up to 512 bits with simple internal structure) may be used. It is noteworthy that most numerical algorithms proceed bit-sequen- tially for which the bit-serial retrieval of storage locations is a sufficient feature.

In some large-scale computer architectures aimed at maximum parallelism and computing power, content-addressability is not the central feature. Accordingly, these solutions fall outside the scope of this book, and they are mentioned only briefly, within a systematics intended to embrace the existing architectures of parallel computers.

Although the presentation of some material in this chapter seems to be somewhat anachronistic in view of its present and future status, it may any- way be informative to proceed systematically, in the order of increasing complexity. At least this then shows what lines of thinking may be followed in the delineation of new ideas.

6.2 Distributed-Logic Memories (DLMs)

The One-Dimensional Distributed-Logic Memory. The processing circuit dubbed *distributed-logic memory* (DLM) has obviously evolved from the CAM array by the addition of local *intercommunication control* (e.g., setting of flags) between the storage locations. In the first of these devices there are a couple of *status flip-flops* and some auxiliary logic associated with each storage location; this combination which forms a circuit module is named *cell*. The status flip-flops at each cell from a simple finite-state machine by which sequential steps of operation can be defined. These flip-flops can be set and reset by external control, possibly conditionally on the matching of the contents of the cell with external variables. Moreover, the status of a location can be announced to the adjacent locations whereby it becomes possible to write programs which in parallel *retrieve variable-length strings, move* data in the array, as well as *edit* them, very much along the same lines as already discussed with the byte-serial CAM in Sect. 3.6.3. An interesting special application of DLMs is *proximity search* as demonstrated in Example 6.3.

In order to implement an instruction set for these operations, the need for circuit logic is appreciable; typically 13 logic gates per bit plus a couple of dozen gates per cell for the sequential control must be included. The real number of active logic circuits may, of course, be much higher in a circuit realization as the case was with a conventional all-parallel CAM bit cell, where the number of logic gates was only seven, but several tens of transistors per bit might have been used.

In its basic form conceived by LEE and PAULL [6.1,2] and LEE [6.3], following a simpler idea of LEE [6.4], and later revised by GAINES and LEE [6.5], the DLM consists of a linear array of identical cells, depicted in Fig. 6.1.

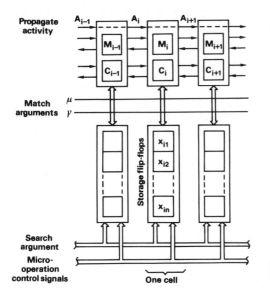

Fig. 6.1. One-dimensional distributed-logic memory

Its design specifications may better be understood if it is restated that this circuit was initially meant only for *information retrieval*, e.g., storage and manipulation of *variable-length character strings*. Each cell was supposed to hold only one character code, whereby no word locations were defined in the array; words composed of characters could be of an arbitrary length and they could be stored contiguously or leaving arbitrary gaps between them. Intercommunication of the cells was needed for successive matching of the characters as well as for justification of the strings, for instance, to close gaps between them. It has later been pointed out, e.g., by GRANE and GITHENS [6.6] that these provisions are also sufficient to perform simultaneous *arithmetic operations* on many sets of binary data, stored in a particular format in the array.

   The micro-operations by which the cell status can be controlled are listed first. After that, it will be exemplified how some of the most important *functional operations*, to be executed by this DLM, are implementable as *microprograms*.

   The set of contents of *storage flip-flops* $X_i = [x_{i1}, \ldots, x_{in}]$ in the cell is called *pattern*. The two status flip-flops in every cell are the *match flip-flop* with value M_i and the *control flip-flop* with value C_i. A cell is said to be *active* if its M_i flip-flop has the value 1. The following micro-operations shall be implementable by the circuit logic embodied in each cell:

Table 6.1. The basic set of micro-operations applied in a DLM

Micro-operation	Definition
match $X_i = \xi$, $C_i = \gamma$	Set $M_i = 1$ in each cell (activate each cell) whose contents match with the search argument given on the input lines, and whose control flip-flop has the value γ
set $X_i = \xi$, $M_i = \mu$, $C_i = \gamma$	Store the pattern ξ and set the values μ and γ, respectively, to the M_i and C_i flip-flops in each cell
store $X_i = \xi$, $C_i = \gamma$	Store the pattern ξ and set the value γ into the C_i flip-flop in each *active* cell (for which $M_i = 1$)
read	Read out the contents of an active cell
mark	Set $M_i = 1$ in all cells to the right of each active cell up to and including the first cell $C_i = 1$

left	Set M_i = 1 in the left neighbor of each cell whose C_i = 1
right	Set M_i = 1 in the right neighbor of each cell whose C_i = 1

Comment 1:

In the first three micro-operations, only part of the specifications may be given: e.g., *match* x_{i1} = 1, x_{in} = 0 means that x_{i2} through $x_{i,n-1}$ are "don't cares" and C_i can have any value.

Comment 2:

For reading, there must be only one active cell. If M_i = 1 for several i, reading must be preceded by a priority-resolution microprogram as explained below. (In practice, a multiple-response resolver might be preferred for this purpose.)

The following *microprograms* exemplify the applications of micro-operations.

Example 6.1:

This microprogram locates all character strings which match with an external search argument, given as a sequence of characters.

Assume that initially each M_i = 0. Denote the string of external characters (search argument) by $\{\xi_1, \ldots, \xi_k, \ldots, \xi_n\}$. The microprogram leaves M_i = 1 in the cells which follow each matching string.

Microprogram	Comments
1) k = 1	
2) *match* X_i = ξ_k	Marks M_i = 1 in all cells that match with ξ_1
3) *set* C_i = 0	
4) *store* C_i = 1	
5) *set* M_i = 0	Marks C_i = 1 in each cell which follows a string
6) *right*	matching so far
7) *set* C_i = 0	
8) *store* C_i = 1	
9) *set* M_i = 0	

10) $k = k + 1$

11) *match* $X_i = \xi_k$, $C_i = 1$ Marks $M_i = 1$ if the string matches so far

12) Repeat from step 3
    until all characters
    of the search argument
    have been exhausted

Example 6.2:

This microprogram singles out the leftmost active cell in the array,
leaving $M_i = 1$ in it and resetting M_i in the others:

1) *set* $C_i = 0$
2) *mark*
3) *store* $C_i = 1$
4) *set* $M_i = 0$
5) *right*
6) *store* $C_i = 0$
7) *set* $M_i = 0$
8) *match* $C_i = 1$

For further illustration, assume that the values of pairs (M_i, C_i) for
$i = 0,1,\ldots,6$ are shown typographically as strings with a space between
the pairs. The transformed contents after each (numbered) microinstruction are shown below.

	(M_i, C_i)						
Initial contents:	01	11	10	00	10	01	00
1)	00	10	10	00	10	00	00
2)	00	10	10	10	10	10	10
3)	00	11	11	11	11	11	11
4)	00	01	01	01	01	01	01
5)	00	01	11	11	11	11	11
6)	00	01	10	10	10	10	10
7)	00	01	00	00	00	00	00
8)	00	11	00	00	00	00	00

Example 6.3 (Proximity search):

This microprogram locates all character strings which match with the search argument exactly or differ from it at most in one character (*replacement error*).

The idea is to reserve two bits, x_{i1} and x_{i2}, in each word as *flags* with the aid of which, when the strings are compared by characters, the degree of matching is indicated. The status of these bits shall be such that if $x_{i1} = x_{i2} = 1$, the string ending with cell i matches exactly so far; if $x_{i1} = 1$, $x_{i2} = 0$, one error has occurred so far, and $x_{i1} = x_{i2} = 0$ means that two or more errors have occurred so far.

Denote the pattern $[x_{i3}, \ldots, x_{in}] = X_i'$. The microprogram reads as follows (cf Example 5.1):

Microprogram	Comment
1) *match* $X_i' = \xi_1$	
2) *set* $C_i = 0$	
3) *store* $C_i = 1$	Sets $x_{i1} = x_{i2} = 1$ in each cell which
4) *set* $M_i = 0$	follows a character matching with ξ_1.
5) *right*	If ξ_1 does not occur in the memory,
6) *set* $x_{i1} = 1$, $x_{i2} = 0$	sets $x_{i1} = 1$, $x_{i2} = 0$ in all cells.
7) *store* $x_{i2} = 1$	
8) *set* $M_i = 0$	
9) k = 2	
10) *match* $X_i' = \xi_k$, $x_{i1} = 1$	Sets $x_{i1} = 1$ in all cells following
11) *set* $C_i = 0$	those in which a match occurred at
12) *store* $C_i = 1$	the last character comparison, and
13) *set* $M_i = 0$	when no more than one error had oc-
14) *right*	curred so far. (Notice that this sub-
15) *set* $x_{i1} = 0$	program is iterated for k = 2...n).
16) *store* $x_{i1} = 1$	Sets $x_{i1} = 0$ in all other cases. Does
17) *set* $M_i = 0$, $C_i = 0$	not yet alter the x_{i2}.
18) *match* $x_{i2} = 1$	Leaves otherwise correct values for
19) *store* $C_i = 1$, $x_{i2} = 0$	x_{i1} and x_{i2} after the last matching
20) *set* $M_i = 0$	operation except sets $x_{i1} = 0$, $x_{i2} = 1$
21) *right*	if the string had been correct so far
22) *store* $x_{i2} = 1$	and if an error occurred at the last
23) *set* $M_i = 0$	character comparison.

24) *match* $x_{i1} = 0$, $x_{i2} = 1$ ⎫
25) *store* $x_{i1} = 1$, $x_{i2} = 0$ ⎬ Corrects the above case.
26) *set* $M_i = 0$ ⎭
27) $k = k + 1$
28) Repeat from step 10 until all characters of the search argument have been exhausted.

This example shows that the microprograms tend to become rather long. It should be realized, however, that each program step is executed in parallel over all cells.

An Example of Bit Control Logic in a Linear DLM. As stated above, the *cell* of a DLM must include two types of logic control: one for the storage, writing, reading, and matching of the *data bits* x_{i1} through x_{in}, and the other for the sequential control of the *status flip-flops* M_i and C_i. From the point of view of production costs, it is the bit logic which is more decisive, and an example of it (according to [6.4]) is shown in Fig. 6.2a. A control circuit for the M_i and C_i flip-flops is delineated in Fig. 6.2b. Since an asynchronous principle of sequential operation was used, a double-rank shifting method had to be applied to propagate the status of the M_i flip-flop into the adjacent cell. It is for this reason that so many micro-operations were necessary in a simple sequential matching (e.g., *set* $M_i = 0$, *match* $X_i = \xi$, *set* $C_i = 0$, *store* $C_i = 1$, *set* $M_i = 0$, *right*) to isolate the partial operations of sending away old information and receiving new one.

The external control of the DLM which interpretes and executes the microprograms must be such that it distributes a sequence of control signals, corresponding to micro-operations listed in Table 6.1, to the respective common control lines. It is hoped that the circuit diagrams of Fig. 6.2 are otherwise self-explanatory.

Microprograms Written for DLMs. To recapitulate, it may be mentioned that at least the following microprograms have been written for DLMs: retrieval, editing, and moving of variable-length character strings [6.1-4]; bulk addition, subtraction, multiplication, and Boolean operations on many sets of operands [6.6,7]; matrix inversion by the Gauss-Jordan elimination procedure [6.6]; multiple-response resolution [6.5,6]; searching for maximum and minimum values [6.6,7]; and magnitude comparison [6.6].

Group-Organized DLM. If the principal mode of use of a DLM is parallel operation on numerical variables, e.g., magnitude search or bulk arithmetic, then

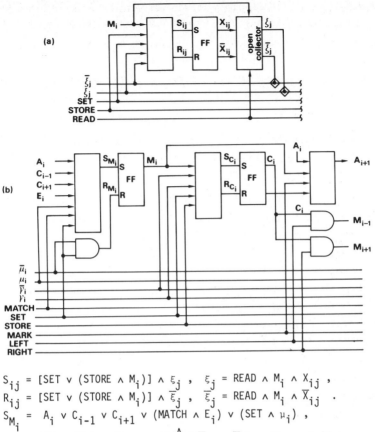

$S_{ij} = [\text{SET} \vee (\text{STORE} \wedge M_i)] \wedge \xi_j \; , \quad \xi_j = \text{READ} \wedge M_i \wedge X_{ij} \; ,$

$R_{ij} = [\text{SET} \vee (\text{STORE} \wedge M_i)] \wedge \bar{\xi}_j \; , \quad \bar{\xi}_j = \text{READ} \wedge M_i \wedge \bar{X}_{ij} \; .$

$S_{M_i} = A_i \vee C_{i-1} \vee C_{i+1} \vee (\text{MATCH} \wedge E_i) \vee (\text{SET} \wedge \mu_i) \; ,$

$E_i = [(\bar{C}_i \wedge \bar{\gamma}_i) \vee (C_i \wedge \gamma_i)] \bigwedge_j [(\bar{X}_{ij} \wedge \bar{\xi}_i) \vee (X_{ij} \wedge \xi_i)] \; ,$

$S_{C_i} = (\text{SET} \wedge \gamma_i) \vee (\text{STORE} \wedge M_i \wedge \gamma_i) \; ,$

$R_{C_i} = (\text{SET} \wedge \bar{\gamma}_i) \vee (\text{STORE} \wedge M_i \wedge \bar{\gamma}_i) \; .$

Fig. 6.2a,b. Circuit logic of the DLM: a) bit cell, b) status flip-flops

it will be profitable to organize the memory array in a slightly different way, as suggested by CRANE and GITHENS [6.6]. The memory is segmented in a number of groups, each one containing m locations named X cells, and one location named Y cell. As seen from Fig. 6.3, the X and Y cells overlap. There are two match flip-flops, M_{ij} in the X cell, and G_i in the Y cell, respectively. The control flip-flop C_i of the previous example has been replaced by circuit logic. The adjacency of the cells is defined in such a way that the left and right neighbor of an X cell can only be found within the same group, whereas the neighbors of the Y cells are the Y cells of adjacent groups.

In accordance with the word-parallel, bit-serial CAM operation, the numerical variables are stored in "bit slices" of the X cells, i.e., horizontally in each group.

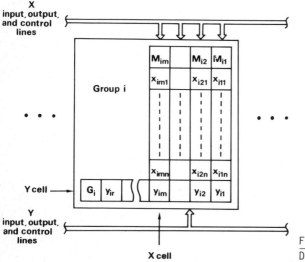

Fig. 6.3. Group-organized DLM, one group

In order to speed up the execution of numerical algorithms, the micro-operations in the group-organized DLM have been made more flexible compared to those of the simple one-dimensional circuit. For instance, the sequence of six instructions used in the former to transfer the matching condition into the M_i flip-flop of the next cell which would be intolerable in bit-serial arithmetics has been replaced by a single instruction *clear match right* $X_{ij} = \xi$. Naturally the internal logic structure of each cell then must be more complex. Table 6.2 lists samples of micro-operations which occur in Example 6.4. For a more complete list, see [6.6].

Table 6.2. Samples of micro-operations of the group-organized DLM

Micro-operation	Definition
match $X_{ij} = \xi$	Set $M_{ij} = 1$ in all X cells whose contents match the input pattern ξ
match $Y_i = \eta$	Set $G_i = 1$ in all Y cells whose contents match the input pattern η
clear match $X_{ij} = \xi \; (Y_i = \eta)$	Similar to above except that in all other cells it is set $M_{ij} = 0 \; (G_i = 0)$
match left (right) $X_{ij} = \xi \; (Y_i = \eta)$	Set $M_{ij} = 1 \; (G_i = 1)$ in the left (right) neighbor of all X (Y) cells whose contents match the input pattern $\xi \; (\eta)$
clear match left (right) $X_{ij} = \xi \; (Y_i = \eta)$	Similar to above except that in all other cells it is set $M_{ij} = 0 \; (G_i = 0)$
propagate left (right) $X_{ij} = \xi \; (Y_i = \eta)$	Set $M_{ij} = 1 \; (G_i = 1)$ in all cells between an already *active* [a] cell and the first cell to its left or right which does not match the input pattern
store $X_{ij} = \xi \; (Y_i = \eta)$	Write the input pattern in all X (Y) cells
store conditionally $X_{ij} = \xi \; (Y_i = \eta)$	Write the input pattern in all *active* X(Y) cells

$\cdots$

[a] An active cell is one with $M_{ij} = 1 \; (Y_i = 1)$. The following microprogram exemplifies the operation.

Example 6.4:

This microprogram adds pairs of binary numbers in parallel over all cell groups. The two operands are stored in the x_{ij2} and x_{ij3} bit slices of the X cells in each group, and the sums are left in the sets of M_i flip-flops, respectively. These additions will be performed only in groups which have $G_i = 1$.

For the carry bits, flip-flops x_{ij1} are used.

Microprogram

1)	*clear match left*	$G_i = 1$, $x_{ij2} = 0$, $x_{ij3} = 0$
2)	*store conditionally*	$x_{ij1} = 1$
3)	*clear match left*	$G_i = 1$, $x_{ij2} = 1$, $x_{ij3} = 1$
4)	*propagate left*	$x_{ij1} = 0$
5)	*store*	$x_{ij1} = 0$
6)	*store conditionally*	$x_{ij1} = 1$
7)	*clear match*	$x_{ij1} = 0$, $x_{ij2} = 0$, $x_{ij3} = 1$
8)	*match*	$x_{ij1} = 0$, $x_{ij2} = 1$, $x_{ij3} = 0$
9)	*match*	$x_{ij1} = 1$, $x_{ij2} = 0$, $x_{ij3} = 0$
10)	*match*	$x_{ij1} = 1$, $x_{ij2} = 1$, $x_{ij3} = 1$

The sum and carry bits are determined in the usual way: if the carry bit is $x_{ij1} = 1$, then the sum (match) bit will become 1 if and only if (x_{ij2}, x_{ij3}) is (0,0) or (1,1) and if $x_{ij1} = 0$, the sum (match) bit becomes 1 if and only if (x_{ij2}, x_{ij3}) is (0,1) or (1,0). This is implemented at steps 6 through 10 of the microprogram. Before that, the correct carry bits had to be computed in the first part of the program. When scanning the bit positions from right to left, one will note that the first carry cannot be generated until a bit combination $(x_{ij2}, x_{ij3}) = (1,1)$ is found after which the carry is propagated to the left until a combination $(x_{ij2}, x_{ij3}) = (0,0)$ is found. When proceeding further left, the generation of a new carry again requires the occurrence of the operand bit combination (1,1).

Further Works on the DLM. A Cryotron implementation of DLM is reported by CRANE and LAANE [6.8]. Suggestions for generalizations and further algorithms have been made by EDWARDS [6.9] and SPIEGELTHAL [6.10]. An algorithmic language for DLM has been presented by TREMBLAY [6.11]. STURMAN [6.12,13] has suggested that a general-purpose computer be implemented by the DLM structure. Cost-effectiveness, due to the high cost of memory modules, is the worst handicap thereby met.

Further DLM applications have been described by SMATHERS [6.14]. Later in Sect. 6.4.4 we shall describe the content-addressable computer named PEPE; its input has been implemented by the DLM principle.

6.3 The Augmented Content-Addressable Memory (ACAM)

The second cellular (iterative) array principle discussed in this section is
the *Augmented CAM* (ACAM) described by KAUTZ et al. [6.15] and KAUTZ [6.16].
It too is a memory provided with distributed logic and local intercommunication between its cells. The main structural differences in comparison with
the DLM are the following: 1) Each cell stores only one bit value (y_{ij}), and
the logic-to-memory ratio is correspondingly larger. 2) While intercommunication in the DLM was implemented by hardware in the horizontal direction
only, and interaction of data elements in the vertical direction had to be
made by pattern matching and subsequent writing (i.e., by software), in the
ACAM the intercommunication and processing of adjacent bit values is directly
made possible by hardware in the two orthogonal directions. 3) One of the
characteristic properties of the DLM not shared by CAM and ACAM is transla-
torial invariance of the stored information, i.e., a string of symbols can
be stored in and retrieved from an arbitrary segment of the memory array; the
ACAM, on the other hand, is word-organized.

   The ACAM is capable of more powerful information processing than the DLM,
and it was, therefore, mainly intended for a programmed array for the realization of arbitrary combinational and sequential logic functions, including
various parallel comparisons, data transfers, and arithmetic operations. It
was anticipated that the ACAM, making use of the advantages offered by large-
scale integration, might have become accepted to perform many CPU functions
in general-purpose computers. It has been demonstrated that the ACAM can be
controlled by external signals to emulate the following hardware functions:
a bank of index registers, page controller, a set of buffer registers such
as stack or queue, encoder, decoder, permutation array, sorting memory,
list memory, and microprogram memory.

   The ACAM may be regarded as belonging to the category of content-addressable memories although it, unlike the CAM and the DLM, is not primarily intended for information retrieval. However, this construct has many features
in common with those of the CAMs: the cells are activated simultaneously and
in parallel by external signals, and information can be written in and read
out in a content-addressed mode.

The Cellular Array and Its Functions. The ACAM is a two-dimensional, iterative,
m by n array of bit cells, each one storing a bit value y_{ij}. Each cell has
eight logic terminals (pins), including one assigned for the clock signal,
and some pins for supply voltages. Two control signals a_i and b_i pass all
cells on each row i, and one control signal c_j passes all cells at each

294

column j. Two sets of intercommunication signals $\{x_{ij}\}$ and $\{z_{ij}\}$ are also defined; transfer of information between the cells occurs in the horizontal direction by means of the z_{ij}, and in the vertical direction by the x_{ij}. The y_{ij} can change their states only in synchronism with a global system clock the signal line of which is threaded through all cells.

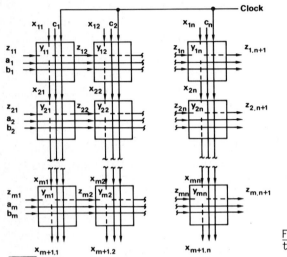

Fig. 6.4. Cellular array of the ACAM

At this point it may be proper to mention that the cells can be controlled to carry out the following modes of logic: 1) Combinational logic operations along the horizontal cascades of cells. 2) Combinational logic operations along the vertical cascades of cells. 3) Sequential logic operations whereby the states of the y_{ij} flip-flops can be made to change individually in synchronism with the system clock.

It may further be pointed out that if the ACAM array shall realize Boolean functions, the stored bit values y_{ij} are used to *program* this logic, i.e., they can be set to define particular functions. It shall further be shown below that *content-addressable search* actually belongs to these functions. Clocking signals are not needed to compute these functions. If, on the other hand, this array is used for the processing of numerical information, the y_{ij}, organized as words by the rows, are used to represent *stored data*. Clocking thereby allows recursive transformation of the contents. When the array is used as a data memory and processor, the various signals then serve the following purposes:

z_{ij} represents the carry bit in arithmetic operations, or propagates data bits horizontally in shifting, writing, and reading operations

x_{ij} propagates stored and transformed data values in the vertical direction

c_j is used for bit masking

a_i, b_i select the operation to be performed on the ith word

Cell Logic. Controlled by the three signals a_i, b_i, and c_j, one of eight processing modes, described by the logic equations in Table 6.3, can be defined for each cell. The result of a cell operation is in general a new value for the stored bit y_{ij}, and new values for the intercommunication variables $x_{i+1,j}$ and $z_{i,j+1}$. It may further be noted that there are control conditions under which the intercommunication signals z_{ij} and x_{ij} are constant along the whole row or column, respectively, and the z_{ij} resp. x_{ij} then can be used as extra control signals. With $c_j = 0$, the cell operates in the *unclocked mode* in which only combinational logic operations are performed. During the *clocked* (arithmetic) operations which in general change the bit value, $c_j = 1$. If it is desirable to disable (mask) a particular column in arithmetic operations, its c_j is made 0.

Table 6.3. The logic of ACAM cells

Control signals c_j b_i a_i	Transformation of the contents y'_{ij} (new)	Intercommunication signals	
		$x_{i+1,j}$	$z_{i,j+1}$
0 0 0	y_{ij}(old)	x_{ij}	$z_{ij} \wedge (x_{ij} \vee \bar{y}_{ij})$
0 0 1	y_{ij}	x_{ij}	$z_{ij} \wedge (x_{ij} \oplus \bar{y}_{ij})$
0 1 0	y_{ij}	$(\bar{y}_{ij} \wedge x_{ij}) \vee (y_{ij} \wedge z_{ij})$	$(\bar{y}_{ij} \wedge z_{ij}) \vee (y_{ij} \wedge x_{ij})$
0 1 1	y_{ij}	$x_{ij} \vee (y_{ij} \wedge z_{ij})$	z_{ij}
1 0 0	$(\bar{z}_{ij} \wedge y_{ij}) \vee (z_{ij} \wedge x_{ij})$	$x_{ij} \oplus (y_{ij} \wedge z_{ij})$	z_{ij}
1 0 1	$(\bar{z}_{ij} \wedge y_{ij}) \vee (z_{ij} \wedge x_{ij})$	y_{ij}	z_{ij}
1 1 0	$(\bar{x}_{ij} \wedge y_{ij}) \vee (x_{ij} \wedge z_{ij})$	x_{ij}	$(\bar{x}_{ij} \wedge z_{ij}) \vee (x_{ij} \wedge y_{ij})$
1 1 1	$x_{ij} \oplus y_{ij} \oplus z_{ij}$	x_{ij}	maj (x_{ij}, y_{ij}, x_{ij})[a]

[a] Majority value of the argument bits

The particular cell logic defined by Table 6.3 represents only one paradigm of ACAM designed to demonstrate the processing ability of such an array. A set of *array functions* implementable by this logic is explained below in more detail. Each bit cell must be realized by about 36 logic gates which is almost three times as many as in the DLM, and about five times as many as in the all-parallel CAM. It is to be noted that the bit storage for y_{ij} must have the structure of a clocked flip-flop to isolate the old and new values from each other. Two-phase (master-slave) flip-flops seem advantageous, whereby in certain electronic circuits two external clocking signals may be used.

Array Functions. A justification for the cell logic defined in Table 6.3 is now given by a detailed discussion of the functions that the array is supposed to implement.

1) *Equality Search:* With $(c_j, b_i, a_i) = (0,0,1)$ the x word is propagated vertically as such through the array. If now $z_{i1} = 1$, the leftmost bit cell on a row forms an output z_{i2} which is the logical equivalence function of x_{i1} and y_{i1}, as immediately verified by a comparison of the truth tables of $(x \overline{\oplus} y)$ and $(x \equiv y)$. The result of this bit comparison (z_{i2}) is propagated to the next cell on the right which performs a similar bit comparison if $z_{i2} = 1$. If, on the other hand, $z_{i2} = 0$, all the further z_{ij} are 0 for $j = 3, \ldots, n$. An induction shows that the final output $z_{i,n+1}$ indicates the *word match*, i.e., $z_{i,n+1} = 1$ if and only if x_{ij} and y_{ij} match at all positions on the row.

Masked equality search is implemented by setting $c_j = 1$ at all masked bit positions. Table 6.3 now shows that for $(c_j, b_i, a_i) = (1,0,1)$ the cell propagates the x_{ij} and z_{ij} values as such, i.e., this bit position cannot produce a mismatch signal.

2) *Searching on the Basis of Logical Implication:* With $(c_j, b_i, a_i) = (0,0,0)$ and $z_{i1} = 1$ another content-addressable searching is performed. A response is obtained at all words which *do not imply* the search argument, i.e., which have $y_{ij} = 1$ at least in one position in which $x_{ij} = 0$. Masking of a bit is again possible by setting $c_j = 1$.

3) *Permutation Switch:* When the control pattern is $(0,1,0)$ and $y_{ij} = 1$, it is seen from Table 6.3 that the cell transmits signals $x_{i+1,j}$ and $z_{i,j+1}$ which are obtained from x_{ij} and z_{ij} by their *interchanging*. If $y_{ij} = 0$, no interchange occurs. By this function, after writing proper values to the cells, any permutation of the input signals becomes possible whereby the set of input signals is also shifted downward by an amount depending on the number of permutations made. (The paths of the signals in the array resemble fletched wires.)

4) *Arithmetic Addition and Subtraction:* When the words represent positive binary numbers, the bits are arranged so that the least significant end is always on the *left*. With the control pattern $(c_j, b_i, a_i) = (1,1,1)$ and $x_{i1} = 0$, the array forms the *arithmetical sum* of the word $(x_{11},...,x_{1n})$ with the stored words $(y_{i1},...,y_{in})$. This is immediately clear from Table 6.3 in which the logic function of y'_{ij} and $z_{i,j+1}$ are found to be those of the sum and carry bits, respectively. A carry in the least significant bit position, obtained by setting $z_{i1} = 1$, is needed in double-precision arithmetic and in arithmetic operations with signed numbers.

Subtraction of the X word from the contents of a word location is performed by first forming the 1's complement of the X word externally, or in some word location (k) above. If the control pattern for such a location is $(1,0,0)$, if the contents of the location are $(1,1,...,1)$, and if $z_{k1} = 1$, the X word retransmitted by this location downward will be complemented. When the complemented word is added to the contents of a location, taking $z_{i1} = 1$ for it, the result is the difference in 2's complement representation.

5) *Magnitude Search:* Since arithmetic operations are already included in the array functions, they can be utilized in content-addressable searches on the basis of magnitude relations. The search argument (an X word which is the 1's complement of the original argument value to be compared) is subtracted from all word locations in the way described above. An overflow $z_{i,n+1}$ is then obtained at all word locations the contents of which are *equal to or greater than* the original search argument value. If now $z_{i1} = 0$ is taken, a response is obtained only if the stored word is *strictly greater* than the search argument.

6) *Horizontal Shifting:* The contents of the array can be shifted to the right at the control pattern $(1,1,0)$. The bits in all cells corresponding to an input $x_{ij} = 1$ are shifted to the next cell on the right. A mask can also be used in shifting by setting $x_{ij} = 0$; the contents of the corresponding cell remain unaltered. An input defined by z_{i1}, is received by the leftmost cell for which $x_{ij} = 1$, and the output signal $z_{i,n+1}$ comes from the rightmost cell for which $x_{ij} = 1$.

7) *Vertical Shifting, Including Readout Operations:* These array functions are defined by the control pattern $(1,0,1)$. Shifting of the contents of all word locations downward occurs when $z_{i1} = 1$ because Table 6.3 shows that such a location transmits its old contents as the X word downward, and accepts the X word from above as its new contents. All words can be made to appear in a sequence at the lowermost location from which they overflow.

If $z_{i1} = 0$, the contents of the location remain unaltered (cf horizontal shifting). This provision can be used for *addressed, nondestructive readout.* Assume first that only one location has to be read out (addressed mode of reading). This location is selected by the control pattern (1,0,1) and holding $z_{i1} = 0$, whereas the control pattern (0,1,1) and the value $z_{i1} = 0$ shall be used for the rest of the locations to make them pass the x_{ij} and z_{ij} signals as such. The $x_{m+1,j}$ signals emerging at the bottom of the array now represent the contents of the location to be read out. Now assume that several locations receive the control pattern (1,0,1). The $x_{m+1,j}$ then represent the contents of the lowermost of such locations. This is equivalent to the *multiple-match resolution* function.

8) *Vertical Processing:* The control patterns (0,1,1) and (1,0,0) define array functions by which various recursive logic operations can be defined on the contents of word locations while the transmitted signals are propagated downward. Their use depends on particular applications.

9) *Masking of Rows:* The control pattern (0,1,1), with $z_{i1} = 0$, can in general be used to mask out word locations that are not involved in array operations. Cells with these control signals simply behave as if they were deleted from the array.

An ACAM Specially Designed for a Kalman Filter. KAUTZ and PEASE [6.17] have designed a content-addressable processor for use in target tracking, optimal filtering, etc. which primarily means an embodiment of the Kalman filter. The array functions were selected by requirements arising in matrix processing, but at little extra cost it was possible to include a number of other functions, to mention the computation of the Fast Fourier Transform. The processor was designed around an ACAM array the bit cells of which are implementable by about 50 logic gates.

Development of this system has mainly been carried out on the software level (microprograms, macros, routines, and application-oriented programs).

Other Iterative-Cell Array Processors. Cell arrays related to that of KAUTZ have been proposed by HOOD et al. [6.18]. A distributed-logic structure has been presented by TREPP [6.19]. Linear arrays that can be chained in many ways have recently been suggested by FINNILA and LOVE [6.20]. Array algorithms have been devised by BERKOVICH et al. [6.21].

6.4 The Association-Storing Processor (ASP)

A common feature of all memory designs discussed so far is that they were intended for the storage and retrieval of *independent items* such as numbers,
strings, and possibly compound identifiers consisting of (name, attribute)-
pairs. While such content addressability may be enough for simple document
retrieval and parallel computation, it does not yet cope with many of the
more complex tasks that occur in problems of artificial intelligence, language understanding, and in general, when dealing with *semantic* expressions
which are stored in memory in the form of *relational structures*. Retrieval
of information from such structures means that all items which occur in a
specified *"context"*, i.e., which are related to a set of other items in a
specified way, must be spotted and read out. To put it in another way, assume
that the elementary items form *relations* which are ordered sets of these
elements. The searching problem may be formulated by expressing a system of
relations in which some elements are left unknown, and the task is to find
all possible values for these unknowns which simultaneously satisfy all stated
relations, i.e., for which a corresponding set of relations can be found in
the data base referred to.

   In Chapter 1, the structure of semantic associative memory was described
briefly, and it was mentioned that the searching problem is usually formulated
and solved using certain high-level computer languages such as LEAP. This
section discusses a special *hardware* memory system named *association-storing
processor* (ASP) which is especially designed for the parallel storage and retrieval of semantic data structures. This design, introduced by SAVITT et al.,
as well as SAVITT, LOVE, RUTMAN etc. [1.14,15, 6.22-28] was preceded by a
careful software study to find out the processing functions to be embodied in
hardware. To the knowledge of this author there exists no hardware implementation of the ASP up to the present time.

   A corollary of the requirements embodied in the ASP design is that a number
of parallel searches, in general involving many different search arguments,
are proceeding simultaneously. This is not possible in the other content-
addressable memories in which the search argument is given externally.

Representation of Relations in the ASP. Semantic data structures can implicitly
be defined and represented by a set of relations or *"associations"* which in
its simplest form consists of ordered triples of the form (A,R,B). Here A and
B are two *items* and R, the link label, specifies the relation between A and B.
A structure resembling a network results when several relations share common
(identical) items. An example of this will be given in Fig. 6.5. The storage

of the data structure itself presents no problems: the relations, the triples, are stored as such. It is in the retrieval where an analysis of the structure must be performed.

The ASP is a regular array of memory cells provided with parallel processing logic. Each cell in it can be used to store any of the following types of information: 1) A single item or link label (e.g., A, B, or R). 2) The coded representation of a relation. 3) The coded representation of a *compound item* which has the form of a relation but can be used instead of an item in another relation.

The array of cells in the ASP is preferably a square. In principle, the ASP could be linear, too. However, in square arrays, the average distance between a pair of cells may be several orders of magnitude shorter than in a linear array, whereby faster communication becomes possible. The two-dimensional geometry is also suitable in large-scale integration of planar structures.

The ASP in fact belongs to the category of distributed logic memories, too, and it has many features in common with the usual DLMs. For instance, in its memory cells, data as well as various flags are stored. The cells can be identified by global content-addressing, and they are locally connected for intercommunication. However, while this latter feature in the DLM originally served the propagation of the matching status to the next cell in order to facilitate sequential matching of characters in connected strings, the objectives of intercommunication in the ASP have another scope. The purpose is to dispatch coded information over longer distances into addressed destinations, very much in the same way as the *pointers* discussed in Chap. 2 defined locations where the relevant items could be found. Although the cells are interconnected only locally, they can be made to pass signals in the same way as the shift registers do.

It is another characteristic feature of the ASP that the same memory array can be used to store "memorized" data and data structures, as well as descriptions of the searching criteria; the latter, named *control structures*, usually consist of a set of relations in some of which one or two of its elements are *unknown* and denoted by special symbols. The control structure is usually written in the form of a data structure which contains unknown items or link labels. For these, all values have to be found such that when many of them are substituted into the control structure this will match with some part of the "memorized" data structure. The values so found then constitute the set of all possible answers to the query. The searching task is somewhat analogous to solving an equation. While an equation, upon substitution of any of its solutions or roots is reduced into an identity, the "satisfied" control condition becomes identical with some part of the data structure.

The ASP memory can also be compared with that of the conventional or von Neumann computer. The latter may store data as well as various representations of program code. In the ASP memory, the control structures are equivalent to instructions or statements of a high-level language, and it is also possible to write *programs* as sequences of control structures, as explained later on.

Some examples of control structures, and an example of the contents of memory with which the control structures match giving values to unknowns are given in Fig. 6.5.

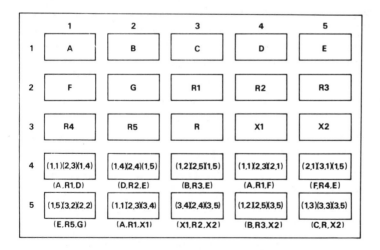

Fig. 6.5. An example of ASP data structures

Description of the ASP Array. The memory structure described below permits completely parallel processing of its contents, i.e., simultaneously over any number of specified relations.

	1	2	3	4	5
1	A	B	C	D	E
2	F	G	R1	R2	R3
3	R4	R5	R	X1	X2
4	(1,1)(2,3)(1,4) (A,R1,D)	(1,4)(2,4)(1,5) (D,R2,E)	(1,2)(2,5)(1,5) (B,R3,E)	(1,1)(2,3)(2,1) (A,R1,F)	(2,1)(3,1)(1,5) (F,R4,E)
5	(1,5)(3,2)(2,2) (E,R5,G)	(1,1)(2,3)(3,4) (A,R1,X1)	(3,4)(2,4)(3,5) (X1,R2,X2)	(1,2)(2,5)(3,5) (B,R3,X2)	(1,3)(3,3)(3,5) (C,R,X2)

Fig. 6.6. An example of contents of the ASP array

Consider Fig. 6.6. which exemplifies a small-size memory array with information, corresponding to that used in Fig. 6.5, stored in it. The cells are designated by their addresses which consist of a pair of coordinates (r, c) with r the row and c the column. Each cell contains three data fields of equal size and some flag fields. If the cell is used to represent a literal item or link label, all of the three fields can be concatenated. If, however, the cell has to hold a triple, i.e., a relation or compound item, its elements are represented *indirectly*, by the address codes of locations at which the literal items and the link label are stored. In this mode of representation, the contents of the above three fields are equivalent to *pointers*. If a triple contains unknown elements, special reserved codes, distinguishable from addresses, must be used for them. The actual values stored in the cells are written as constants in Fig. 6.6; for clarity, symbolic descriptions of the contents are written beneath the cells.

Each cell, in addition to the contents of its three data fields, can express its own address code in the form of wired-in signals. Any of these four addresses can be switched and transmitted into the communication lines, using global control signals. The switching and transmission is activated only in cells which have their match flip-flop set; however, the field to be switched can be defined individually in each cell, using two control bits reserved in the cells for this purpose. These bits can again be set in a separate processing step by global control, with the aid of match flip-flops for the location of the corresponding cells. It may thus be obvious that a great number of cells can simultaneously transmit information into the communication lines where they are propagated like bits in a shift register, under the timing control of a common system clock.

The cells can be made to pass address codes to each other in the upward (south-north) direction as well as to the left (from the west to the east). Codes transmitted to the west are replicas of those entering the cell either from the east or from the south. If the codes come from the east, the propagation is automatically continued unless the received code matches with the cell address, in which case the propagation is stopped and the match flip-flop is set in that cell. If the code was received from the south, it will automatically turn to the west if the *row address* part of it matches with the row address of the cell. Otherwise the code will continue travelling to the north.

Codes transmitted to the north may be replicas of codes entering the cell either from the south or from the east, or they may originate within the cell (being the contents of one of the three data fields or the cell address).

Which one of these alternatives is selected is defined in the following way. The case in which the code originates within the cell must be determined by global control (corresponding to a special microinstruction) whereby a special *transmit flag* within the cell must be set. The case in which the code came from the south was already discussed above. The remaining alternative, deviating westbound signals to the north, is discussed below in connection with the writing and reading (I/O) functions.

The cells which are at the edges of the array need special considerations. They are connected to the cells of the opposite edge cyclically so that the array can topologically be regarded as a torus; thus, signals supposed to be transmitted to the north by the uppermost cells actually enter the lowermost cells from the south, and a similar connectivity between the leftmost and rightmost cells exists.

The rather peculiar switching modes discussed above are necessary for the implementation of the functions described below.

Blockage. Since many codes are simultaneously propagated in the array, there is a nonzero probability for codes coming from the south and from the east to meet at a cell simultaneously whereby a conflict arises if the northbound code tries to turn to the west; the cell can pass only one of the codes. This instance is called *blockage* and it is solved by giving priority to the code coming from the east. The signal coming from the south is transmitted further to the north and it will cyclically return to this row later on. If the array is not quite full, the extra propagation cycles do not significantly reduce the average processing speed.

The Memory Functions. To accomplish the parallel searching on the basis of control structures, the ASP can execute microprogrammed sequences of the following eight *memory functions*:

1) *Content-Addressable Search:* This is the normal CAM operation in which a search argument, or a masked version of it, is broadcast to all cells, and it is called *search* for brevity. Upon agreement, the corresponding match flip-flops are set. In order to avoid erroneous matches with literal items or link labels, a special flag must have been turned on in cells which contain a relation (not containing unknowns), and a corresponding bit must be defined in the search argument. It is not specified in which way the broadcasting is made; notice that the intercommunication paths are only one item wide, but if the search argument is longer, it might be multiplexed on these lines in parts, in a few subsequent steps. A more straightforward method is to use global lines for content-addressable search. In any case, for the definition

of the search argument, the ASP needs an external argument register and a mask register.

It seems that the original reports and subsequent reviews of the ASP overlook the simple fact that the content-addressable search function can be applied only to cells in which *items* or *link labels* are stored (in literal form); the elements in relations, on the other hand, are *address codes* which are not directly known and thus cannot be used as (masked) search argument. For the content-addressable search of *relations* which contain the address code of a particular item or link label, it seems necessary to apply the special "box-car" function discussed below.

2) *Write:* This function writes the contents of the search argument register, corresponding to unmasked fields, into all cells whose match flip-flop is set.

In particular, the write function can be used to set some of the *flags* (transmit flag, box-car flag, relation identifier, etc.) described below in proper context, in all cells whose match flip-flop is set.

3) *Reset:* This function resets the match flip-flop in each cell, and in addition, sets the *sequence flip-flop*, similar to that of the multiple-response storage described in Sect. 3.3.2, to the value 1. The latter is also used to collect intermediate results from several passes as explained in the next function.

4) *Pulse:* The purpose of this function is to form intersections of the sets of responses obtained in subsequent passes. The pulse function resets the sequence flip-flop in each cell whose match flip-flop was not set, and resets the match flip-flop. Thus, the sequence flip-flop forms the logical AND of its previous contents and of the match flip-flop value.

5) *Context Addressing:* This memory function, being one partial step in context-addressable retrieval, aims at the location of stored items or link labels which could be possible solutions for a relation that occurs in a control structure. As a control structure is often complex, this function is usually applied in several passes as explained below.

Assume first that the transmit flags have been set in all cells which contain relations of the form (A,R,X), with A and R specified and X arbitrary; such cells must be located by the special "box-car" function explained below in paragraph 6. The context-addressing function transmits all the values corresponding to X, from cells whose transmit flag has been set, into the intercommunication lines. Notice that many codes which are mutually different may

be sent away. When these codes meet cells with matching addresses, they set
the match flip-flop in these cells.

It is obvious that the context-addressing function can be applied to relations in which the unknown occurs in any field.

The context-addressing function is applied in several passes in the case
in which an unknown is shared by several relations. (Such a control structure
may be regarded as a system of equations all of which must simultaneously be
satisfied. Assume, for instance, that the control structure is of the form
$A \xrightarrow{R1} X \xrightarrow{R2} B$ which is equivalent to a system of two relations, (A,R1,X) and
(X,R2,B). A preparatory operation is to reset the match flip-flops, and to
set all sequence flip-flops to 1 by the reset function. The retrieval commences with a set of operations (described in paragraph 6 in connection with
the box-car function) such that the match flip-flops will become set in all
relation cells whose left and middle fields contain addresses of A and R1,
respectively. Then, a write function is applied to simultaneously set the
transmit flags in these cells. It is to be noted that this first processing
phase may select a number of relations which do not contain solutions for the
whole control structure but only for one relation, whereas all final solutions are certainly contained in the set of selected cells. A context-addressing function executed next masks off the A and R1 fields and sends the
contents of the third fields into the intercommunication lines. When these
codes hit cells with matching addresses, they set the match flip-flops in
them. In this way, a set of candidates for solutions are located. A pulse
function applied next resets the match flip-flops and leaves the value 1 in
sequence flip-flops, thereby identifying the candidates obtained at the first
pass. The next part of the retrieval is similar to the phases described above.
It begins with a set of operations which locate all relation cells whose
middle and right fields contain addresses of R2 and B, respectively. The
transmit flags in these cells are set. The contents of the fields corresponding
to X are dispatched into the intercommunication lines, and they set the match
flip-flops in all cells with matching address. A pulse function executed next
resets the sequence flip-flop if the match flip-flop at the second pass was
not set, and resets the match flip-flop. Thus, the value 1 remains in the sequence flip-flop if and only if this cell responded to both context-addressing
functions described above, and this value signifies all final solutions, i.e.,
those cells which contain a value of X which satisfies the complete control
structure.

It may be clear that more complicated control structures are handled in a
similar way in several passes.

6) *The Box-Car Function:* The central application of this function seems to be in selecting all cells which contain a relation with the address code of a particular item or link label stored in it. When this function is called out, all cells whose transmit tags are set transmit the code in their specified field (e.g., the left one) to the north. In addition, the address of the transmitting cell is made to follow the first code, like a box-car follows a locomotive. This is physically possible since the intercommunication lines act like shift registers, and the transmission is made at two subsequent clock phases. After that, these two codes must follow the same routes which means that the cells passing them must be able to identify their role. It may be necessary to provide the box-car with some identifier, e.g., a special bit, and to make the switching control logic at each cell capable of perceiving it. Now assume that address codes from some field have been dispatched, and they meet cells with matching addresses. At this point it shall be mentioned that every cell contains a further flag named the *box-car flag*. If this flag is reset in the above cells, the propagation of the incoming codes stops and nothing further happens. However, if the box-car flag is set, the box-car code is retransmitted, and its propagation in the array continues until it meets the cell at which it originated. Then it sets the match flip-flop at that cell.

Assume that all relations of the form (A,X1,X2) are to be located, with the A item as "search argument". The box-car flag is first set in all cells (usually only one) where the literal value of item A is stored; this can be done by the search and write functions. Consider next cells which contain *any* relation; such cells can have a special marker, and they are located by a search with this marker as search argument, followed by a write function to set their transmit flags. If a box-car function is issued to all these cells, with the middle and rightmost fields marked off, many address codes from leftmost fields followed by the cell addresses will be sent on their way. Sooner or later they find cells with matching addresses. However, only those few cells (possibly only one) which have their box-car flag set will respond by sending the box-car back. When the latter returns to the address where it originated, it will set the match flip-flop in it. In this way the selection of the relations has been accomplished. This also yields all solutions to a control structure which is a relation with two unknowns. The *reading* of pairs of related items, however, is another problem discussed later on.

For the location of relations with two specified fields, e.g., those of the form (A,R,X), the execution of two box-car functions is needed. This is done by continuing the previous example. Recall that by one box-car function, all relations of the form (A,X1,X) were found; a writing function can be used to set the transmit flags in this subset. Next the literal items R are searched and their box-car flags are set; these flags are reset in other cells. Another box-car function, with the leftmost and rightmost fields masked off, dispatches the addresses from the middle fields of all relations found in the first pass. The box-cars can only be received by such cells which have the address code of R in the middle field. These cells constitute the solution to the above problem.

7) *Read:* This function initiates a simultaneous transmission of information from all cells whose match flip-flop is set. The intercommunication lines are used which means that only one specified field can be read out at a time. These signals are made to travel towards the cell in the *upper left corner* with address (1,1) which is attached to the output port. This is a box-car type transmission with address (1,1) as the "locomotive", and the specified fields as the box-car. Whenever a match occurs, the box-car only is sent out. Notice that any number of cells may be outputting their contents producing a stream of data, and if there was conflict at cell (1,1), it is handled in the normal way. There thus does not exist any problem from multiple responses with the ASP.

8) *Mass Write:* This is a function for the automatic construction and storage of a set of relations which have a regular structure. For instance, if $\{A_k\}$ is a set of items stored in arbitrary cells, and the elements of this set can be located by some characteristics which facilitate content-addressable search, then the mass write function can be used to create a set of relations of the type (A_k,R,B) and to store their representations in some set of cells which happen to be empty. As in the CAM, information in the ASP can be stored in arbitrary cells.

It will be necessary to introduce a further marker in each cell, named *usage flag*, to indicate its vacancy. Let this flag have the value 1 for an empty cell. A signal line which passes all cells on each row may be used to form the Wired OR-function of the corresponding usage flags, indicating whether any empty cells exist on that row.

For the mass write function, all cells A_k have to be searched and their transmit flags set. When the mass write function is called out, these cells simultaneously send their addresses to the west, provided that the signal line described above announces that there is vacancy on the same row. When during the execution of this function an address signal meets a cell which is empty, it writes its value into *all the three fields* of the cell, sets the match flip-flop, and turns the usage flag off. It will be necessary to write the information into all fields because it is not yet specified at this stage which field shall be reserved for the A_k. As soon as all vacancies on a row are filled, the address codes being propagated on that row must be deviated to the north; when they meet a row with vacancy, they are made to turn to the west. Conflicts are handled as described earlier. After the address codes of all the A_k have been written, the write function is applied to store the common address codes of R and B in proper fields. If R and B are stored in the memory, their addresses can be found out by a search, with R or B as the search argument, followed by a read function with the cell address specified as the data to be read out.

Figure 6.7 illustrates the signal trajectories in six of the memory functions. The cells with their transmit flags set are denoted by T; other flags turned on are M = match flip-flop, B = box-car flag, and U = usage flag.

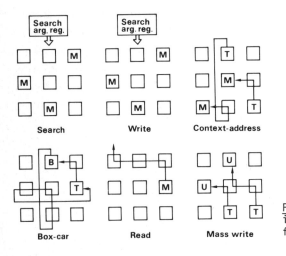

Fig. 6.7. Signal trajectories in the ASP for six memory functions

It will be obvious that for the implementation of all memory functions discussed above, the cell structure of the ASP has to be yet more complex than that of the CAMs and DLMs. The number of control and communication lines is also appreciable. It is striking, however, what an enormous amount of parallelism in global as well as local operations is thereby achievable; as the internal structure moreover is regular and thus suitable for large-scale integration, the principles of this architecture ought to be considered as an alternative for control principles applied in present computers.

Routing Cells. In order to speed up propagation, a further feature might be added to the ASP. A small number of cells ($\sqrt{N}$ if N is their total number), distributed over the memory as a superarray, can be used to pass signals directly to each other, i.e., to skip the intervening cells. It is necessary only to deduce from the address codes how far in the superarray it is necessary to propagate the signals in order to approach the correct row or column. For details, see [1.15].

An Example of More Complicated Control Structures. It occurs often that a control structure contains relations with one as well as *two* unknowns which may be shared by several relations. All of the above-discussed memory functions are then needed to yield values for the unknowns.

Assume that the control structure is

$$A \xrightarrow{R1} X1 \xrightarrow{R2} X2 \xleftarrow{R3} B \ ,$$

with X1 and X2 unknown. The direction of the last arrow is to be noticed; the corresponding relation is (B,R3,X2), i.e., with B defined to be on the left and X2 on the right, respectively. The solution proceeds in several phases. First, a number of candidates for the values of X1 are located with $A \xrightarrow{R1} X1$ as the partial control structure. It takes two searches, two box-car functions, and a few associated flag-setting operations to locate the relations. A context-address function executed next locates the first set of candidates for X1, and the sequence flip-flop in these cells is set. A similar series of operations is performed to find candidates for X2, using $X2 \xleftarrow{R3} B$ as the partial control structure, and the sequence flip-flop in these cells is set. The last phase, with $X1 \xrightarrow{R2} X2$ as the partial control structure, begins by location of cells with contents of the form (X1,R2,X2) for which the box-car function must be used as explained earlier. The transmit flag is set in these cells. The addresses of X1 are dispatched from these cells, and when they meet the proper items, the match flip-flops are set. When now a *pulse* function is applied, in the sequence flip-flops the value 1 is left for only such items X1 which satisfy both (A,R1,X1) and (X1,R2,X2). Next the addresses of X2 are dispatched from the (X1,R2,X2) cells and similarly they screen out only those candidate items for X2 which satisfy both (X1,R2,X2) and (B,R3,X2). Two independent sets of candidates, one for X1 and the other for X2, have now been found. The last task is to apply the constraint that the X1 are related to the X2 by R2. To do this, the box-car flags in the last-found X1 and X2 candidates are set, the transmit flag is set in *all* relations having the address code R3 in the middle (irrespective of X1 and X2), and two box-car functions are issued, one with the addresses of the leftmost fields, the other with those of the rightmost fields as box-cars. Using the match and sequence flip-flops of the transmitting cells, it is then possible to screen out only those transmitting cells which receive the box-car in both of these latter box-car functions. It may be clear that these are the cells which have the address code of the final solution in their X1 and X2 fields.

The final task is to read out the *pairs* of lateral items (X1,X2). Because the read function is able to output only one item at a time, reading of pairs is a bit lengthy, although straightforward task. For instance, the X1 items can readily be located on the basis of the address codes found in the relations, and read out. The address codes of the X1 can similarly be read. It is now necessary to make an external list of the X1 and their address codes. To find the related items X2, the *address code* of each X1 in turn is used as search argument to locate the corresponding relations, after which reading of the related X2 is a straightforward task.

Programs for the ASP. Let it be restated that all the design features of the ASP were aimed at the location of items or link labels which occur within a syntactic structure; this may be followed by processing steps in which a new value is given to these items or labels, new relations are created and connected to the items so found, or relations connected to these items are deleted. In general, thus, the syntactic data structure stored in the memory is changed by processing. This kind of change means *updating* of the stored knowledge. A simpler *query*, however, aims only at retrieval of specified items, without any changes made in the memory.

A typical processing step which also might be a simple retrieval can be defined by an *instruction* which is represented graphically; an instruction consists of a *control structure* described earlier, and a *replacement structure*. The latter shall be substituted for the control structure everywhere in the stored data structures. Assume that the control structure matches with the data structure; this is signified as "success" (S), and the opposite is "failure" (F). When the instructions are provided with *labels* for their identification, S or F defines which instruction shall be executed next. This is equivalent to *conditional branching* in usual computers. An example of a complete instruction is given in Fig. 6.8.

The bookkeeping of instructions, and the detailed execution of memory functions contained in them, can be controlled by a simple microprogram held in an external microprogram memory. The structure of the instructions, in the form of a set of relations having special symbols for their unknowns, however, is held in the ASP memory.

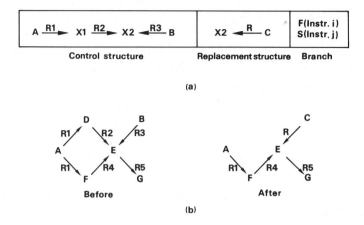

(a)

(b)

Fig. 6.8a,b. An ASP instruction: a) its format, b) its effect upon a data structure

6.5 Content-Addressable Processors with High-Level Processing Elements

If consideration is again directed toward numerical computation, it may be stated that parallelism in the DLM and the ACAM was implemented at the bit level of operation for which the external global control had to define long sequences of rather trivial micro-operations. This section discusses another approach in which a higher degree of autonomy of the cell operation is a characteristic feature. A parallel processor would then consist of an array of cells each of which is a complete subprocessor capable of storing a small number (possibly a few hundred) operands in its local memory, and equipped with a versatile arithmetic-logic unit. From the technological point of view the fundamental difference betweeen these approaches is that control logic in the ACAM is integrated around every bit storage with the result that the cell modules can exchange only elementary bit-oriented information; in the alternative types of cell discussed in this section, on the other hand, each integrated module is supposed to take care of circuit-level arithmetic and logic operations using its own automatic sequential control. Accordingly, intercommunication between the cells of the latter type is made by machine instruction codes and representations of numerical variables. This, of course, is a

very desirable feature from the programming point of view, especially when
dealing with complex problems.

6.5.1 The Basic Array Processor Architecture

Figure 6.9 schematically shows the simplest system organization of an *array
processor*. The array can be nonstructured, linear, rectangular, etc., specified by the interconnection network. If there are no interconnections between the cells except through the control or I/O bus, the system may be
named *ensemble processor*. Each subprocessor consists of a *processing element*
(PE) and a *processing element memory* (PEM). It will be necessary to point
out that all subprocessors, or an active subset of them, operate under a
common stored program control; the only programmed control in the PEs is that
defined in their fixed microprograms. On the other hand, the only storages
for data needed during computations exist in the PEMs. Instead of a special
control unit and program memory, the operation sequences may be controlled
by a general-purpose host computer which treats the array processor as a
special peripheral device.

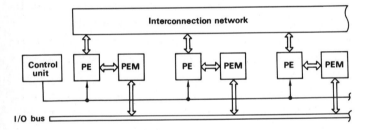

Fig. 6.9. The basic system organization of an array processor

   In the above architecture there exists still no operation which would be
signified as content-addressable search. The simplest, although a bit facti-
tious way to incorporate this feature in the system would be to have a small
CAM at every PE, with a purpose of responding to symbolically presented instructions or data.

Interconnectivity of the Cells. In some applications, the cells may operate
independently and intercommunicate with the host processor only, through the
input-output bus. More often, however, parallel processors are intended to
handle spatial problems or problems involving differential equations or multiple integrals, whereby the variables are defined in a grid, and all compu-

tations are oriented towards the description of interactions between neighboring nodes. For instance, in a simple relaxation problem, each node in a
planar grid has to accumulate an increment which is a weighted sum of values
at the four neighboring nodes. In special problems, the topology of the array
can be fixed, each cell being interconnected to four, six, eight, etc. neighbors, or in a way which corresponds to a shuffle of indices ($1 \to 1$, $n/2 + 1 \to 2$,
$2 \to 3$, ...). If the computer is intended to handle a wide range of problems,
interconnectivity ought to be programmable as in the RADCAP processor discussed
in Sect. 6.5.3. The cells can be interconnected directly, or through signal
paths generated in a network of multiplexer switches.

HOLLAND [6.29] has introduced a general systematics for the definition of
interconnectivity and functional specialization of the subprocessors. It seems
that the structures of the processor arrays often comply with the topology of
the computational problem.

The Tree-Channel Processor (TCP). As demonstrated in connection with the DLM,
ACAM, and ASP, intercommunication between distant cells is possible using
iterative transfer operations, although interconnectivity is provided only
between adjacent cells. However, in large arrays the signal delays may then
become appreciable. For this reason, LIPOVSKI [6.30] devised a tree-structure
interconnectivity pattern which can reduce propagation delays by several
orders of magnitude. By means of the intercommunication logic it is also possible to segment the array into arbitrary subsets, each one performing a
particular computational task.

The processor cells in the *tree-channel processor* (TCP), as this structure
is called, may be as simple as those of the DLM, or as complex as CPUs. In
any case this structure is recommendable with processors forming very large
arrays. Each cell must have a comparand register, comparison logic, and a
match flip-flop for content-addressable selection; other status flip-flops
are needed to determine which cells are transmitting, receiving, or processing
information, and which ones shall be idle.

For global communication such as broadcasting a search argument simultaneously to all cells, each cell is connected to a *channel*. Two auxiliary paths
named *rails* are provided for propagation of intermediate results, and local
interconnectivity is defined through the coupling of cells by the rails. All
cells are thought to be ordered in accordance with a single index, and they
will be organized in a tree structure (cf Fig. 6.10) in which the root has the
highest index value; of two cells the one nearer to the root has always a
higher index, and of two cells on the same level with respect to the root, the

314

one on the left has always the higher index. The cell with the lowest index value dircectly communicates with the root.

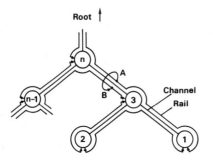

Fig. 6.10. Interconnectivity of the tree channel processor (TCP)

The tree can be a binary one like in Fig. 6.10, or any cell may be connected to more neighbors. The rails always connect bilaterally a pair of cells which have consecutive indices. If this were the only mode of intercommunication, the cells then could be drawn as a linear array. However, when an array is conceived as a tree, it is possible to define *shortcut paths* for signals such as that which would connect the points A and B in Fig. 6.10; this will decrease the propagation time, especially if the whole array is partitioned into small subtrees.

6.5.2 The Associative Control Switch (ACS) Architecture

In large-scale computer systems, to guarantee a high degree of utilization of the equipment, there exist many occasions on which a set of parallel units may have to be interconnected into another set of parallel units. Such "crossbar switching" is commonplace in multiprocessor systems between a set of CPUs and another set of memory banks. The organization presented in Fig. 6.11 is intended to increase the efficiency of an array processor. Without this provision, if a subset of processors were working at a particular program step, the rest would be idle. In the solution designed by SCHMITZ [6.31] for bulk filtering of radar data, each of the PEs can be connected to one of a set of independent programmed control units. The system appears as a set of independent array processors.

The connections are defined by content addressing from which the name *associative control switch* (ACS) for this architecture is derived. One of the

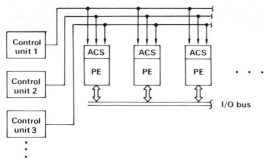

Fig. 6.11. The associative control switch (ACS) architecture

registers of each PE assumes the role of a *compare register* in which intermediate results are displayed; its contents are automatically compared with
the search arguments issued by the various control units. The result determines the switching of the PE into one particular control unit.

6.5.3 An Example of Control-Addressable Array Processors: RADCAP

A real content-addressable processor with a great number (1024) of high-level
processing elements was designed by Texas Instruments Corporation for parallel
computing experiments performed at the Rome Air Development Center (RADC).
This system, originally named RADCAP but later redubbed SIMDA (cf the classification of computers in Sect. 6.6.1) has been reported in many papers of the
1972 Sagamore Computer Conference [6.32]. The whole system is in fact rather
complicated with all its control structures and software [6.32-34]. Let it
suffice here to mention its most central specifications.

   The basic cell module of the RADCAP consists of a 256-word by 5-bit local
random-access memory (PEM) and a processing element (PE) with 4-bit arithmetic-logic unit. One bit in each memory word is reserved for parity checking.
Each cell further contains two 4-bit arithmetic registers, two other registers
for microprogram control, and two status registers. Intercommunication between the cells and at the I/O is made in terms of 4-bit "bytes".

   The instruction set of each PE consists of 48 machine instructions, including arithmetic, logic, and register transfer operations. Arithmetic is
based on 16-bit words, which means that all operands must be partitioned in
4-bit groups. Activation of the cells can be made by usual addressing or
content addressing; indexing of the words in the PEMs is made by global addressing. The status registers are used to indicate the following information:
PE activity, comparison results (less than, equal to, or greater than), overflow, fault, and microprogram status. A built-in logic locates the maximum

(minimum) value contained at a particular memory address of all active cells; the support logic performs many other operations such as priority selection and counting the active elements.

The array is divided into 32 blocks, each with 32 PEs. Each block is provided with a "dual-control multiplexer" switch which can connect all the cells of a block either to a program-control unit or to the I/O. Any subset of blocks can be connected to the control while the rest simultaneously perform I/O operations.

Intercommunications between PEs is a feature which is implemented in RADCAP in a very elaborate way. Corresponding to each cell i there exist three multiplexers X_i, Y_i, and Z_i. Multiplexer Z_i directly connects cell i with cells i-8 to i+7; multiplexer Y_i connects cell i with multiplexers Z_i, Z_{i+16}, Z_{i+32}, ..., and multiplexer X_i connects cell i to multiplexers Y_i, Y_{i+128}, ... All connections are mod 1024.

6.5.4 An Example of Content-Addressable Ensemble Processors: PEPE

Another real parallel processor with content-addressable features is PEPE (Parallel-Element Processing Ensemble) [6.35-41] designed for ballistic missile radar track data processing. Its current version consists of a 288-element ensemble of processing elements, each one intended to store and handle with one reference track. Each processing element contains a *correlation unit* which performs the preprocessing and matching of phased-array radar signals against its stored tracks, an *arithmetic unit*, a *memory*, and an *"associative" output unit*. The internal structure of the processing elements is rather complex: they contain about 8800 gates *in addition to* the memory module, and they are intended to perform 32-bit floating point arithmetic operations with single or double precision (one single-precision floating-point addition taking about 800 ns). Part of the complexity of a processing element comes from an architecture oriented towards an extended FORTRAN language which is capable of treating the processing-element variables in parallel.

The processing elements are not directly interconnected so this machine must be regarded as an ensemble processor. Its operation is content addressable in two respects: first, the operation of the correlation units is controlled by the group-organized DLM principle (cf Sect. 5.1.2) which allows content-addressable selection of the processing elements, as well as highly parallel processing of their contents. Secondly, the associative output unit issues radar control commands in parallel.

6.6 Bit-Slice Content-Addressable Processors

While the array and ensemble processor architectures described above are very
effective for the handling of bulk computations in special applications such
as filtering of radar data, it has been proven that computer systems built
around a word-parallel, bit-serial CAM are the most cost-effective and flexible ones over a range of diversified parallel computing and searching problems. Such computer systems will hereupon be named *bit-slice content-addressable processors*. The structures and operations of their central parts, the
CAM array and the results storage, were already discussed in much detail in
Sects. 3.4 and 4.3.

It may be illustrative to first compare the organization of a "highly
parallel" processor, say, a group-oriented DLM, and that of a simple word-
parallel, bit-serial CAM system. The memory array of the latter is here visualized using shift registers.

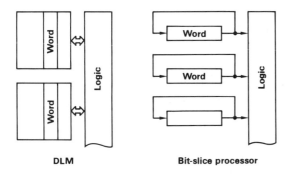

DLM   **Bit-slice processor**

Fig. 6.12. Comparison of a
DLM and a bit-slice processor

Storage of operands as shown on the left in Fig. 6.12 directly implies that if
arithmetic operations are to be performed, some intercommunication logic must
be provided between the X cells to take care of the carry signals. On the
other hand, if the storage of operands is made as shown on the right, and the
operands are rotated through the results storage by shifting, the iterative
intercommunication logic is replaced by a single sequential logic circuit
per word location. Thus, if these systems are used mainly for arithmetic
operations, there is no significant difference in speed between them since
bit-serial steps of operation must anyway be executed in both. On the other
hand, it is clear that the memory hardware of the latter is very much cheaper
per bit, especially with long operands; the memory itself has to perform nothing else but shifting.

An extension in the parallel processing capabilities over those of the DLM is obtained if the response storage of the CAM system is provided with intercommunication between the word locations. This is usually done by programming; the machine instructions which define the detailed data transfer operations may contain a complicated principle of addressing the bit slices and word locations. This is exemplified in the following.

6.6.1 The STARAN Computer

One of the well-known content-addressable parallel computers is the STARAN of Goodyear Aerospace Corporation [6.42-51]. This design, completed around 1972, was preceded by a simpler machine named the "RADC 2048 word memory" [6.52-55]. While the CAM array in the latter was implemented by plated-wire technology (cf Sect. 4.3.2), in the former the RAM memory module hardware with the EXOR skew addressing principle (cf Sect. 3.4.2) is used.

The purpose here is not so much to yield an overall view of STARAN but rather to pick up those features from its operation which describe programmed, parallel intercommunication between the storage locations.

A block diagram of the most important parts, primarily intended to aid their description, is given in Fig. 6.13.

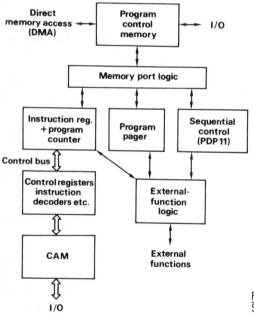

Fig. 6.13. Block diagram of the STARAN computer

The CAM Array. In STARAN, addressed reading and writing of words is made in parallel. It is mainly for this purpose that the skew addressing principle was introduced. The EXOR skew network for diagonal addressing, built of commercially available EXOR chips, is named *flip network* (FN) in this system. It is to be noted, however, that a programmer does not "see" the permutation of data in skew addressing; the array normally appears to him as simply being addressable by word locations or bit slices. The EXOR skew network could, however, be programmed to pick up also other types of slices from the CAM array, by the application of a particular argument in the address mode register (cf Fig. 3.16).

Because of the skew addressing principle, words and bit slices of a memory unit have to be of equal length. The CAM array of STARAN, therefore, consists of square modules called blocks; a size of 256-word by 256-bit for them was chosen, mainly along with the size of the most usual commercial RAM modules. Selection of words, bit slices, or other bit patterns in a block is made by an 8-bit address and an 8-bit address-mode argument. The data read out appear in parallel at a 256-bit I/O port, or they communicate via a 256-bit wide path with the results storage. The array may contain 1 to 32 blocks, the maximum number being determined by the indexing capacity of the machine instruction words which control the overall system operation. Bit slices over sets of blocks can be defined by programming.

Since the writing and reading principles as well as the logic of the results storage were already discussed in Sects. 3.4.2,4 there is not much to be added here to the description of the CAM hardware.

The Results Storage Hardware. The results storage of STARAN (cf Sect. 3.4.4) contains for each block three 256-bit registers X, Y, and M, respectively. The X register is generally used to store temporary results in recursive operations, the Y register acts as results register for searches and arithmetic and logic operations, and the M register is used as a mask register for words, and a word-select register in the writing of bit-slices into the CAM. In the control circuitry of the CAM array there is a 256-bit argument register, also named *common register* F which holds either the *search argument*, or serves as the buffer register in addressed writing and reading of data by words or by bit slices.

As mentioned earlier, each bit cell of the X and Y registers is equipped with logic which allows various sequential operations to be performed on their contents. This provision compensates for the need for having special arithmetic circuits for addition and subtraction.

3) *CAM Reference Instructions:* Addressing of the CAM array can be defined by several registers situated in the control block. These registers can be loaded from the control memory, and in conjunction with the instruction word, then can be multiplexed to the common control bus.

The *array select register*, with one bit for every block, indicates by bit values 1 which blocks are to participate in searching.

For addressed writing and reading referring to the CAM array, the F register acts as the source or destination of data, respectively. The contents of F may be subdivided into eight fields of 32 bits each, whereby only a specified field may be written or read out. An address mode defined in the instruction determines whether a word or a bit slice in a block is addressed. The most usual addressing uses two field pointers, e.g., FP1 to define the block and FP2 the word or bit slice. Other options exist, too, for which the control block contains two additional field pointers. All types of instruction defined in Table 6.5 are possible.

Table 6.5. CAM reference instructions

Source	Destination	Comments
CAM	F	All or only some fields of F may acquire new contents
F	CAM	All or only some fields of the CAM word may acquire new contents
CAM	X, Y, or M	
X, Y, or M	CAM	Masking of X or Y by M may be used

Between a pair of reading and writing instructions, the field pointers FP1, FP2, etc. may be modified, e.g., by incrementing or decrementing them. This allows various shifting and permutation operations to be performed.

Program Control. All of the above instructions dealt solely with the CAM system. The stream of instructions which defines the information process, however, must also contain instructions which are connected with the *stored program control* and which refer to the *control memory*. These include instructions for unconditional and conditional branching, control of program loops of specified length, handling of priority interrupts, communication with the

external devices, as well as arithmetic operations which change fields in
other instructions associated with the program control. It may not be necessary to give a complete list of all of these, since many instructions are
similar to those used in general-purpose computers. Let a few features, characteristic of the system in presentation, be pointed out.

A general-purpose computer usually has very few parts in its arithmetic-
logic unit, whereby the branching conditions are also few and simple; jumps
are usually conditional on the status of a few flags which indicate overflow,
zero result, or sign of the result. Contrary to this, many different types of
status conditions must be taken into account in parallel computers. Jumps in
the programs can be conditional , for instance, on the contents of the following registers which may be changed by the computing results: 1) Field pointers
FP1, FP2, FP3, and FPE. 2) Field length registers FL1 and FL2 which are used
as counters in program loops. 3) Branch and link registers which are used to
store program counters in the interrupt mode.

Other Control Conditions. As illustrated in Fig. 6.13, there are other units
in the STARAN system named *program pager, external function logic,* and *sequential control* which operate more or less autonomously, making use of priority interrupts. They perform various kinds of scheduling jobs, move pages
in the control memory, take care of interlocking conditions, and indicate
error conditions. As the sequential control unit, a PDP-11 host computer is
used, and its main responsibility is communication with the operator, including
monitoring of error conditions.

The I/O of STARAN is only loosely defined and could have many options.

Software. The assembler language of STARAN is named APPLE and it includes
instructions for register operations, loading, storing, and branching, procedures to perform equality and magnitude searching, parallel arithmetic,
as well as instructions for all necessary control and test operations. Assembly programs are defineable in terms of macrodefinitions.

6.6.2 Orthogonal Computers

The computing operations in STARAN and similar computers were mainly restricted to bit slices, whereas the word-addressing provision was primarily used
to guarantee fast I/O. In other words, the computational variables which
were stored in word locations were processed *independently,* by simultaneous
execution of similar program steps on all or a selected subset of them (bulk
processing). However, for interrelated variables further dimension in parallel

computations is necessary. These frequently raise problems, e.g., in *vector arithmetics* and various *transformations* in which sequences of operations must be performed over the sets of stored variables like elements of vectors or matrices, or variables in functions and transforms. The computing principle thereby applied, introduced by SHOOMAN [6.56-58], is named *orthogonal processing*.

The Basic Orthogonal Computer. In order to retain a possibility to compute with words (horizontally) as in a general-purpose computer, but to take a speed advantage from bit-slice operations (vertical processing) which refer to a large number of data simultaneously, two arithmetic units can be used. One of them, the *horizontal arithmetic unit* is of the conventional type, and a host computer can be used to provide it. In addition, a *vertical arithmetic unit* in many respects similar to the results storage described above, has to be constructed. In order to facilitate flexible operation in the applications for which this design was intended, the vertical arithmetic unit ought to contain several arithmetic registers. Figure 6.14 illustrates the system. Because of conventional computer operation with horizontal operands, the words can be rather short (e.g., 32 bits as shown).

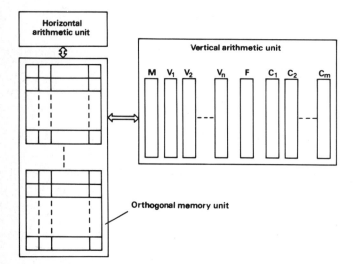

Fig. 6.14. The orthogonal computer. M = mask register; V_1 through V_n = vertical arithmetic buffer registers; F = function generator; C_1 through C_m = carry buffer registers

The block in the vertical arithmetic unit named *function generator* has a
logic circuit at every bit position, capable of forming any of the 16 Boolean
functions on two variables which are the corresponding bits in registers V_i.
Alternatively, one of the bits may be that found in a register, and the second one is read from the memory. The result can be subsituted into any
vertical register or bit slice. In order to speed up arithmetic operations,
each bit position in the vertical arithmetic unit is further provided with
a full adder network and a carry bit storage; there are several *carry registers* C_j to allow several arithmetic operations to be performed simultaneously.

The OMEN Computer. An outgrowth of the basic orthogonal computer of SHOOMAN
is the OMEN (Orthogonal Mini EmbedmeNt) system produced by Sanders Associates
[6.59]. Instead of bit slices, its vertical arithmetic operates on byte slices
(two bytes per horizontal word); there are eight buffer registers for the
operands. The CAM array is built of Intel 1103 memory chips, and skew addressing is used. The CAM capacity is up to 128 K words, 16 bits each. The
horizontal arithmetic unit is provided by a PDP-11 host computer which can
perform all usual 16-bit operations treating the CAM as its memory unit.

Computations particularly suitable to orthogonal computers are matrix
multiplications, evaluation of discrete transformations such as the Fast
Fourier Transform, and various tasks in signal processing.

6.7 An Overview of Parallel Processors

6.7.1 Categorizations of Computer Architectures

Several attempts have been made to represent general systematics for parallel
computers, and for content-addressable parallel processors in particular
[6.60-69]. In view of the fact that most existing parallel processors comprise unique solutions for special purposes, it may be more advisable to
restrict this discussion to the main lines along which different configurations
can be identified; the number of individual systems is too large to be reviewed here.

One of the first attempts to classify existing computer systems is due to
FLYNN [6.62]. He divided computers in four categories:

1) SISD (Single Instruction Stream, Single Data Stream) machines, also named
   uniprocessors; this is the fundamental type of general-purpose computers.
2) MISD (Multiple Instruction Stream, Single Data Stream) machines, also named
   pipeline processors. These could be particular general-purpose computers,

but more often a pipeline processor consists of a series of computing units which pass results to each other. Pipeline processors are used in picture processing.

3) SIMD (Single Instruction Stream, Multiple Data Stream) machines. This category contains array and ensemble processors with or without content-addressable selection.

4) MIMD (Multiple Instruction Stream, Multiple Data Stream) machines, also named *multiprocessors*. Architectures of large computer systems such as Univac 1108 are often built according to the MIMD principle.

A distinction between MIMD machines and computer networks should be made. While in MIMDs several CPUs and memory banks are switched to each other and intercommunicate mutually under a common control, using fast parallel switching networks, the computers in a computer network are almost autonomous units, usually located some distance apart, and they intercommunicate through transmission lines according to a "handshaking protocol". It has become popular to exchange information between the computers in terms of transmissional units named *packets*; the operation is dubbed *packet switching* [6.70].

The processors which have been discussed earlier in Sect. 6.6 all belong to the SIMD category; however, there also exist SIMD machines which are not content-addressable.

Another categorization is due to SHORE [6.66,67]. He divides the computers into six types named Machine I through Machine VI, depicted in Fig. 6.15.

If it is assumed that each *processing unit* (PU) also has its control memory, then it may be recognized that Machine I is a SISD processor, Machines II and III are SIMD processors capable of operating bit-serially on data, Machine IV describes SIMD constructs presented in Sect. 6.4, Machine V represents MIMD architectures of which ILLIAC IV [6.71-77] is the most powerful and best known example, and Machine VI distributes control logic with the memory hardware, and it embraces DLM, ASP, ACAM, and similar solutions. Often computers of the type of Machine VI are named "highly parallel computers" although this usually means only that parallelism has been implemented at a low organizational level throughout the system. If complexity were the most important attribute to characterize a computer system, then "highly parallel" actually should mean that the operations are parallel at all levels of hierarchical organization.

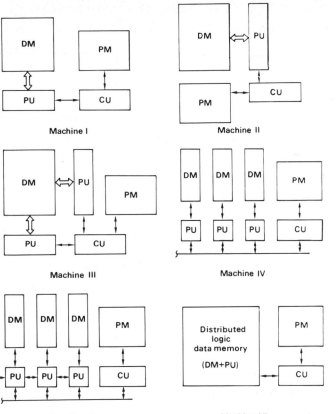

Fig. 6.15. Six categories of computers

6.7.2 Survey of Additional Literature on Content-Addressable and Parallel Processing

Books. Development of the state-of-art of content-addressable and parallel processing may be followed from the following publications: the books of HOBBS et al. [6.64], JACKS [6.78], FOSTER [6.68], and THURBER [6.69], Proceedings of the Sagamore Computer Conference 1972 through 1975 [6.32, 79-81], Proceedings of the IMACS symposium [6.82], and the Proceedings of the 1976 and 1977 International Conferences on Parallel Processing [6.83,84].

General Review Articles. The following reviews, presented below in an ap-proximately chronological order, shall be mentioned: CONTROL ENGINEERING

[6.85], FULLER [6.86-89], FULLER et al. [6.90,91], WESTINGHOUSE [6.92],
FELDMAN [6.93], GENERAL PRECISION [6.94], REICH [6.95], DUGAN et al. [6.96],
KNIGHT [6.97], KNAPP [6.98], JAUVITS [6.99], RUDOLPH [6.100], CANNELL et al.
[6.101], HOBBS and THEIS [6.102], MURTHA [6.103], MEILANDER and GALL [6.104],
MINSKY and PAPERT [6.105], THURBER and BERG [6.106], PARHAMI [6.107], THURBER
and PATTON [6.108], LEA [6.109], HIGBIE [6.110,111], PRENTICE et al. [6.112],
THURBER and WALD [6.113], LEWIN [6.114], SUMMERS [6.115], BAER [6.116],
INFOTECH INTERNATIONAL [6.117], YAU and FUNG [6.118], and ZIMMERMAN and
SIPS [6.119].

Processor Systems. Descriptions of architectures of parallel processor systems in addition to those already mentioned can be found in the following
articles: on the SOLOMON computer by SLOTNICK et al. [6.120] and WESTINGHOUSE
[6.121]; on the NEBULA computer by WEINGARTEN et al. [6.122] and BOLES [6.123];
on the IBM "Project Lightning" in [6.124,124]; and on other systems by UNGER
[6.126], HOLLAND [6.127], COMFORT [6.128], FEIGENBAUM and SIMON [6.129],
DAVIES [6.130-132], EVREINOV and KOSAREV [6.133], EWING and DAVIES [6.134],
HASBROUCK et al. [6.135], GALL [6.136,137], GALL and BROTHERTON [6.138],
ROHRBACHER [6.139], CASS [6.140], FENG [6.141-144], AUERBACH [6.145], LEA
[6.146-148], MIT [6.149], LOVE [6.150,151], MELLIAR-SMITH [6.152], SHORE and
POLKINGHORN [6.153], TUMA [6.154], KRESSLER [6.155], BERG and THURBER [6.156],
SHORE and COLLINS [6.157], URBAN [6.158], ARNOLD [6.159], BALDAUF [6.160],
GAMBINO [6.161], and LIPOVSKI [6.162].

Software. BERKOVICH et al. [6.163] have devised algorithms for group processing of data; RESNICK [6.164] describes an extension of COBOL for associative arrays, KERSCHBERG et al. [6.165] have designed an English-type query
language for relational data structures, and HICKS [6.166] describes algorithms
for syntactic and semantic analysis.

Special Questions. Some details and special algorithms have been discussed
in the following articles: logic circuit structures by SEEBER and LINDQUIST
[6.167], interconnectivity of the processing element by SQUIRE and PALAIS
[6.168], ENTNER [6.169], KOCZELA and WANG [6.170], and LIPOVSKI [6.171];
input-output problems by FOSTER [6.172] and KROEGER [6.173]; multifunctional
memory systems by GLANZ et al. [6.174], fault-tolerant processors by PARHAMI
and AVIZIENIS [6.175]; hardware floating-point arithmetic and algorithms by
THURBER and PATTON [6.176], and an analysis of the allocation of resources
in the processor system by NUTT [6.177].

Applications. The following listing tries to cover the most important applications of the content-addressable processors:

Data Management and Information Retrieval

KISYLIA [6.178], LINDE et al. [6.179], DeFIORE [6.180], DeFIORE et al. [6.181],
ARLAZAROV et al. [6.182], SALTON [6.183], DeFIORE and BERRA [6.184-185],
MOULDER [6.186], OZKARAHAN et al. [6.187,188], CHARNAYA [6.189], ASRATYAN
and LYSIKOV [6.190], BEAUFILS and SANSONNET [6.191], BIRNEY et al. [6.192],
ISHIKAWA [6.193], LANGDON [6.194], and SCHUSTER et al. [6.195].

Text Handling

DYKE and LEA [6.196], PRONINA and CHUDIN [6.197], LOVE and BAER [6.198],
and LEA [6.199].

Pattern Recognition and Picture Processing

FULLER and BIRD [6.200], YANG [6.201,202], YAU and YANG [6.203,204], STILLMAN
et al. [6.205], KRUSE [6.206].

Signal Processing (Radar, Sonar, Voice, Etc.)

Works on these areas are reported by JOSEPH and KAPLAN [6.207], LIBRASCOPE
[6.208], EDDEY [6.209,210], MEILANDER [6.211], CANNON [6.212], COSTANZO and
GARRETT [6.213], BIRD [6.214], THURBER [6.215], WALD [6.216,217], EDDEY et al.
[6.218], MOREFIELD [6.219], SCHMITZ [6.220], SINGHANIA [6.221], and LAMB and
VANDERSLICE [6.222].

Mathematical Problems

Parallel processing is needed in many mathematical tasks of which the following are examples: matrix computations [6.223-226]; Fast Fourier Transform
[6.227-228]; Hadamard transform [6.229]; solution of differential equations
[6.230]; linear decision processes [6.231]; multidomain algorithm evaluation
[6.232,233]; statistical processing [6.234], and spatial problems [6.235].

Miscellaneous

Further examples of applications of parallel processors are: space applications
[6.236]; machine translation [6.237]; data collection and display [6.238];
network problems [6.239]; evaluation of various arithmetic, logic and searching algorithms [6.240]; control of communication multiplexing [6.241]; weather

forecasting [6.242-244]; searching from decision trees [6.245] and evaluation
of the function of electrical power systems [6.246].

Simulation Studies. The following studies of the performance of hardware
content-addressable processing schemes by simulation shall be mentioned:
FINDLER [6.247,248], FINDLER and McKINZIE [6.248], FOSTER [6.249], SHORE
[6.250], and MEYERS [6.251]. It may also be recalled that content-addressable
processing is implementable by pure software, examples of which (in addition
to those mentioned in Chaps. 1,2) are the works of ASH and SIBLEY [6.252,253],
ASH [6.254], SIBLEY et al. [6.255], ROVNER and FELDMAN [6.256,257], FELDMAN
and ROVNER [6.258], ROVNER and HENDERSON [6.259], FELDMAN et al. [6.260],
and FELDMAN [6.261].

Chapter 7 Review of Research Since 1979

The first edition of this book contained literature references up to the year 1979. The purpose of the present chapter is to review more recent works on content-addressable memories, relating to Chaps. 2 through 6.

7.1 Research on Hash Coding

The basic principles and most of the practical solutions explained in Chap. 2 have remained almost as such in later implementations. The works reviewed in this section mainly contain new analyses, refined details, and applications. There is one novel idea, though, the *linear hashing* (Sect. 7.1.5) which seems to constitute the most important advance in hash coding in the 1980'ies. It neatly continues the original philosophy of locating the item by a few arithmetic mappings. (This principle should not be confused with linear probing, or with the linear quotient method, discussed in Sect. 2.3.2).

7.1.1 Review Articles

We start out with a listing of more recent reviews or tutorial articles on the hash-coding techniques: COWAN et al. [7.1], DEVILLERS and LOUCHARD [7.2], HEMENWAY and TEJA [7.3], GUNJI and GOTO [7.4], GOTO et al. [7.5], GOTO and TERASHIMA [7.6], KAPECKI [7.7], LYON [7.8], NISHIHARA [7.9], VIZZONE [7.10], GALL and NAGL [7.11], IDA [7.12], and TAMMINEN [7.13].
   Comparative surveys and evaluations of various file-access methods, with special mention of hash coding, have been presented by ANCONA and ANTOY [7.14], BLANK [7.15], INCE [7.16], MENDELSON and YECHIALI [7.17], KOCHIN [7.18], STANDISH [7.19], TRIFONOV [7.20], LEWIS [7.21], MITCHELL [7.22], JENKINS et al. [7.23], REGNIER [7.24], LITWIN and BELL [7.25], and ELLIS [7.26].

Simple hash-coding implementations, especially relating to small computers or special data banks, can be found in (anon.) [7.27], and in the
works of HEWLETT-PACKARD [7.28], LABEK [7.29], BILLIONNET [7.30], OSBORN
[7.31], LOMET [7.32], and DEEN and BELL [7.33].

7.1.2 Hashing Functions

This section contains some newer studies on hashing functions. Collections
of functions have been published by CANNELL [7.34], CARTER and WEGMAN [7.35],
KINZER [7.36], SORENSON et al. [7.37], WEGMAN and CARTER [7.38], and WEGMAN
et al. [7.39].

The following specific types of hashing functions have been suggested.
"Generalized" hashing functions: LITVINOV and IVANENKO [7.40]; "H-trees":
MALY et al. [7.41].

A random-number generator for hashing has been proposed in (anon.)
[7.42].

The performance of hashing when one aims at a balanced or uniform distribution has been analyzed by PAPADIMITRIOU and BERNSTEIN [7.43], LARSON
[7.44], and YAO [7.45]. Geometric considerations for analysis have been
used by COMER and O'DONNELL [7.46].

"Perfect Hashing Functions". The ideal hashing function —perfect or optimal
hashing (Sect. 2.2), for which no collisions occur —has still occupied a
few researchers: e.g., ANDERSON and ANDERSON [7.47], as well as CORMACK et
al. [7.48]. Especially the problem of the so-called minimum perfect hashing
has been discussed by CICHELLI [7.49], JAESCHKE [7.50], COOK et al. [7.51],
CHANG et al. [7.52], CHANG [7.53-54], CHANG and LEE [7.55], CHANG and SHIEH
[7.56], BELL and FLOYD [7.57], and DU et al. [7.58]. A collection of various perfect hashing functions has been presented by BERMAN et al. [7.59].
"Optimality" of hashing has been discussed by YUBA [7.60] and KRICHEVSKY
[7.61].

One has to realize, however, that the speed of searching is the only
reason for the introduction of any hashing techniques, and therefore the
hashing algorithm ought to be as simple as possible. If then, with an im-
perfect but quickly computable hashing function, the average number of
searches per item does not significantly exceed unity, there is no benefit
from using a slower but perfect (single-access) hashing scheme (see also
Sect. 7.1.5).

7.1.3 Handling of Collisions

Open Addressing. For the construction of an open hash table, an improved
program has been devised by SCHMIDT and SHAMIR [7.62]. Considerations of
the effect of unequal frequencies of the keys and its utilization have been
given by LARSON [7.63] and GONNET [7.64].

   Some new analyses of probing lengths can be found in [7.65] by MENDELSON
and YECHIALI, [7.66] by MURAO, and [7.67] by GONNET.

Open Addressing with Chaining (Coalesced Chaining). The basic method of
direct chaining in the open hash table has been discussed by TAI and THARP
[7.68,69] and VITTER [7.70]. Optimization of chaining can be found in VITTER
[7.71,72]. Deletion algorithms that preserve randomness are published in
VITTER [7.73,74]. For a shared-memory scheme, see [7.75].

   The efficiency of coalesced chaining has been analyzed by VITTER [7.76],
CHEN and VITTER [7.77,78], and LARSON [7.79].

Chaining Through Overflow Area. A general description has been presented by
TÖRN [7.80].

7.1.4 Hash Table Organization

One of the traditional features of hash tables, the bucket organization, has
further been developed by LYON [7.81,82]. A comparison of various bucket-
organization methods can be found in QUITTNER et al. [7.83].

   For indirect addressing of data spaces in general, see [7.84] by CREMERS
and HIBBARD.

   A hybrid approach for overflow handling has been presented by SCHEUERMANN
[7.85]. Another efficient combination of index tables and hashing has been
suggested by QUITTNER [7.86].

   Retrieval from hash tables can be reduced by several methods: ordering of
the tables (GONNET and MUNRO [7.87]), using predictors (NISHIHARA and IKEDA
[7.88]), split sequence search (LODI and LUCCIO [7.89]), self-organizing
search (BURKHARD [7.90]), direct rehashing (MADDISON [7.91]), repeated hashing (LARSON [7.92]), and reorganization of the table (SCHOLL [7.93]).
Another organizational idea can be found in [7.94] by ASTAKHOV.

   Criteria for efficient packing of hash tables have been presented by
LYON [7.95] and MUEHLBACHER [7.96].

7.1.5 Linear Hashing

It was stated above that most of the fundamental ideas in hash coding have remained almost as such since their introduction in the 1950'ies. There is, however, one significant exception. The concept of *linear hashing* (actually, *linear virtual hashing*) was developed around 1980 by LITWIN [7.97,98] for the management of files that can expand dynamically during use. The average number of accesses to the table or file stays reasonably close to unity at rather high loading of the table, and even a great number of insertions can be made without heavy reorganization.

The most serious disadvantage of the usual hashing methods is that if the load factor approaches unity, the average number of accesses for to search an item grows rapidly. This handicap is worse with open addressing, whereas in chaining through an overflow area, each item in the chain may add an extra access to memory, too. The rehashing method discussed in Sect. 2.3.5 solved this problem by moving all the items into a new, bigger table when the old one became filled up, whereby a completely new hashing function had to be chosen.

It would be more reasonable, however, that additional address space to the old hash table were allocated gradually, according to need; modification of the hashing function can thereby not be avoided, but it turns out that this kind of *dynamic hashing function* is derivable from the static one in a simple way.

The bucket organization (Sect. 2.4.2) will now be assumed. Let the address space of the hash table first be 0...N-1, with a bucket containing a certain number of slots for items at each address. Any number of new addresses $N, N+1,...$ shall be appendable upon need. Denote the key by K. The original hashing function, $h_0(K)$, shall hash uniformly over $0...N-1$. When any of the buckets is filled, the items overflowing are first chained to it.

The original idea of LITWIN is that every time when an overflow from *any bucket* occurs, the new item is chained to it, but, in addition, a new bucket is appended to the memory, and one of the old buckets *in the numerical order* (i.e., not necessarily where the overflow occurred) is rehashed. Rehashing shall split the contents of the old bucket between it and the bucket appended last. If there was a chain appended, all items in it are rehashed, too. Assume that in total $p+1$ overflows have occurred, whereby address p is, in turn, to be rehashed; then its contents are randomly split between the addresses p and $p+N$. A rehashing function $h_1(K')$ which does this can be, e.g., of the form

$$h_1(K') = h_0(K') + b_0(K') \cdot N \tag{7.1}$$

where $b_0(K')$ is an extra hash bit computed at the same time as $h_0(K')$. Notice that K' does not only stand for the key which caused the overflow, but any of the keywords at address p. It may be obvious that with growing p most of the buckets and chains appended to them will be rehashed. After N overflows, the address space will be 0...2N - 1. The next-level hashing function shall then be

$$h_2(K') = h_1(K') + b_1(K') \cdot 2N \quad , \tag{7.2}$$

and the process shall continue with p = 0 again. At level i, the hashing function is

$$h_i(K') = h_{i-1}(K') + b_{i-1}(K') \cdot 2^{i-1}N \quad . \tag{7.3}$$

For *searching* with key K it will be necessary and sufficient to know the current value of p and the current highest level i. Assume that key K shall be located. In the following, K is assumed to exist in the table. The searching starts with the hashing function $h_i(K)$. Let m be the correct bucket to be found. The following procedure can easily be deduced to compute m (although the complete argumentation shall be abandoned here):

begin

> *if*   p = 0   *then*   m = $h_i(K)$
>
> *else*   m = $h_{i-1}(K)$;
>
> *if*   m < p   *then*   m = $h_i(K)$;

end

Notice that no accesses to the memory are necessary until the correct address is known. (It can be shown that if K is not found at the address m defined by this procedure, it does not exist in the table). The total average number of accesses is slightly greater than unity, because some buckets may still have chains although most of them will have been rehashed away.

For the many details, performance, and variants of this method, see [7.99] by MULLIN, [7.100-103] by LARSON, and [7.104] by RAMAMOHANARAO and SACKS-DAVIS.

7.1.6 Dynamic, Extendible, and External Hashing

The linear hashing scheme was preceded by the ideas of *dynamic hashing* and *extendible hashing* (Sect. 2.9). Their principles were analogous, namely, allocation of new buckets upon need and splitting overflowing buckets. However, because the systematic linear order, characteristic of linear hashing, was not applied, the dynamic hashing function had to be defined using auxiliary index tables or directories. The address structure thereby became a tree of TRIE (Sect. 2.8). The following works, in addition to the original ones of LARSON [2.135] and FAGIN et al. [2.136] shall be mentioned: SCHOLL [7.105-107], REGNIER [7.108], FROST [7.109], FROST and PETERSON [7.110], RAMAMOHANARAO and LLOYD [7.111], MULLIN [7.112], VEKLEROV [7.113], KAWAGOE [7.114], FAGIN et al. [7.115], TAMMINEN [7.116-123], YAO [7.124], LLOYD and RAMAMOHANARAO [7.125], MENDELSON [7.126], FLAJOLET [7.127], BECHTHOLD and KÖPERT [7.128], BRYANT [7.129], and HUANG [7.130].

In *external hashing*, in general, a small internal table is used to direct accesses to an external storage. This method facilitates addressing of collections of data structures: LIPTON et al. [7.131], GONNET and LARSON [7.132], BELL and DEEN [7.133], and LARSON and RAMAKRISHNA [7.134].

7.1.7 Multiple-Key and Partial-Match Hashing

The following new papers on multiple-key searching methods shall be mentioned: BOLOUR [7.135,136], DORNG and CHANG [7.137], CHANG [7.138], OTOO [7.139], and VALDURIEZ and VIEMONT [7.140].

By its nature, a hashing function does not usually tolerate errors in the keyword; if the items from the proximity of the keywords have to be searched, either multiple functions or multiple searches are necessary (Sects. 2.6,7). For newer studies on partial-match operations, see BURKHARD [7.141], ROBERTS [7.142], RAMAMOHANARAO et al. [7.143], RAMAMOHANARAO and SACKS-DAVIS [7.144], COLOMB [7.145], and KOHONEN et al. [7.146].

7.1.8 Hash-Coding Applications

In addition to the traditional symbol lists and data base management applications, it will be interesting to mention the following new applications of hash-coding methods: high-performance memory management (THAKKAR and KNOWLES [7.147]), sorting (DUCOIN [7.148]), storing a sparse table (TARJAN and CHICHIH YAO [7.149], FREDMAN et al. [7.150]), storage structures for a

DBTG data system (BLANKEN et al. [7.151]), document retrieval (WILLET
[7.152]), bibliographic data and INSPEC codes (COOPER et al. [7.153,154]),
index-maintenance (BURKHARD [7.155]), dictionaries (DODDS [7.156], COMER
et al. [7.157], RADUE [7.158]), hyphenation (BARTH and NIRSCHL [7.159]),
spelling-error detection and correction (ZAMORA [7.160] and MOR and FRAENKEL
[7.161]), erroneous sentence networks (VENTÄ [7.162,163], VENTÄ and KOHONEN
[7.164]), priority queues (AJTAI et al. [7.165]), special key words (DOSTER
[7.166]), extracting statistical data from text (HILL and ZEIN [7.167]),
text signatures (THARP and TAI 7.168), passwords (PORTER 7.169), large-
scale mathematical programming system (SUHL [7.170]), virtual-to-real address
translation (COCKE and WORLEY [7.171]), medical imaging (FRIEDER et al.
[7.172]), and management of an automated stereo-warehouse (HEJUN and YUEFANG
[7.173]).

7.1.9 Hash-Coding Hardware

Special hardware must be used if there is a need to perform several searches
in parallel, e.g., in the control of parallel processors. The following
works have been published: memory with non-changeable address block (BRAIDT
and TAYLOR [7.174], associative mass storage (WOLF [7.175], HIRAKI et al.
[7.176]), hashing addresses to a cache (ROBINSON and TAYLER [7.177]), database machine with hashing hardware (DOHI et al. [7.178]), resetting storage
unit directories (BENHASE [7.179]), pseudoassociative store (DA SILVA and
WATSON [7.180]), symbol manipulation (IDA [7.181]), multiterm string com-
parator (BURKOWSKI [7.182]), address translation (RAMAMOHANARAO and SACKS-
DAVIS [7.183], THAKKAR and KNOWLES [7.184]), and distributed data-driven
processor (KISHI et al. [7.185]).

7.2 CAM Hardware

7.2.1 CAM Cells

Surveys of semiconductor memories, with special emphasis on CAM, can be
found in [7.186] by BARRY and GEORGE, as well as in [7.187] by CIROVIC. The
following implementations of CAM cells have been developed recently: with
TTL circuits (three switches) (HOWARD [7.188]), with several bipolar transistors and diodes (DENIS [7.189]), with six FET components (AIPPERSPACH and
WU [7.190]), and with five resistor-load type MOS transistors (YOSHIDA et

al. [7.191,192]). A hybrid memory cell has been suggested by HOU [7.193]. For high-performance complementary CAM/RAM circuits, see HARASZTI [7.194].

7.2.2 CAM Arrays

The following LSI CAM arrays have been reported: 1-Kbit array (NIKAIDO et al. [7.195]), 1.5-Kbit array (VERNAY et al. [7.196]), and 4-Kbit array (OGURA et al. [7.197]). An 8-Kbit content-addressable and re-entrant memory has been published by KADOTA et al. [7.198]. A CAM with 256-byte "superwords" has been reported by LAMB [7.199].

A re-orderable-content RAM array has been used in CAM arrays by PUCKNELL and RAYMOND [7.200,201]. Description of a parallel cellular memory has been given by PEREZ [7.202].

A methodology for the testing of CAMs has been developed by GILES and HUNTER [7.203].

7.2.3 Dynamic Memories

A dynamic semiconductor CAM with refresh feature has been described by SCHUSTER [7.204]. Bubble memories for CAM and array logic have been reported by LEE and CHANG [7.205]. Some comments on bubble memory architectures have been made by STRADER [7.206].

7.2.4 CAM Systems

The "associative random access memory" has been described by TAVANGARIAN [7.207]. SHIRAKAWA and KUMAGAI [7.208] describe the implementation of a content-addressable memory using three-dimensional addressing. NODES et al. [7.209] have introduced a fuzzy associative memory module for signal processing applications.

7.3 CAM Applications

A CAM chip for virtual memory has been designed by HAMAZAKI [7.210]. The implementation of a paged-memory management unit is described by COHEN and McGARITY [7.211]. The realization of the LRU algorithm is discussed by SCHUBERT [7.212]. Parallel garbage collection can be implemented using an associative tag (SHIN et al. [7.213]). Memory defects can be corrected by spare components, using a CAM for control (anon.) [7.214]; similarly, fault-tolerant MOS RAMs can be implemented (HARASZTI [7.215]). Reference data structures using CAM have been organized by GEKHT and FROLOV [7.216].

The fifth-generation computers have called for new architectures. Content addressing for their information management has been discussed in MALLER and SCARROT [7.217]. Performance enhancement of computation in general has been published by MALMS et al. [7.218].

Logical minimization of multilevel coding functions using optical CAM has been devised by MIRSALEHI and GAYLORD [7.219]. Truth-table lookup, especially residue numbers, and its optical CAM implementation has been suggested by GUEST and GAYLORD [7.220], GUEST et al. [7.221], MIRSALEHI and GAYLORD [7.222], GAYLORD et al. [7.223], as well as GAYLORD and MIRSALEHI [7.224]. Multivalued logic seems to be another useful application area for the CAM (PAPACHRISTOU [7.225]. The number of storage locations thereby needed has been estimated by BUTLER [7.226].

Arithmetic operations are implementable by content addressing (NIKITIN et al. [7.227], PAPACHRISTOU and KAI HWANG [7.228]).

Content-addressable memory systems have further been suggested for the following special applications: pattern recognition (MALMS [7.229], BHAVSAR et al. [7.230], BADI'I and MAJD [7.231]), image analysis (SNYDER and SAVAGE [7.232], SNYDER and COWART [7.233]), designing assemblers (SINHA and SRIMANI [7.234], BADI'I and JAYAWARDENA [7.235]), evaluation of logic programs (NAKAMURA [7.236], OLDFIELD [7.237]), for a time-division switch cell (DEMANGE [7.238]), learning controller of an unstable system (BOZINOVSKI et al. [7.239]), branch path prediction ((anon.) [7.240]), burst-error correction (YAMAMOTO and FURUKAWA [7.241]), determining the state of a multistage switching circuit (KACZAMAREK and GOFTA [7.242]), as well as architectures for artificial intelligence (DEERING [7.243]).

7.4 Content-Addressable Parallel Processors

The following survey articles on the content-addressable or associative parallel processors have appeared: SCHUEGRAF [7.244], ICHIKAWA and KAMIBAYASHI [7.245], KOLLER [7.246], as well as DURHAM [7.247].

7.4.1 Architectures for Content-Addressable Processors

A special mention is due to a three-volume book series that describes the
implementation of LUCAS, the Lund University of Technology associative array
processor: SVENSSON [7.248], FERNSTRÖM [7.249], and KRUSELA [7.250].

A new content-addressable parallel processor design, especially for picture processing, has been developed by FORSTER [7.251].

The following particular types of processors have been described: Titanic,
a VLSI based content-addressable parallel array processor (WEEMS et al.
[7.252]), vector associative processor (BERKOVICH and PULLEN [7.253]), bit-
serial processing system (BATCHER [7.254]), multiple-instruction associative
processor ((anon.) [7.255]), and holographic processor (GERASIMOVA and
ZAKHARCHENKO [7.256]). A multilevel architecture has been described by
KUMAR et al. [7.257].

The performance of a highly parallel system has been measured by PARKIN-
SON and LIDDEL [7.258].

Systolic arrays for content addressing have been studied by WING [7.259]
and WALLIS [7.260].

Memory activation for the associative processor NOAH is described by
SHIMAZU and TAMATI [7.261].

7.4.2 Data Base Machines

After submission of the manuscript for the first edition of this book, a
special issue on data base machines was published in *Computer* [7.262]. The
reader is adviced to look at this 73-page edition because many of the ideas
of Chapter 6 appear there in slightly updated form.

Hardware systems for text information retrieval can be found in HOLLAAR
and KUEHN [7.263]. An associative backend processor for data base management
is described by HURSON [7.264]. Performance analysis of several alternative
data base machines has been given by HAWTHORN and De WITT [7.265]. A multiprocessor system which supports relational database systems has been presented by De WITT [7.266]. Memory allocation for multiprocessor systems
with content-addressable memories is discussed in KARTASHEV and KARTASHEV
[7.267].

A very ambitious objective of encoding and retrieving large amounts of
text by a special high-density MOS circuitry has been envisioned by GOSER
et al. [7.268]. They suggest an adaptive, distributed character cell which
is able to tolerate fabrication errors, bound to occur at extremely high
component densities.

7.4.3 Applications of Content-Addressable Processors

The following is a list of applications solved by content-addressable processors: iconic and symbolic processing (LAWTON et al. [7.269], WEEMS et
al. [7.270]), image analysis and computer vision (VERONIS [7.271], WEEMS
and LAWTON [7.272], POTTER [7.273], LEA [7.274]), determination of the components of a flow field (STEENSTRUP et al. [7.275]), knowledge-based systems
(McGREGOR [7.276]), and intelligence information retrieval (CYRE [7.277]).

7.5 Optical Associative Memories

7.5.1 General

To the last the recent work on optical associative memories should be mentioned, especially on systems storing items in *distributed* form, as superimposed memory traces. Many of them aim at the implementation of the optimal associative mappings described in Sect. 1.4.4 and in [1.1] and [7.278].

There exist in fact two types of these devices. In one of them, the matrix
operations for associative recall are performed by multiplying light intensities using transmission filters, and summing up convergent light beams
locally, cf. PSALTIS and FARHAT [7.279]. The second type is based on holography.

7.5.2 Holographic Content-Addressable Memories

A special issue on optical computing, with mention of optical "associative"
memories, was published by the IEEE Spectrum [7.280].

The simple Fourier-holography type of associative memory (Sect. 4.4.2)
has further been suggested by MADA [7.281]. An extension of this principle
for a multifocus hologram, for a larger memory capacity, has been implemented in practice by GREGORY and LIU [7.282].

Research in new optical materials has had a great impact on optical computing. Several optically bistable media have been known for some time, and
the associative processing power has significantly been enhanced by the introduction of the so-called *phase-conjugation mirrors* (PCM) which facilitate
a two-way transmission of phased wavefronts. Optical architectures which
involve PCM devices have been published by CAULFIELD [7.283], FISHER and
GILES [7.284], and SOFFER et al. [7.285].

Information-retrieving optical systems which, however, need a spatial
beam switch and are thus not genuine associative memories are described by
BECKER et al. [7.286], GERASIMOVA and ZAKHARCHENKO [7.287], VERBOVETSKII
[7.288], as well as WARDE and KOTTAS [7.289].

A novel approach to distributed memories is one in which optical modes
or eigenstates of a resonator, eventually based on a PCM, are utilized as
memory states: ANDERSON [7.290] and YARIV and KWONG [7.291].

More references can be found, e.g., from the Technical Digest of the 1985
Annual Meeting of the Optical Society of America [7.292] p. 51 and pp. 57-58,
and from the Proceedings of the SPIE Special Institute on Optical and
Hybrid Computing [7.293].

References

1.1   T. Kohonen: *Associative Memory – A System-Theoretical Approach*, Com-
      munication and Cybernetics, Vol. 17 (Springer, Berlin, Heidelberg,
      New York 1978)
1.2   J.R. Anderson, C.H. Bower: *Human Associative Memory* (Winston & Sons,
      Washington, D.C. 1973)
1.3   A.C. Hanlon: IEEE Trans. EC-*15*, 509-521 (1966)
1.4   J. Minker: Comput. Rev. *12*, 453-504 (1971)
1.5   B. Parhami: Proc. IEEE *61*, 722-730 (1973)
1.6   V. Bush: Atl. Mon. *176*, 101 (1945)
1.7   F. Rosenblatt: *Principles of Neurodynamics: Perceptrons and the Theory
      of Brain Mechanisms* (Spartan Books, Washington, D.C. 1961)
1.8   M.H. Lewin: RCA Rev. *23*, 215-229 (1962)
1.9   V.L. Newhouse, R.E. Fruin: Electronics *35*, 31-36 (1962)
1.10 A.E. Slade, H.O. McMahon: Proc. EJCC *10*, 115-120 (1956)
1.11 J.D. Noe: Curr. Res. Devel. Sci. Doc. *9* (1956)
1.12 E.H. Frei, J. Goldberg: IEEE Trans. EC-*10*, 718-722 (1961)
1.13 A.D. Falkoff: J. ACM *9*, 488-511 (1962)
1.14 D.A. Savitt, H.H. Love, R.E. Troop: "Association Storing Processor",
      Vol. I (AD 818 529), Vol. II (AD 818 530) (Hughes Aircraft Co. 1967)
1.15 D.A. Savitt, H.H. Love, R.E. Troop: AFIPS Proc. SJCC *31*, 87 (1967)
1.16 J.A. Feldman, P.D. Rovner: Commun. ACM *12*, 439-449 (1969)
1.17 P.D. Rovner, J.A. Feldman: In *Information Processing 68* (North Holland,
      Amsterdam 1969) pp. 579-585
1.18 J. McCarthy, P.W. Abrahams, D.J. Edwards, T.P. Hart, M.I. Levin: *LISP
      1.5 Programmer's Manual* (MIT Press, Cambridge, Mass. 1962)
1.19 H.A. Simon, A. Newell: "Information-Processing in Computer and Man".
      A Sigma Xi-RESA National Lecture 1964. In *Perspectives on the Computer
      Revolution*, ed. by Z.W. Pylyshyn (Prentice-Hall, Englewood Cliffs 1970)
1.20 P.D. Rovner: *LEAP User's Manual* (MIT Lincoln Laboratory, Mass. 1968)
1.21 K. Van Lehn: *SAIL User Manual*, Stanford Computer Science Report
      STAN-CS-73-373 (1973)
1.22 L.A. Zadeh: Inf. Control *8*, 338 (1965)
1.23 B. Widrow: "Generalization of Information Storage in Networks of Adaline
      Neurons", in *Self Organizing Systems 1962*, ed. by G.T. Yovits (Spartan
      Books, Washington, D.C. 1962)
1.24 K. Steinbuch: *Automat und Mensch* (Springer, Berlin, Heidelberg, New York
      1963)
1.25 P.J. van Heerden: *The Foundation of Empirical Knowledge with a Theory
      of Artificial Intelligence* (Wistik, Wassenaar, Netherlands 1968)
1.26 D. Gabor: IBM J. Res. Dev. *13*, 156 (1969)
1.27 M.D. Levine: Proc. IEEE *57*, 1391 (1969)
1.28 D. Marr: J. Physiol. London *202*, 437 (1969)
1.29 D.J. Willshaw, H.C. Longuet-Higgins: "Associative Memory Models", in
      Machine Intelligence, Vol. 5, ed. by B. Meltzer and D. Michie (Edinburgh
      University Press, Edinburgh 1970)

344

1.30 K. Nakano, J. Nagumo: In *Advance Papers of the Conference, 2nd Intern. Joint Conf. Artificial Intelligence* (The British Computer Society, London 1971)

1.31 T. Kohonen: A class of randomly organized associative memories. Acta Polytech. Scand., Electr. Eng. Ser. No *El 29* (1971)

1.32 J.A. Anderson: Math. Biosci. *14*, 197 (1972)

1.33 T. Kohonen: IEEE Trans. C-*21*, 353 (1972)

1.34 T. Kohonen, M. Ruohonen: IEEE Trans. C-*22*, 701 (1973)

1.35 L.N. Cooper: "A Possible Organization of Animal Memory and Learning". in *Proc. Nobel Symp. Collective Properties of Physical Systems*, ed. by B. Lundquist, S. Lundquist (Academic Press, New York 1974)

1.36 M.L. Yevick: Pattern Recognition *7*, 197 (1975)

1.37 R. Sorabji: *Aristotle on Memory* (Brown University Press, Providence, R.I. 1972)

1.38 R.W. Hamming: Bell Syst. Tech. J. *29*, 147 (1950)

1.39 D.J. Rogers, T.T. Tanimoto: Science *132* (1960)

1.40 L. Ornstein: J. M. Sinai Hosp. *32*, 437 (1965)

1.41 K. Sparck-Jones: *Automatic Keyword Classification and Information Retrieval* (Butterworth's, London 1971)

1.42 J. Minker, E. Peltola, G.A. Wilson: Tech. Report 201, University of Maryland, Computer Science Center (1972)

1.43 J.S. Liénard, M. Mlouka, J.J. Mariani, J. Sapaly: "Real-Time Segmentation of Speech", in *Preprints of the Speech Communication Seminar*, Vol. 3 (Almqvist & Wiksell, Uppsala 1975) p. 183

1.44 T. Tanimoto: *An Elementary Mathematical Theory of Classification and Prediction* (IBM Corp., 1958)

1.45 J. Łukasiewicz: Ruch Filos. *5*, 169 (1920)

1.46 E.L. Post: Am. J. Math. *43*, 163 (1921)

1.47 L.A. Zadeh: IEEE Trans. SMC-*3*, 28 (1973)

1.48 L.A. Zadeh, K.S. Fu, K. Tanaka, M. Shimura (eds.): *Fuzzy Sets and Their Applications to Cognitive and Decision Processes* (Academic Press, New York 1975)

1.49 V.M. Velichko, N.G. Zagoruyko: Int. J. Man Mach. Stud. *2*, 223 (1970)

1.50 K. Abe: *Technical Report of the Professional Group on Pattern Recognition of IECEJ*, PRL 74-5 (1974) (In Japanese)

1.51 V.I. Levenshtein: Sov. Phys. Dokl. *10*, 707 (1966)

1.52 T. Okuda, E. Tanaka, T. Kasai: IEEE Trans. C-*25*, 172 (1976)

1.53 K.S. Fu: *Syntactic Methods in Pattern Recognition* (Academic Press, New York 1974)

1.54 K.S. Lashley: In *The Neurophysiology of Lashley; Selected Papers of K.S. Lashley*, ed. by F.A. Beach (McGraw-Hill, New York 1960)

1.55 T. Poggio: Biol. Cybern. *19*, 201 (1975)

1.56 A.E. Albert, L.A. Gardner, Jr.: *Stochastic Approximation and Nonlinear Regression* (MIT Press, Cambridge, Mass. 1967)

1.57 M.L. Minsky: *Computation: Finite and Infinite Machines* (Prentice-Hall, Englewood Cliffs, N.J. 1967)

1.58 G. Bohn: Biol. Cybern. *29*, 193 (1978)

2.1 D. Knuth: *The Art of Computer Programming. Vol. 3: Sorting and Searching* (Addison-Wesley, Reading, Mass. 1973)

2.2 J. Martin: *Computer Data-Base Organization*. 2nd printing (Prentice-Hall, Englewood Cliffs, N.J. 1977)

2.3 I. Flores: *Data Structure and Management* (Prentice-Hall, Englewood Cliffs, N.J. 1970)

2.4 W.H. Desmonde: *Real-Time Data Processing Systems* (Prentice-Hall, London 1964)

2.5 D. Lefkovitz: *File Structures for On-Line Systems* (Hayden, New York 1969)

2.6 C.W. Bachman: Commun. ACM *15*, 628-634 (1972)

2.7   N. Chapin: Proc. FJCC 1969, pp. 413-422
2.8   G.G. Dodd: Comput. Surv. *1*, 117-139 (1969)
2.9   R.F. Schubert: Datamation (July 1972) pp. 42-47
2.10 D.K. Chow: Inf. Control *15*, 377-396 (1969)
2.11 S.P. Ghosh: Inf. Sci. *1*, 363-380 (1969)
2.12 S.P. Ghosh, M.E. Senko: J. ACM *16*, 569-579 (1969)
2.13 S.P. Ghosh: Commun. ACM *15*, 802-808 (1972)
2.14 S.P. Ghosh: Inf. Sci. *6*, 1-9 (1973)
2.15 S.P. Ghosh: Inf. Control *25*, 145-169 (1974)
2.16 H. Hellerman: Commun. ACM *5*, 205-207 (1962)
2.17 V.Y. Lum: Commun. ACM *13*, 660-665 (1970)
2.18 D.R. Morrison: J. ACM *15*, 514-534 (1968)
2.19 E. Wong, T.C. Chiang: Commun. ACM *14*, 593-597 (1971)
2.20 CODASYL Data Base Task Group: April 1971 report (Amsterdam 1971)
2.21 CODASYL Systems Committee: *Feature Analysis of Generalized Data Management Systems* (Amsterdam 1971)
2.22 R. Morris: Commun. ACM *11*, 38-44 (1968)
2.23 A. Gill: *Linear Sequential Circuits: Analysis, Synthesis, and Applications* (McGraw-Hill, New York 1966)
2.24 T. Kohonen: *Digital Circuits and Devices* (Prentice-Hall, Englewood Cliffs, N.J. 1972)
2.25 A.D. Lin: Proc. AFIPS 1963 SJCC *24* (Spartan Books, New York) pp. 355-366
2.26 M. Hanan, F.P. Palermo: IBM J. Res. Dev. *7*, 127 (1963)
2.27 G. Schay, N. Raver: IBM J. Res. Dev. *7*, 121 (1963)
2.28 E.R. Berlekamp: *Algebraic Coding Theory* (McGraw-Hill, New York 1968)
2.29 V.Y. Lum, P.S.T. Yuen, M. Dodd: Commun. ACM *14*, 228-239 (1971)
2.30 R. Sprugnoli: Commun. ACM *20*, 841-850 (1977)
2.31 D. Mitra: Inf. Control *23*, 205-220 (1973)
2.32 B.H. Bloom: Commun. ACM *13*, 422-426 (1970)
2.33 W.D. Maurer: Commun. ACM *11*, 35-38 (1968)
2.34 J.D. Beyer: Commun. ACM *11*, 378 (1968)
2.35 C.E. Radke: Commun. ACM *13*, 103-107 (1970)
2.36 A.C. Day: Commun. ACM *13*, 481-482 (1970)
2.37 F.R.A. Hopgood, J. Davenport: Comput. J. *15*, 314-315 (1972)
2.38 A.F. Ackerman: Commun. ACM *17*, 164 (1974)
2.39 V. Batagelj: Commun. ACM *18*, 216-217 (1975)
2.40 A. Ecker: Comput. J. *17*, 340-343 (1974)
2.41 J.R. Bell: Commun. ACM *13*, 107-109 (1970)
2.42 L. Lamport: Commun. ACM *13*, 573-574 (1970)
2.43 J.R. Bell, C.H. Kaman: Commun. ACM *13*, 675-677 (1970)
2.44 F. Luccio: Commun. ACM *15*, 1045-1047 (1972)
2.45 A.J.D. Pawson: Comput. J. *16*, 285 (1973)
2.46 S.K. Bandyopadhyay: Commun. ACM *20*, 262-263 (1977)
2.47 W.D. Maurer: Commun. ACM *11*, 378 (1968)
2.48 R.P. Brent: Commun. ACM *16*, 105-109 (1973)
2.49 C. Halatsis, G. Philokyprou: Commun. ACM *21*, 554-557 (1978)
2.50 J.A. Feldman, J.R. Low: Commun. ACM *16*, 703 (1973)
2.51 F.R.A. Hopgood: Comp. Bull. *11*, 297-300 (1968)
2.52 F.R.A. Hopgood: *Compiling Techniques* (Elsevier Publishing Co., New York 1969)
2.53 C. Bays: Comput. J. *16*, 126-131 (1973)
2.54 C. Bays: Commun. ACM *16*, 11-14 (1973)
2.55 D. Severance, R. Duhne: Commun. ACM *19*, 314-326 (1976)
2.56 K. Furukawa: Inf. Process. Japan *13*, 13-18 (1973)
2.57 O. Amble, D.E. Knuth: Comput. J. *17*, 135-142 (1974)
2.58 D.G. Bobrow: Commun. ACM *18*, 413-415 (1975)
2.59 L.D. Higgins, F.J. Smith: Comput. J. *14*, 249-253 (1971)
2.60 W.W. Peterson: IBM J. Res. Dev. *1*, 130 (1957)

346

2.61 W. Buchholz: IBM Syst. J. *2*, 86 (1963)
2.62 M.H. McKinney: Proc. Nat. Comput. Conf., 1977, pp. 371-377
2.63 J.A. Feldman, P.D. Rovner: Stanford Artificial Intelligence Project Memo AI-66 (1968)
2.64 J.A. Feldman, P.D. Rovner: Commun. ACM *12*, 439-449 (1969)
2.65 P.D. Rovner, J.A. Feldman: MIT Lincoln Laboratory, Technical Note 1967-19 (1967)
2.66 P.D. Rovner, J.A. Feldman: In *Information Processing 68* (North-Holland, Amsterdam 1969) pp. 579-585
2.67 M.D. McIlroy: Commun. ACM *6*, 101 (1963)
2.68 A.P. Ershov: Doklady Akad. Nauk SSSR *118*, 427-430 (1958)
2.69 G. Schay, Jr., W.G. Spruth: Commun. ACM *5*, 459-462 (1962)
2.70 J. Král: Comput. J. *14*, 145-149 (1971)
2.71 L.R. Johnson: Commun. ACM *4*, 218-222 (1961)
2.72 W.P. Heising: IBM Syst. J. *2*, 112 (1963)
2.73 C.A. Olson: Proc. 1969 ACM Nat. Conf., San Francisco, pp. 539-549
2.74 J.A. van der Pool: IBM J. Res. Dev. *16*, 579 (1972)
2.75 J.A. van der Pool: IBM J. Res. Dev. *17*, 27 (1973)
2.76 M. Tainiter: J. ACM *10*, 307-315 (1963)
2.77 J.G. Williams: Commun. ACM *14*, 172-175 (1971)
2.78 V.Y. Lum, P.S.T. Yuen: Commun. ACM *15*, 996-997 (1972)
2.79 V.Y. Lum: Commun. ACM *16*, 603-612 (1973)
2.80 J.D. Ullman: J. ACM *19*, 569-575 (1972)
2.81 D. Knuth: *The Art of Computer Programming*, Vol. 1: Fundamental Algorithms (Addison-Wesley, Reading, Mass. 1968)
2.82 B.F. Cheydleur: "Dimension: An Associative Memory", Philco Computer Division, Dec. 1962
2.83 B.F. Cheydleur: *Vistas in Information Handling*, Vol.1 (Spartan Books, Washington, D.C. 1963) p. 55
2.84 B.F. Cheydleur: Am. Doc. *14*, 56 (1963)
2.85 P. Weston, S.M. Taylor: Coordinated Sci. Lab. Rept. R-393, Sept. 1968 (AD-679 948)
2.86 P.C. Patton: Computer *3*, 19 (1970)
2.87 N.S. Prywes, H.J. Gray, W.I. Landauer, D. Lefkowitz, S. Litwin: U. Pennsylvania, The Moore School of El. Eng., Tech. Rept. No. 1 (AD-270 573) (1961)
2.88 N.S. Prywes, H.J. Gray: In *Information Processing 1962* (North-Holland 1962) pp. 273-278
2.89 N.S. Prywes, H.J. Gray: AIEE Special Publication S 136 (1962) pp. 87-101
2.90 N.S. Prywes, H.J. Gray: IEEE Trans. Commun. and Electron. *82*, 488-492 (1963)
2.91 N.S. Prywes: Proc. IEEE *54*, 1788-1966 (1966)
2.92 R.L. Rivest: SIAM J. Comput. *5*, 19-50 (1976)
2.93 R.A. Gustafson: Proc. Symp. on Information and Retrieval, ACM, New York, April, 1971, pp. 163-174
2.94 T. Kohonen: *Associative Memory - A System-Theoretical Approach*, Communication and Cybernetics, Vol.17 (Springer, Berlin, Heidelberg, New York 1978)
2.95 T. Kohonen, E. Reuhkala: Helsinki University of Technology, Report TKK-F-A335, 1978
2.96 T. Kohonen, E. Reuhkala: Proc. 4th Intern. Joint Conf. on Pattern Recognition, Kyoto (1978) pp. 807-809
2.97 G. Dewey: *Relative Frequency of English Speech Sounds* (Harvard University Press, Cambridge, MA 1923)
2.98 A.G. Debus, Ed.: *World Who's Who in Science*, 1st ed. (Marquis Who's Who Inc., Chicago 1968)

2.99   M.C. Harrison: Commun. ACM *14*, 777-779 (1971)
2.100 C.E. Goble: Comput. J. *18*, 18-20 (1975)
2.101 A.V. Aho, M.J. Corasick: Commun. ACM *18*, 333-340 (1975)
2.102 C.N. Alberga: Commun. ACM *10*, 302-313 (1967)
2.103 F.J. Damerau: Commun. ACM *7*, 171-176 (1964)
2.104 W. Doster: IEEE Trans. Comput. C-*26*, 1090-1101 (1977)
2.105 V.I. Levenshtein: Sov. Phys. Dokl. *10*, 707-710 (1966)
2.106 H.L. Morgan: Commun. ACM *13*, 90-94 (1970)
2.107 T. Okuda, E. Tanaka, T. Kasai: IEEE Trans. C-*25*, 172-178 (1976)
2.108 E.M. Riseman, A.R. Hanson: IEEE Trans. C-*23*, 480-493 (1974)
2.109 J.R. Ullman: Comput. J. *20*, 141-147 (1977)
2.110 E. Fredkin: Commun. ACM *3*, 490-499 (1960)
2.111 K. Maly: Commun. ACM *19*, 409-415 (1976)
2.112 W.A. Burkhard: J. Comp. Syst. Sci. *15*, 280-299 (1977)
2.113 E.G. Coffman, Jr., J. Eve: Commun. ACM *13*, 427-436 (1970)
2.114 D.G. Severance: Comput. Surv. *6*, 175-194 (1974)
2.115 E.G. Mallach: Comput. J. *20*, 137-140 (1977)
2.116 A.I. Dumey: Comput. and Autom. *5*, 6-9 (1956)
2.117 W.D. Maurer, T.G. Lewis: Computing Surv. *7*, 5-19 (1975)
2.118 P.G. Sorenson, J.P. Tremblay, R.F. Deutscher: Inf. Sci. Can. *16*, 1 (1978)
2.119 G.D. Knott: Comput. J. *18*, 265 (1975)
2.120 A. Bookstein: J. Amer. Soc. Inf. Sci. *25*, 232 (1974)
2.121 W. Doster: Wiss. Ber. A. *51*, 104 (1978)
2.122 W.B. Samson, R.H. Davis: Comput. J. *21*, 210 (1977)
2.123 D.G. Severance, J.V. Carlis: Minnesota Univ. Man. Inf. Syst. Res. Center Rept. MISRC-TR-77-06 (1977)
2.124 T. Yuba, M. Hoshi: Trans. Inst. Electron. Commun. Eng. (Japan) E*61*, 52 (1978)
2.125 R.L. Rivest: J. ACM *25*, 200 (1978)
2.126 W. Littwin: CR Acad. Sci. Ser. A (France) *286*, A695 (1978)
2.127 W. Littwin: In *Proc. 4th Int. Conf. on Very Large Data Bases* (IEEE, New York 1978)
2.128 M. Ajtai, J. Komlos, E. Szemeredi: Inf. Proc. Lett. *7*, 270 (1978)
2.129 L.J. Guibas: J. ACM *25*, 544 (1978)
2.130 L.J. Guibas, E. Szemeredi: J. Comp. Syst. Sci. *16*, 226 (1978)
2.131 G. Lyon: Proc. COMPCON Fall '78, p. 378 (1978)
2.132 A. Kramli, J. Pergel: Probl. Control Inf. Theory *6*, 207 (1977)
2.133 S. Fortune, J. Hopcroft: Inf. Proc. Lett. *8*, 20 (1979)
2.134 A.L. Tharp: Inf. Syst. *4*, 55 (1979)
2.135 P.-A. Larson: BIT (Sweden) *18*, 184 (1978)
2.136 R. Fagin, N.J. Pippenger, H.R. Strong: IBM Tech. Discl. Bull. *21*, 809 (1972)
2.137 W.A. Burkhard: Proc. 11th Hawaii Int. Conf. on Syst. Sci. (1978) p. 99
2.138 R.C. Lee, Y.H. Chin, S.C. Chang: IEEE Trans. SE-*2*, 185 (1976)
2.139 E. Hill, Jr.: "A Comparative Study of Very Large Data Bases", in *Lecture Notes on Computer Science*, Vol. 59 (Springer, Berlin, Heidelberg, New York 1978)
2.140 M.L. Griss: Proc. 10th Hawaii Int. Conf. on Syst. Sci. (1977) p. 169
2.141 M. Imai, T. Fukumura, Y. Yoshida: Inf. Process. Soc. Jpn. (Joho Shori) *18*, 639 (1977)
2.142 W.T. Wipke, S. Krishnan, G.I. Ouchi: J. Chem. Inf. Comput. Sci. *18*, 32 (1978)
2.143 L. Hodes, A. Feldman: J. Chem. Inf. Comput. Sci. *18*, 96 (1978)
2.144 T.G. Lewis: "A Hashed-Array Database Technique for Minicomputers", in *Microprocessors, Microprogramming, and Minicomputers*, ed. by R. Chattergy, U.W. Pooch (Western Periodicals, N. Hollywood 1977)

348

2.145 S. Nachmens, S. Berild: Data (Sweden), No. 6, 41 (1976)
2.146 P. Heckel: Commun. ACM *21*, 264 (1978)
2.147 L.H. Groner, A.L. Goel: "Concurrency in Hashed File Access", in
      Information Processing 74 (IFIP, North-Holland, Amsterdam 1974)
2.148 E. Goto, T. Ida, T. Gunji: Inf. Proc. Lett. *6*, 8 (1977)
2.149 T. Ida, E. Goto: "Performance of a Parallel Hash Hardware with Key
      Deletion", in *Information Processing 77* (IFIP, North-Holland,
      Amsterdam 1977)
2.150 R.S. Fabry: Commun. ACM *17*, 403 (1974)
3.1   R.F. Rosin: Proc. AFIPS 1962 SJCC, p. 203
3.2   C.C. Foster: IEEE Trans. C-*17*, 788 (1968)
3.3   G.A. Anderson: IEEE Trans. C-*23*, 1317 (1974)
3.4   J.T. Koo: IEEE Trans. SC-*5*, 208 (1970)
3.5   F.A. Behnke: U.S. Patent No. 3,195,109, July 13 (1965)
3.6   W. Hilberg: IEEE Trans. EC-*15*, 117 (1966)
3.7   W. Hilberg: U.S. Patent No. 3,706,078, Dec. 12 (1972)
3.8   V.F. Rudakov, Yu.I.Il'yashenko: Star *5*, 1420 (1967)
3.9   A.V. Campi, B.H. Gray: U.S. Patent No. 3,634,829, Jan. 11 (1972)
3.10 E.E. Davidson: IBM Tech. Discl. Bull. *17*, 855 (1974)
3.11 E.H. Frei, J. Goldberg: IRE Trans. EC-*10*, 718 (1961)
3.12 J. Favor: Goodyear Aerospace Corp., AP-111770, Oct. 1964
3.13 C.A. Hill: Goodyear Aerospace Corp., GER-12181, May 1965
3.14 C.C. Foster, F. Stockton: IEEE Trans. C-*20*, 1580 (1971)
3.15 Y. Chu: IEEE Trans. EC-*14*, 600 (1965)
3.16 H.S. Stone: Proc. AFIPS 1968 FJCC, p. 949 (1968)
3.17 K.E. Batcher: U.S. Patent No. 3,800,289, March (1974)
3.18 K.E. Batcher: Proc. 1974 Nat. Comput. Conf., p. 405
3.19 M.H. Lewin: RCA Rev. *23*, 215 (1962)
3.20 R.R. Seeber, A.B. Lindqvist: IBM J. Res. Dev. *6*, 126 (1962)
3.21 H. Weinstein: IEEE Trans. EC-*12*, 564 (1963)
3.22 L. Johnson, M. McAndrew: IBM J. Res. Dev. *8*, 189 (1964)
3.23 H.S. Miller: IEEE Trans. EC-*13*, 614 (1964)
3.24 V. Chlouba: Inf. Process. Mach. *13*, 139 (1967)
3.25 R.R. Seeber: Proc. EJCC *18*, 179 (1960)
3.26 A. Wolinsky: Commun. ACM *11*, 488 (1968)
3.27 C.V. Ramamoorthy, J.L. Turner, B.W. Bah: IEEE Trans. C-*27*, 800 (1978)
3.28 K.E. Batcher: Proc. AFIPS 1968 SJCC, p. 307 (1968)
3.29 E.J. Gauss: J. ACM *8*, 418 (1961)
3.30 J.L. Anderson: IBM Tech. Discl. Bull. *4*, 28 (1962)
3.31 G. Estrin, R.H. Fuller: Proc. IEEE Pacific Comput. Conf., p. 118 (1963)
3.32 A. Kaplan: Proc. FJCC *24*, 193 (1963)
3.33 A. Wolinsky: IEEE Symp. on Search Memory, May 1964 (IEEE New York
      1964)
3.34 A. Wolinsky: IEEE Trans. C-*18*, 899 (1969)
3.35 A.B. Lindqvist: IBM Tech. Discl. Bull., August 1965, p. 372
3.36 R.M. Bird, J.L. Cass, R.H. Fuller: Rome Air Dev. Center, Rept. No.
      TR-66-209, Sept. 1966, Vol 1, AD-800 387
3.37 R.M. Bird, S.L. Cabs, R.H. Fuller: Rome Air Dev. Center, Rept. No.
      TR-66-209, Sept. 1966, Vol. 2, AD-376 572
3.38 R.R. Seeber, A.B. Lindqvist: U.S. Patent No. 3,430,205, Feb. (1969)
3.39 T.Y. Feng: Proc. 4th Annu. Princeton Conf. on Inf. Sci. and Syst.,
      Princeton U., March 1970, p. 442
3.40 C.C. Foster: Univ. Massachusetts, Comput. Inf. Sci. Dept., Tech. Note
      CS-00016, July 1970
3.41 D.P. Agrawal: Proc. 1974 Conf. on Comput. Syst. and Technol.,
      Oct. 29-Nov. 1, 1974, p. 180
3.42 W.A. Crofut, M.R. Sottile: IEEE Trans. EC-*15*, 529 (1966)

3.43   D.C. Alexander, R.H. Dennard, F.L. Post: IBM Advanced Systems, 17.022, May 1961
3.44   F.H. Young: Oregon State Univ., Dept. Math., In-House Doc. 1962
3.45   P.T. Rux: Oregon State Univ., July 1967 (AD-660 792)
3.46   P.T. Rux: Oregon State Univ., Feb. 1968 (AD-671 910)
3.47   P.T. Rux: IEEE Trans. C-*18*, 512 (1969)
3.48   P.T. Rux, F.W. Weingarten, F.H. Young: IEEE Comput. Group Repository, No. 67-72, March 1967
3.49   B. Parhami: Tech. Report UCLA-ENG-7213, Univ. California LA (1972)
3.50   B. Parhami: Proc. AFIPS 1972 FJCC, p. 681 (1972)
3.51   G.L. Hollander: Proc. 1956 JCC, p. 128 (1956)
3.52   D. Warren: IEEE Symp. on Search Memory, May 1964 (IEEE,New York 1964)
3.53   R.I. Roth: U.S. Patent No. 3,257,646, June 21 (1966)
3.54   D.L. Slotnick: Adv. in Comput. *10*, 291 (1970
3.55   L.D. Healy, G.J. Lipovski, K.L. Doty: Proc. AFIPS 1972 FJCC, p. 691 (1972)
3.56   N. Minsky: Proc. AFIPS 1972 FJCC, p. 587 (1972)
3.57   G.B. Houston, R.H. Simonsen: U.S. Patent No. 3,906,455, Sep. 16 (1975)
3.58   Chyuan Shiun Lin, D.C.P. Smith: ACM Trans. Database Syst. *1*, 53 (1976)
3.59   M. Flinders, P.L. Gardner, R.J. Llewelyn, J.F. Minshull: Proc.1970 IEEE Int. Comput. Group Conf., p. 314
3.60   P.L. Gardner: IEEE Trans. C-*20*, 764 (1971)
3.61   P.L. Gardner: U.K. Patent No. 1 281 387, July 12 (1972)
3.62   T. Kohonen: *Digital Circuits and Devices* (Prentice-Hall, Englewood Cliffs, N.J. 1972)
3.63   J.R. Brown, Jr.: Proc. Spec. Tech. Conf. on Nonlinear Magnetics, Los Angeles, Cal., Nov. 1961
3.64   M.H. Lewin, H.R. Beelitz, J.A. Rajchman: Proc. AFIPS 1963 FJCC *24*, 101 (1963)
3.65   G.G. Pick, D.B. Brick: Am. Doc. Inst., 26th Annu. Meeting, p. 245, Oct. 1963
3.66   G.G. Pick: Proc. AFIPS 1964 FJCC *26*, 107 (1964)
3.67   E.L. Younker, C.H. Heckler, D.P. Masher, J.M. Yarborough: Stanford Res. Inst., Oct. 1964 (AD-609 126)
3.68   E.L. Younker, C.H. Heckler, D.P. Masher, J.M. Yarborough: Proc. SJCC *25*, 515 (1964)
3.69   S.T.C.: B.P. 1013241, Dec. 1965
3.70   M.H. Lewin: U.S. Patent No. 3,245,052, Apr. 5 (1966)
3.71   R.A. Henle, I.T. Ho, G.A. Maley, R. Waxman: Proc. FJCC 1969, p. 61 (1969)
3.72   D.C. Wyland: Comput. Des. No. 9, p. 61 (1971)
3.73   K.E. Iverson: *A Programming Language* (Wiley, New York 1962)
3.74   A.D. Falkoff: J. ACM *9*, 488 (1962)
3.75   G. Estrin: Proc. WJCC, May 1960, p. 33
3.76   J. Ausley: Moore School of Electr. Eng., M. Sc. Thesis, 1961
3.77   M.J. Flynn: Purdue Univ., Ph. D. Thesis, BTP-62-1782, Jun 1961
3.78   R.H. Fuller: Disser. Absts. *24*, 1960 (1963)
3.79   R.H. Fuller: UCLA, Dept. of Eng., Rept. No. 63-25 (1963)
3.80   Computer Command and Control: Rept. No. 5-101-5, Jan. 1964
3.81   J.E. McAteer, J.A. Capobianco, R.L. Koppel: Proc. 1964 FJCC, p. 81 (1964)
3.82   A.E. Slade: Am. Doc. Inst. 27th Annu. Meeting (1964)
3.83   S. Sohara: IEEE Symp. on Search Memory (1964)
3.84   A.V. Campi, R.M. Dunn, B.H. Gray: IEEE Trans. AES-*1*, 168 (1965)
3.85   W.F. Chow: Univac, Quart. Prog. Rept., Oct. 1965 (AD-477 446)
3.86   W.F. Chow: Sperry Rand Corp. Quart. Rept. 1965 (AD-472 571)
3.87   W.F. Chow: Sperry Rand Corp. Quart. Rept. 1966 (AD-804 628)
3.88   R.W. Haas, E.H. Blevis: Marquardt Corp., July 1965 (AD-620 915)
3.89   R.R. Seeber, A.B. Lindqvist: Proc. IFIP Cong. *2*, 479 (1965)

3.90   R. Haas, E. Blewis, S. Requa, I. Hanlet: Marquardt Corp., AF 30(602)-
        3709, June 1966 (AD-488 453)
3.91   R.W. Haas, J.M. Hanlet: Marquardt Corp., Final Rept. Dec. 1967
        (AD-825 274)
3.92   R.C.M. Barnes, I.N. Hooton: H.M. Stationary Office, England 1966
3.93   C.F. Chong, P.A. Rivelli, J.S. Mathias, P. Geaneotes: Sperry Rand
        Corp. Quart. Rept., June 1967 (AD-815 774L)
3.94   J.P. Bartlett: Proc. IEEE Int. Comput. Group Conf., Wash.
        June 16-18, 1970, p. 299
3.95   J. Bartlett, J. Mudge, J. Springer: Electronics 43, 96 (1970)
3.96   J. Cashera: Air Force Avionics Lab. Rept. No. AFAL-TR-70-71,
        Aug. 1970 (AD-876 691/7SL)
3.97   Electron 13, 53 (1972)
3.98   A.B.E. Ellis: The Marconi Review XXXV, 42 (1972)
3.99   J. Minker: Univ. of Maryland, Tech. Rept. TR-195, July 1972
3.100   H.-O. Leilich: "Assoziative Speicher", in *Taschenbuch der Informatik*,
        Vol.I (Springer, Berlin, Heidelberg, New York 1974) p. 479
3.101   H.-O. Leilich: "Access Methods and Associative Memories", Proc.
        Digitale Speicher (NTG-Fachber. Germany) 58, 328 (1977)
3.102   G. Wolf: Elektron. Rechenanlagen 17, 264 (1975)
3.103   G. Wolf: Data Report 11, 29 (1976)
3.104   W. Motsch: Electron. Rechenanlagen 19, 274 (1977)
3.105   W. Motsch: Proc. Digitale Speicher, Stuttgart, Germany, 22-24 March
        1977 (NTG-Fachber. Germany) 58, 339 (1977)
3.106   Electronics 51, 63 (1978)
3.107   E.S. Lee: Proc. AFIPS 1963 SJCC 23, 381 (1963)
3.108   E.G. Wagner, J. McCarthy: U.S. Patent No. 3,093,814, June (1963)
3.109   R.D. Ross: IBM Tech. Discl. Bull. No. 10, p.561 (1967)
3.110   A.M. Peskin: Proc. FJCC 1970, p. 615 (1970)
3.111   D.E. Davis, C.W. Hannaford: IBM Tech. Discl. Bull. 15, 719 (1972)
3.112   P.A. DiGleria, M.H. Hallett: U.K. Patent No. 1,265,645, March 1 (1972)
3.113   R.B. Derickson: Comput. Des. 7, 60 (1968)
3.114   J.D. Erwin, E.D. Jensen: Proc. AFIPS 1970 FJCC, p. 621 (1970)
3.115   W.K. King: IEEE Trans. C-20, 671 (1971)
3.116   A. Weinberger: Comput. Des. 10, 77 (1971)
3.117   A. Weinberger: U.K. Patent No. 1 280 753, July 5 (1972)
3.118   M. Handrich: Informatique et Gestion, No. 49, p. 95 (1973)
3.119   N.K. Natarajan, P.A.V. Thomas: IEEE Trans. C-18, 424 (1969)
3.120   B. Parhami: Asilomar Conf. on Circuits, Syst. and Comput.,
        7th Annu. Conf. Rec. Pap., p. 439 (1973)
3.121   J.F. Minshull, A.S. Murphy: U.K. Patent No. 1 289 249, Sep. 3 (1972)
3.122   D.W. Digby: IEEE Trans. C-22, 768 (1973)
4.1   J. Bartlett, J. Mudge, J. Springer: Electronics 43, 96 (1970)
4.2   D. Aspinall, D.J. Kinniment, D.B.G. Edwards: *Information Processing 68*
        (North-Holland, Amsterdam 1969) p. 800
4.3   D.J. Kinniment, A.E. Knowles, D.B.G. Edwards: Proc. Int. Conf. on
        Microelectronics (IEEE, London 1969) p. 37
4.4   D.W. Hillis, T.W. Hart: U.K. Patent No. 1 255 206, Dec. 1 (1971)
4.5   J.B. Hughes: U.S. Patent No. 3,704,456, Nov. 28 (1972)
4.6   SERT J. 8, 78 (1974)
4.7   Motorola Semiconductor Products, Inc.: In *Developments in Large-Scale
        Integration (Engineering Edition)*. (Phoenix, Ariz. 1968)
4.8   S. Matsue: U.S. Patent No. 3,703,709, Nov. 21 (1972)
4.9   R.W. Murphy: IBM Tech. Discl. Bull. 14, 1669 (1971)
4.10   H.H. Berger, C.L. Schuenemann, S.K. Wiedmann: IBM Tech. Discl. Bull.
        14, 3088 (1972)
4.11   A.W. Bidwell, W.D. Pricer: Proc. Int. Solid-State Circuits Conf.
        (1967) p. 78

351

4.12 D.P. Repchick: IBM Discl. Bull. *10*, 502 (1967)
4.13 W. Hilberg: U.S. Patent No. 3,706,078, Sep. 8 (1971)
4.14 W.D. Pricer: IBM Tech. Discl. Bull *16*, 3160 (1974)
4.15 A.A. Orlikovskii: Sov. Microelectron. (USA) *6*, 355 (1977)
4.16 J.R. Burns, J.H. Scott: Proc. AFIPS 1969 FJCC, p. 469 (1969)
4.17 L.D. Wald: Proc. 1970 NAECON, p. 277
4.18 R.M. Lea: Electron. Lett. *8*, 391 (1972)
4.19 C.J. Shead: GEC J. Sci. Tech. *40*, 119 (1973)
4.20 I.V. Prangishvili, V.A. Dementujev, M.S. Sonin: Mikroelektronika (USSR) *4*, 497 (1975)
4.21 R. Igarashi, T. Kurosawa, T. Yaita: ISSCC Digest Tech. Papers, (1966) p. 104
4.22 R. Igarashi, T. Yaita: Proc. AFIPS 1967 FJCC *22*, 499 (1967)
4.23 R.M. Lea: Datafair 73. *II*, 418 (1973)
4.24 R.M. Lea: Radio Electron. Eng. *45*, 177 (1975)
4.25 W.F. Bankowski, Jr.: IBM Tech. Discl. Bull. *18*, 477 (1975)
4.26 W.F. Bankowski, Jr., K. Simanavicius: IBM Tech. Discl. Bull. *18*, 475 (1975)
4.27 J.L. Mundy, R.E. Joynson: Proc. IEEE Comput. Soc. Conf., (1971) p. 189
4.28 J.L. Mundy: U.S. Patent No. 3,701,980, Oct. 31 (1972)
4.29 J.L. Mundy, J.F. Burgess, R.E. Joynson, C. Neugebauer: IEEE JSSC *7*, 364 (1972)
4.30 R.M. Lea: IEEE JSSC *10*, 179 (1975)
4.31 R.M. Lea: Electr. Eng. *49*, 77 (1977)
4.32 J.R. Burns, J.J. Gibson, A. Harel, K. Hu, R.A. Powlus: RCA Labs. Final Rept. June 1967 (AD-828 570)
4.33 M.L. MacKnight: U.S. Patent No. 3,696,174, Sept. 19 (1972)
4.34 G. Carlstedt, G.P. Petersson, K.O. Jeppson: IEEE JSSC *8*, 338 (1973)
4.35 A. Corneretto: Electron. Des. *2*, 40 (1963)
4.36 C.P. Wang, A.E. Ruehli: IBM Confidential, 65C-061359-MFO03 (1965)
4.37 R.J. Koerner, S. Nissim: U.S. Patent No. 3,284,775, Nov. 8 (1966)
4.38 W.K. French: U.S. Patent No. 3,123,706, March 1964
4.39 W.K. French: U.S. Patent No. 3,131,291, April 1964
4.40 A.E. Slade, C.R. Smallman: Solid-State Electr. *11*, 357 (1960)
4.41 C.R. Smallman, A.E. Slade, M.L. Cohen: Proc. IRE *48*, 1562 (1960)
4.42 R.W. Ahrons: RCA Rev. *24*, 325 (1963)
4.43 R.W. Ahrons: RCA Rev. *26*, 557 (1965)
4.44 R.W. Ahrons: IEEE Trans. EC-*14*, 267 (1965)
4.45 R.W. Ahrons, L.L. Burns: RCA Final Rept., May 1964 (AD-448 504)
4.46 R.W. Ahrons, L.L. Burns: Comput. Des. *3*, 12 (1964)
4.47 J.L. Anderson: U.S. Patent No. 3,229,255, Jan. 11 (1966)
4.48 M. Asher: Electron. News *6*, 59 (1962)
4.49 J.D. Barnard, F.A. Behnke, A.B. Lindqvist, R.R. Seeber: Proc. IEEE *52*, 1182 (1964)
4.50 F.A. Behnke: U.S. Patent No. 3,184,717, May 18 (1965)
4.51 F.A. Behnke, G.B. Rosenberger: IBM Final Rept. Sept. 1963, (AD-423 492)
4.52 J.W. Bremer, D.W. Doss, B.T. McKeever: U.S. Patent No. 3,311,898 March 28 (1967)
4.53 J.W. Bremer, D.W. Doss, B.T. McKeever: U.S. Patent No. 3,312,956 April 4 (1969)
4.54 P.M. Davies: Proc. AFIPS 1962 SJCC *21*, 79 (1962)
4.55 R.J. Ferris: RADC Rept. No. TDR 64-457, Dec. 1964 (AD-610 131)
4.56 H. Fleischer, R.I. Roth: U.S. Patent No. 3,221,157, Nov. 30 (1965)
4.57 General Electric Co.: U.S. Patent No. 3,321,746, May (1967)
4.58 J. Goldberg, M.W. Green: Stanford Res. Inst. RADC-TR-61-233, Aug. 1961 (AD-266 169)

4.59 J. Goldberg, M.W. Green: "Large Files for Information Retrieval Base on Simultaneous Interrogation of All Items", in *Large-Capacity Memory Techniques for Computing Systems*, ed. by M.C. Yovits (MacMillan Co., New York 1962) p. 63

4.60 K. Goser, H.G. Kadereit: Proc. IEEE *56*, 121 (1968)

4.61 C.C. Green, B. Raphael: Stanford Res. Inst. May 1967 (AD-656 789)

4.62 M.W. Green: Suppl. C. Quart. Rept. 2, AF-30(602)-2142, RADC July 1960

4.63 M.W. Green: U.S. Patent No. 3,243,785, March 29 (1966)

4.64 R.S. Green, J. Minker, W.E. Shindle: Auerbach Corp., Management Rept., Vol I, Final Rept. July 1966 (AD-489 660)

4.65 R.S. Green, J. Minker, W.E. Shindle: Auerbach Corp., Tech. Disc. Final Rept., July 1966 Vol II (AD-489 661)

4.66 C.H. Heckler, Jr.: In *Multiple Instantaneous Response File*, p. 195, ed. by J. Goldberg, RADC-TR-61-233, 1961 (AD-266 169)

4.67 H.G. Kadereit, K. Goser: Proc. IEEE *56*, 121 (1968)

4.68 A.B. Lindqvist: IBM Tech. Discl. Bull. *7*, 1115 (1965)

4.69 H.T. Mann, J.L. Rogers: Proc. Nat. Aerosp. Electron. Conv. (1962) p. 359

4.70 W.L. McDermid, R.I. Roth: U.S. Patent No. 3,242,468, March 22 (1966)

4.71 B.T. McKeever: Proc. AFIPS 1965 FJCC *28*, 371 (1965)

4.72 V.L. Newhouse, R.E. Fruin: Proc. AFIPS 1962 SJCC *21*, 89 (1962)

4.73 V.L. Newhouse, R.E. Fruin: Electronics *35*, 31 (1962)

4.74 J.P. Pritchard: Texas Instruments F.T. Rept. 100CT66, RADC-TR-66-775 (AD-811 983)

4.75 J.P. Pritchard: Texas Instruments, Final Rept. May 1965 (AD-618 491)

4.76 J.P. Pritchard: IEEE Spectrum *3*, 46 (1966)

4.77 J.P. Pritchard: IEEE Comput. Group News *2*, 25 (1968)

4.78 J.P. Pritchard, L.D. Wald: Proc. Int. Conf. Nonlinear Magn. (1964) p. 2-5-1

4.79 J.P. Pritchard, L.D. Wald: IEEE Trans. MAG-*1*, 68 (1965)

4.80 J.A. Rajchman: ONR Rept. ACR-97, Inform. Syst. Summaries, July 1964

4.81 G. Retiz: IEEE Symp. on Search Memory, May 1964

4.82 J.L. Rogers: TRW Space Tech. Lab., Quart. Rept. Apr. 1963

4.83 J.L. Rogers: TRW Space Tech. Lab., Quart. Rept. Aug. 1963

4.84 J.L. Rogers: ONR Rept. ACR-97, Task No. NR-348-002, RR 003-10-02, 1964

4.85 J.L. Rogers, A. Wolinsky: TRW Space Tech. Labs., Final Rept. No. NR 3839 (1001), May 1964

4.86 J.L. Rogers, A. Wolinsky: U.S. Gov. Res. Repts. *39*, 166 (1964)

4.87 H. Rosenberg: U.S. Patent No. 3,235,839, Feb. 15 (1966)

4.88 G.B. Rosenberger: IBM Data Syst. Div. Final Tech. Rept. 1964 (AD-602 067)

4.89 R.F. Rosin: Proc. AFIPS 1962 SJCC *21*, 203 (1962)

4.90 P. Schupp, T. Singer: Mitre Corp., Aug. 1963 (AD-416 301)

4.91 R.R. Seeber, Jr.: Proc. EJCC *18*, 179 (1960)

4.92 R.R. Seeber, Jr.: IBM Data Systems, TR-00,756, Nov. 1960

4.93 R.R. Seeber, Jr.: Proc. Nat. Conf. ACM *14* (1960)

4.94 R.R. Seeber, Jr., A.J. Scriver, Jr.: U.S. Patent No. 3,191,155, June 22 (1965)

4.95 A.E. Slade: Proc. Int. Symp. Theory Switching, (1959) p. 326

4.96 A.E. Slade: Proc. IRE *50*, 81 (1962)

4.97 Space Technology Laboratories, Inc.: Rept. Proposal 0739.00, July 1961

4.98 E.D. Van De Rift: In *Multiple Instantaneous Response File*, p.158, ed. by J. Goldberg, RADC-TR-61-233, 1961 (AD-266 169)

4.99 I.D. Voitovich: Rept. No. FTD-HT-23-942-68, May 5, 1969 (AD-695 318)

4.100 C. Yang: "A Study of Cryotron Associative Memory in Digital Systems", M.Sc. Thesis, Northwestern Univ. (1964)

4.101 C.C. Yang, J.T. Tou: J. Franklin Inst. *284*, 109 (1967)

4.102 S.S. Yau, C.C. Yang: Proc. Nat. Electronic Conf. *22*, 764 (1966)

4.103 S.S. Yau, C.C. Yang: Northwestern Univ. Tech. Rept., Nov. 1966 (AD-644 439)

4.104 J. Matisoo: Proc. IEEE *55*, 172 (1967)
4.105 J. Matisoo: IEEE Trans. MAG-*5*, 848 (1969)
4.106 W. Anacker: IEEE Trans. MAG-*5*, 968 (1969)
4.107 H.H. Zappe: IEEE Trans. MAG-*13*, 41 (1977)
4.108 W.Y.Lum, H.W. Chan, T. Van Duzer: IEEE Trans. MAG-*13*, 48 (1977)
4.109 P. Wolf: *Proc. Int. Conf. SQUID 1976* (de Gruyter, Berlin, New York 1977) p. 519
4.110 H.W. Chan, B.T. Ulrich, T. Van Duzer: *Proc. Int. Conf. SQUID 1976* (de Gruyter, Berlin, New York 1977) p. 555
4.111 T.A. Fulton, L.N. Dunkleberger, R.C. Dynes: Phys. Rev. B *6*, 855 (1972)
4.112 D.J. Herrell: IEEE JSSC *9*, 277 (1974)
4.113 H.H. Zappe: Appl. Phys. Lett. *25*, 424 (1974)
4.114 P. Guéret: IEEE Trans. MAG-*11*, 751 (1975)
4.115 P. Guéret, Th. O. Mohr, P. Wolf: IEEE Trans. MAG-*13*, 52 (1977)
4.116 S. Hasuo, T. Imamura, K. Dazai: *Proc. Int. Conf. on SQUID 1976* (de Gruyter, Berlin, New York 1977) p. 541
4.117 W.F. Bankowski, H.C. Hamel: IBM Tech. Discl. Bull. *16*, 3321 (1974)
4.118 E.I.Il'yashenko, V.F. Rudakov: *Assotsiativnye zapominayuzie ustroistva na magnitnykh elementakh* (Energiya, Moscow 1975)
4.119 R.T. Hunt, D.L. Snider, J. Surprise, H.N. Boyd: U.S. Gov. Res. Repts. *39*, 188 (1964)
4.120 H.M. Beisner: IBM Tech. Discl. Bull *8*, 445 (1965)
4.121 G.T. Tuttle: U.S. Patent No. 3,300,761, Jan. 24 (1967)
4.122 Y. Chu: *Digital Computer Design Fundamentals* (McGraw-Hill, New York 1962)
4.123 R.C. Corbell: UCLA M.Sc. Thesis (1962)
4.124 R.R. Seeber, F.B. Hartman: IBM Tech. Discl. Bull. *4*, 73 (1962)
4.125 L.L. Lussier, R.P. Schneider: Electron. Ind. *22*, 92 (1963)
4.126 G.T. Tuttle: Electronics *36*, 43 (1963)
4.127 J.A. Capobianco, J.E. McAteer, R.L. Koppel: Proc. AFIPS 1964 FJCC *26*, 27 (1964)
4.128 R.J. Koerner, A. Scarborough: IEEE Symp. Search Memory, May 1964
4.129 R.J. Koener, A.D. Scarborough: U.S. Patent No. 3,297,995, Jan. 10 (1967)
4.130 J. McAteer: IEEE Symp. on Search Memory, May 1964
4.131 A.D. Robbi, R. Ricci: Proc. Int. Conf. Nonlinear Magn. (1964) p. 8-3-1
4.132 E.S. Lee III: U.S. Patent No. 3,206,735, Sep. 14 (1965)
4.133 V. Chlouba: Inf. Process. Mach. *13*, 113 (1967)
4.134 J.T. Franks, G.T. Tuttle: U.S. Patent No. 3,300,760, Jan. 24 (1967)
4.135 R.G. Ofengenden, F.N. Berezin: Prib. Tekh. Eksp. No. 2, p. 5 (1967)
4.136 E.I. Il'yashenko: Autom. Remote Control USSR *34*, 1164 (1973)
4.137 J.A. Rudolph: *The Associative Processor, a New Computer Resource*, GER-14087 (Goodyear Aerospace Corp., Akron, Ohio 1969)
4.138 J.A. Rudolph, L.C. Fulmer, W.C. Meilander: Electronics *43*, 96 (1970)
4.139 J.A. Rudolph, L.C. Fulmer, W.C. Meilander: Electronics *44*, 91 (1971)
4.140 R.H. Fuller, J.D. Tu, R.M. Bird: Proc. Nat. Aerosp. Electron. Conf. (1965) p. 1
4.141 R.J. Koerner: U.S. Patent No. 3,257,650, June 21 (1966)
4.142 W.F. Chow: IEEE Trans EC-*16*, 642 (1967)
4.143 W.F. Chow, L.M. Spandorfer: Proc. AFIPS 1967 SJCC *28* (1967)
4.144 M. Bialer, J. Garrett, W.C. Meilander: Symp. Parallel Processor Systems, Tech. & Appl., June 1969
4.145 J.P. Mc Callister, C.F. Chong: Proc. AFIPS 1966 FJCC *29* (1966)
4.146 A. Corneretto: Electron. Des. *10*, 8 (1962)
4.147 C.A. Rowland, W.O. Berge: Proc. AFIPS 1963 FJCC *24*, 59 (1963)
4.148 J.I. Raffel, T.S. Crowther: IEEE Trans. EC-*13*, 611 (1964)
4.149 M. Naiman: "Content-Addressed Memory Using Magneto-Resistive Readout of Magnetic Thin Films", IEEE Intermag. 1965
4.150 C.P. Wang: J. Appl. Phys. *39*, 1220 (1968)
4.151 C.P. Wang: U.S. Patent No. 3,466,631, Sept. 9 (1969)

4.152   C.P. Wang: U.S. Patent No. 3,466,632, Sept. 9 (1969)
4.153   P.A. Lord, M.P. Marcus: U.S. Patent No. 3,426,335, Feb.(1969)
4.154   J. Kiseda, H. Petersen, W. Seelbach, M. Teig: IBM J. Res. Dev. 5,
         106 (1961)
4.155   J.R. Kiseda: "A 128-Word, 36-Bit Magnetic Associative Memory". IBM
         confidential, NC-358, March 1964
4.156   W.I. McDermid, J.E. Petersen: IBM J. Res. Dev. 5, 59 (1961)
4.157   R.G. Gall: Proc. AFIPS 1964 FJCC, p. 159 (1964)
4.158   A. Apicella, J. Franks: "BILOC - A High Speed NDRO One-Core-Per-Bit
         Associative Element", IEEE Intermag. 1965
4.159   Tse-Yun Feng: Proc. AFIPS 1968 SJCC 33, 275 (1968)
4.160   C.F. Pulvari, M. Szabo, M.J. Walsh, A. De La Paz, W.B. Penzes:
         RADC-TR-68-105, April 1968 (AD-668 475)
4.161   C.F. Pulvari: IEEE Trans. ED-16, 580 (1969)
4.162   R.F. Herlein, A.V. Thompson: Proc. Int. Solid State Circuits Conf.
         (1969) p. 42
4.163a   D. Toombs: IEEE Spectrum 15, 22 (1978)
4.163b   D.F. Barbe: Charge-Coupled Devices, Topics in Applied Physics,
         Vol. 38 (Springer, Berlin, Heidelberg, New York 1980)
4.164   C. Kooy, U. Enz: Philips Res. Rep. 15, 7 (1960)
4.165   A.H. Bobeck: Bell Syst. Tech. J. 46, 1901 (1967)
4.166   H. Chang: Magnetic-Bubble Memory Technology (Dekker, New York, Basel
         1978)
4.167   A.M. Bobeck, H.E.D. Scovil, W. Shockley: U.S. Patent 3,541,522
         (1967, Pat. 1970)
4.168   H. Murakami: U.S. Patent 3,760,390 (1972, Pat. 1973)
4.169   S.Y. Lee, H. Chang: IEEE Comp. Soc. Conf. COMPCON 75 1, 91 (1975)
4.170   W. Kluge: Proc. IEEE 120, 1308 (1973)
4.171   I.G. Avaeva, E.I. Il'yashenko, V.G. Kleparskii, Yu.L. Kopylov,
         V.B. Kravchenko, A.T. Sobolev: Mikroelektronika (USSR) 5, 188 (1976)
4.172   I.G. Avaeva, E.I. Il'yashenko, Yu.L. Kopylov, V.B. Kravchenko,
         F.V Lisovskii, S.N. Matveev, A.T. Sobolev, V.I. Shapovalov:
         Mikroelektronika (USSR) 5, 500 (1976)
4.173   R. Allan: IEEE Spectrum 12, 49 (1975)
4.174   D.M. Lee, R.A. Naden: NAECON '76 Record, p. 724 (1976)
4.175   E.I. Il'yashenko, I.G. Avaeva, V.B. Kravchenko, S.N. Matveev:
         Phys. Status Solidi A: 36, K1 (1976)
4.176   E.I. Il'yashenko, V.G. Kleparskii, S.E. Yurchenko: Phys. Status
         Solidi A: 28, K153 (1975)
4.177   E.I. Il'yashenko, S.N. Matveev, N.I. Karmatzky: IEEE Trans.
         MAG-12, 663 (1976)
4.178   V.G. Kleparsky, E.I. Il'yashenko, S.N. Matveev: IEEE Trans.
         MAG-12, 700 (1976)
4.179   E.I. Il'yashenko, S.N. Matveev: Zh. Tekh. Fiz. 3, 137 (1977)
4.180a   R.A. Naden: U.S. Pat. Appl. 809 729 (AD-D004 142/6S) (1977)
4.180b   A.H. Eschenfelder: Magnetic Bubble Technology, Springer Series in Solid
         State Sciences, Vol.14 (Springer, Berlin, Heidelberg, New York 1980)
4.181   D.O. Smith, K.J. Harte: IEEE Trans. EC-15, 123 (1966)
4.182   D.O. Smith, K.J. Harte: IEEE Trans. EC-16, 372 (1967)
4.183   K.Y. Ahn, Y.S. Lin: IBM Tech. Discl. Bull. 15, 2017 (1972)
4.184   R.J. Collier: IEEE Spectrum 3, 67 (1966)
4.185   G.W. Stroke: Appl. Phys. Letters 6, 201 (1965)
4.186   G.W. Stroke: An Introduction to Coherent Optics and Holography
         (Academic Press, New York 1966)
4.187   W. Kulcke, K. Kosanke, E. Max, M.A. Habegger, T.J. Harris, H. Fleischer:
         Proc. IEEE 54, 1419 (1966)
4.188   H. Kogelnik: Microwaves 6, 68 (1967)
4.189   D.R. Bosomworth, H.J. Gerritsen: Appl. Opt. 7, 95 (1968)

4.190 M. Sakaguchi, N. Nishida: U.S. Patent 3,704,929 (1970, Pat. 1972)
4.191 G.R. Knight: Appl. Opt. *13*, 904 (1974)
4.192 P.P. Sorokin: IBM J. Res. Dev. *8*, 182 (1964)
4.193 M. Sakaguchi, N. Nishida, T. Nemoto: IEEE Trans. C-*19*, 1174 (1970)
4.194 G.R. Knight: Appl. Opt. *14*, 1088 (1975)
4.195 J.T. LaMacchia, D.L. White: Appl. Opt. *7*, 91 (1968)
4.196 J. Thire: L'Onde Electr. *52*, 452 (1972)
4.197 V. Malina, M. Chomat: Slaboproudy (Czechoslovakia) *37*, 572 (1976)
4.198 N. Minnaja, M. Nobile: Rend. Riunione Annu. Assoc. Elettrotec. Ital.
       49, B3/1 (1974)
4.199 N. Minnaja, M. Nobile: Elettrotecnica *61*, 674 (1974)
4.200 N. Minnaja: U.S. Patent No. 3,887,906, June 3 (1975)
4.201 D. Garelli: Elektrotechnik *57*, 30 (1975)
4.202 N. Nishida, M. Sakaguchi, F. Saito: Appl. Opt. *12*, 1663 (1973)
4.203 G.N. Aleksakov, Yu.A. Bykhovskii, A.I. Larkin, A.A. Markilov:
       Prib. Tekh. Eksp. (USSR) *17*, 203 (1974)
4.204 M.T. Fatehi, S.A. Collins Jr.: Int. Opt. Comput. Conf. (Digest of
       Papers), Apr. 23-25, 1975, p. 105
4.205 V.K. Bykhovsky, A.K. Glotov, A.E. Krasnov: Int. Opt. Comput. Conf.,
       Capri, Italy, Aug. 28 - Sep. 2, 1976
4.206 E. Mühlenfeld: Int. Opt. Comput. Conf., Capri, Italy, Aug. 28 - Sep. 2,
       1976 (Digest of Papers) p. 111
4.207 E. Mühlenfeld: Complete manuscript of [4.206]
4.208 E. Mühlenfeld, W. Kistner, A. Keller: Elektron. Rechenanlagen *18*,
       224 (1976)
4.209 H. Akahori: Res. Electrotech. Lab. (Japan), No. 771, p. 1, Oct. 1977
4.210 I.V. Prangishvili, A.K. Glotov, A.E. Krasnov, V.K. Bykhovskii: Proc.
       2nd USA-USSR Seminar Opt. Inf. Processing Novosibirsk, July 1976
       (Plenum Press, New York 1977)
4.211 V.N. Morozov: Sov. J. Quantum Electron. (USA) *8*, 1 (1978)
4.212 A.A. Vasiliev, I.N. Kompanet, S.P. Kotova, V.N. Morozov: Kvantovaya
       Elektron. *5*, 1298 (1978)
4.213 A.A. Vasiliev: Proc. 1st Eur. Conf. Opt. Syst. and Appl. 4-6 Apr.
       1978, Brighton, England, Vol. 6 (1978)
5.1   C.J. Conti: IEEE Comput. Group News *2*, 9 (1969)
5.2   R.M. Meade: Electronics *45*, 58 (1972)
5.3   R.M. Meade: IFIPS 1970 FJCC, p. 33
5.4   L.A. Belady: IBM Syst. J. *5*, 78 (1966)
5.5   R.M. Meade: Comput. Des. *10*, 87 (1971)
5.6   I.S. Reed: Computer *5*, 47 (1972)
5.7   D.M. Nessett: Aust. Comput. J. *7*, 33 (1975)
5.8   O.P. Agrawal: Proc. 1977 Nat. Comp. Conf., 13-16 Jun. (1977)
5.9   O.P. Agrawal, R.J. Zingg, A.V. Pohm: Proc. 14th IEEE Comput. Soc.
       Int. Conf. (IEEE, New York 1977) p. 74
5.10   J.D. Jones, D.M. Junod, R.L. Partridge, B.L. Shawley: IBM Tech. Discl.
       Bull. *19*, 594 (1976)
5.11   J.D. Jones, D.M. Junod: IBM Tech. Discl. Bull. *20*, 295 (1977)
5.12   T. Kilburn, D. Edwards, N. Lanigan, F.H. Summer: IEEE Trans. EC-*11*,
       233 (1962)
5.13   S.G. Campbell: Proc. AFIPS 1963 FJCC *24*, 473 (1963)
5.14   M.F. Wolff: Electronics *36*, 35 (1963)
5.15   G.G. Scarrot: Proc. IFIP Cong. 1965, *1*, 137
5.16   L.C. Hobbs: IEEE Trans. EC-*15*, 534 (1966)
5.17   A.B. Lindqvist, R.R. Seeber, L.W. Comeau: Proc. IEEE *54*, 1774 (1966)
5.18   W. Anacker, C.P. Wang: IEEE Trans. EC-*16*, 764 (1967)
5.19   E.J. Joseph: Comput. Des. *8*, 165 (1969)
5.20   R.L. Mattson, J. Gecsei, D.R. Slutz, I.L. Traiger: IBM Syst. J. *2*,
       78 (1970)

5.21 H. Katzan, Jr.: Proc. 1971 SJCC, p. 325
5.22 J.G. Williams: Commun. ACM *14*, 172 (1971)
5.23 A. Bensoussan, C.T. Clingern, R.C. Daley: Commun. ACM *15*, 308 (1972)
5.24 R.P. Parmelee, T.I. Peterson, C.C. Tillman, D.J. Hatfield: IBM Syst.
      J. *11*, 99 (1972)
5.25 R.E. Brundage, A.P. Batson: Proc. 2nd Annu. Symp. Comput. Archit.,
      Jan. 20-22, 1975 (IEEE, New York 1975) p. 85
5.26 R.F. Brundage, A.P. Batson: Rev. Fr. Automat. Inf. Rech. Oper. *10*, 47
      (1976)
5.27 S.Ya. Berkovich, Yu.Ya. Kochin, Yu.N. Khrebtov: Program. and Comput.
      Software (USA) *2*, 455 (1976)
5.28 C. Schuenemann: IBM Tech. Discl. Bull. *21*, 663 (1978)
5.29 T.D. Chase, R.M. Glorioso: Proc. 25th Anniv. Conf. *1*, 6 (1972)
5.30 M.V. Wilkes: IEEE Trans. EC-*14*, 270 (1965)
5.31 M.V. Wilkes: IEEE Trans. C-*20*, 674 (1971)
5.32 D.H. Gibson, W.L. Shevel: Electronics *21*, 105 (1969)
5.33 F.F. Lee: IEEE Trans. C-*18*, 1062 (1969)
5.34 D.C. Gunderson: Computer *3*, 7 (1970)
5.35 H.S. Stone: IEEE Trans. C-*19*, 73 (1970)
5.36 J.H. Kroeger, R.M. Meade: Proc. 1971 Comput. Des. Conf., Vol. 1 (1971)
      p. 252
5.37 H. Barmasian, A.L. DeCegama: Digest of Papers 6th Annu. IEEE Comput.
      Soc. Int. Conf., 1972, p. 107
5.38 K.R. Kaplan, R.O. Winder: Computer *6*, 30 (1973)
5.39 J. Niederreichholz: Elektron. Rechenanlagen *18*, 122 (1976)
5.40 B.D. Ackland, D.A. Pucknell: Electron. Lett. *11*, 588 (1975)
5.41 B.D. Ackland, D.A. Pucknell: N76/5 For. Meet.: Comput. in Eng., Perth
      Sept. 6-7, 1976, p. 120
5.42 A.J. Smith: Proc. 2nd Int. Conf. Software Eng. (IEEE, New York 1976)
      p. 286
5.43 A.J. Smith: IEEE Trans. SE-*4*, 121 (1978)
5.44 H. Iizuka, T. Terui: J. Inf. Proc. Soc. Jpn. *14*, 669 (1973)
5.45 M. Bennett, P. Berard, C. Boksenbaum, M. Veran: Rev. Fr. Automat. Inf.
      Rech. Oper., 10 May,1976, p. 1
5.46 C.K. Chow: IBM Tech. Discl. Bull. *17*, 3163 (1975)
5.47 C.K. Chow: IBM Tech. Discl. Bull. *18*, 1643 (1975)
5.48 C.K. Chow: IEEE Trans. C-*25*, 157 (1976)
5.49 C.G. Bell, D.P. Casasent: Comput. Des. No. 11, Nov. 1971, p. 83
5.50 J. Bell, D.P. Casasent, C.G. Bell: IEEE Trans. C-*23*, 346 (1974)
5.51 R. Monroe: Digest of Papers 10th IEEE Comp. Soc. Int. Conf. *3*, 1034
      (1975)
5.52 W.D. Strecker: Proc. 3rd Annu. Symp. Comput. Archit. (IEEE New York
      1976) p. 155
5.53 A. Weinberger: U.K. Patent No. 1 280 753, July 5 (1972)
5.54 D.K. Chia: IBM Tech. Discl. Bull. *17*, 3361 (1975)
5.55 A.M. Weiner: IBM Tech. Discl. Bull. *19*, 4697 (1977)
5.56 C.J. Conti, D.H. Gibson, S.H. Pitkowsky: IBM Syst. J. *7*, 2 (1968)
5.57 J.S. Liptay: IBM Syst. J. *7*, 15 (1968)
5.58 R.A. McLaughlin: Datamation, Sept. 1972, p. 58
5.59 Comp. Des. *15*, 26 (1976)
5.60 J.F. Mastranadi: IBM Tech. Discl. Bull. *14*, 3760 (1972)
5.61 M.J. Haims: IBM Tech. Discl. Bull. *18*, 278 (1975)
5.62 H. Gelernter: Proc. Symp. on Large Capacity Memory Tech., May 1961
5.63 H.E. Petersen: Proc. IFIPS Cong., Aug. 1962
5.64 H. Hellerman, IBM Watson Res. Center Rept. No. RC-1095, Oct. 1963
5.65 F.J. Hilbing: Ph. D. Dissert., Stanford Univ. (1968)
5.66 J.B. Rothnie, T. Lozano: Commun. ACM *17*, 63 (1974)
5.67 G.M. Furxhi: Riv. Inf. (Italy) *7*, 15 (1977)

5.68   P.B. Berra: Proc. COMPSAC 1978 (IEEE, New York 1978) p. 698

5.69   A.J. Symonds: IBM Syst. J. *7*, 229 (1968)

5.70   J. Fotheringham: Commun. ACM *4*, 435 (1961)

5.71   D. Aspinall, D.J. Kinniment, D.B.G. Edwards: IFIP Edinburgh, Aug. 1968, p. D81

5.72   H.R. Holt, J.A. Timmons, D.C. Gunderson: Honeywell Inc., Final Tech. Rept. 12099-FR1, Sept. 1968

5.73   M.H.J. Baylis, D.G. Fletcher, D.J. Howarth: *Inform. Proc. 68* (North-Holland, Amsterdam 1969) p. 831

5.74   Y. Chu: *Computer Organization and Microprogramming* (Prentice-Hall, Englewood Cliffs, N.J. 1972)

5.75   R. Moulder: Proc. AFIPS Nat. Conf. Comput. Composition and Expo. *42*, 171 (1973)

5.76   W.B. Riley: Electronics *45*, 91 (1972)

5.77   J.L. Gertz: Infotech. State-of-the-Art Rept. (Maidenhead, England 1976) p. 273

5.78   K. Koch: IBM Tech. Discl. Bull. *15*, 3088 (1973)

5.79   J.R. Carlberg: Taylor Naval Ship Res. and Dev. Center, Rept. No. DTNSRDC-77-0083, Aug. 1977

5.80   M. Takesue: Inform. Process. Soc. Jpn. *19*, 158 (1978)

5.81   C.V.W. Armstrong: Proc. 2nd Annu. Symp. Comput. Archit. (IEEE, New York 1975) p. 34

5.82   C.E. Shannon: Bell Syst. Tech. J. *28*, 59 (1949)

5.83   T.F. Tabloski, F.J. Mowle: IEEE Trans. C-*25*, 684 (1976)

5.84   W.E. Donath: IBM J. Res. Dev. *18*, 401 (1974)

5.85   B.A. Holum: IBM Confidential, SRI Term Paper, No. 11-31, April 1964

5.86   F.T. Baker, W.E. Triest, C.H. Forbes, N. Jacobs, J. Schenken: IBM, Final Rept. May 1966, AF-30(602)-3573

5.87   D.C. Gunderson, J.P. Francis, W.L. Heimerdinger: Honeywell Inc., Rept. No. 12029, Dec. 1966 (RADC TR-66-573)

5.88   D.C. Gunderson, W.L. Heimerdinger, J.P. Francis: "A Multiprocessor with Associative Control", in *Prospects for Simulation and Simulators of Dynamic Systems* (Spartan Books, New York 1967) p. 183

5.89   R. Gonzales, D.C. Gunderson, J.A. Timmons: Honeywell Inc., Final Rept. Nov. 1967 (AD-662 361)

5.90   R.P. Bair: Moore School of Electr. Eng., May 1968 (AD-674 199)

5.91   L.D. Wald, G.A. Anderson: Final Rept. NAS 12-2087, Sept. 1971

5.92   F. Tsui: IBM Tech. Discl. Bull. *15*, 2342 (1972)

5.93   L. Hellerman, G.E. Hoernes: IEEE Trans. C-*17*, 1144 (1968)

5.94   I.N. Hooton: In *Automatic Acquisition and Reduction of Nuclear Data* (Ges. Für Kernforschung G.m.b.H., Karlsruhe 1964) p. 338

5.95   I.N. Hooton: In Ref. 5.94, p. 349

5.96   H. Meyer, W. Stuber: In Ref. 5.94, p. 357

5.97   E. Blanca, A. Carriere: CEA-R-3394, Dec. 1967

5.98   M.D. Johnson, D.C. Gunderson: Proc. 1970 Int. Telemetry Conf., April 1970, p. 109

5.99   L. Rettelbusch, H. Pfahlbusch: Nachrichtentech. Elektron. *24*, 340 (1974)

5.100 T.L. Saxton, C.-C. Huang: IEEE Trans. C-*26*, 170 (1977)

5.101 R.R. Seeber, Jr.: Commun. ACM *4*, 301 (1961)

5.102 E.S. Gershuny, O.L. Lamb: IBM Tech. Discl. Bull. *15*, 1109 (1972)

5.103 B.A. Crane: IEEE Trans. C-*17*, 691 (1968)

5.104 B.H. Scheff: Electron. Prog. *10*, 31 (1966)

5.105 C. Peters: NTIS AD-824 213

5.106 S.N. Porter: J. ACM *13*, 369 (1966)

5.107 R.C. Minnick: IEEE Trans. EC-*13*, 685 (1964)

5.108 C.C. Yang, S.S. Yau: IEEE Trans. EC-*15*, 522 (1966)

5.109 S.S. Yau, M. Orsic: IEEE Trans. C-*19*, 259 (1970)

358

5.110 S.S. Yau, C.K. Tang: IEEE Trans. C-*19*, 141 (1970)
5.111 C. Barre: Electron. Appl. Ind. (France) *250*, 21 (1978)
6.1   C.Y. Lee, M.C. Paull: Proc. IEEE *51*, 924 (1963)
6.2   C. Lee, M. Paull: Proc. IEEE *52*, 312 (1964)
6.3   C.Y. Lee: "Content-Addressable and Distributed Logic Memories", in
       Applied Automata Theory, ed. by J.T. Tou (Academic Press, New York
       1968)
6.4   E.S. Lee: Proc. AFIPS 1963 SJCC, p. 381 (1963)
6.5   R.S. Gaines, C.Y. Lee: IEEE Trans. EC-*14*, 72 (1965)
6.6   B.A. Crane, J.A. Githens: IEEE Trans. EC-*14*, 186 (1965)
6.7   G. Németh: Helsinki U.Tech., Dept. Tech. Phys. Report TKK-F-A347
       (1978)
6.8   B.A. Crane, R.R. Laane: Proc. AFIPS 1967 SJCC, p. 517 (1967)
6.9   R.P. Edwards: Proc. IEEE *52*, 83 (1964)
6.10   E.S. Spiegelthal: Proc. IEEE *52*, 74 (1964)
6.11   A. Tremblay: Cybernetics *XIX*, 105 (1976)
6.12   J.N. Sturman: IEEE Trans. C-*17*, 2 (1968)
6.13   J.N. Sturman: IEEE Trans. C-*17*, 10 (1968)
6.14   J.E. Smathers: Ph. D. Dissert., Oregon State Univ. (1969)
6.15   W.H. Kautz, K.N. Levitt, A. Waksman: IEEE Trans. C-*17*, 443 (1968)
6.16   W.H. Kautz: J. ACM *18*, 19 (1971)
6.17   W.H. Kautz, M.C. Pease III: AD 763 710 (1971)
6.18   J. Hood, M. Mark, J. Cotton: Proc. 1976 Int. Conf. Parallel Processing,
       Aug. 24-27, 1976, p. 168
6.19   R. Trepp: RADC-TR-66-182, June 1966
6.20   C.A. Finnila, H.H. Love, Jr.: IEEE Trans. C-*26*, 112 (1977)
6.21   S. Ya. Berkovich, Ya.Ya. Kochin, G.M. Lapir: Autom. Remote Control
       35, 1342 (1974)
6.22   D.A. Savitt, H.H. Love, R.E. Troop: AD 488 538 (1966)
6.23   D.A. Savitt, H.H. Love, R.E. Troop, R.A. Rutman: *Association Storing
       Processor Interpretive Program - Program Logic Manual*. Final Report,
       Hughes Aircraft Co., FR-11-558 (1968)
6.24   H.H. Love, D.A. Savitt: RADC-TR-65-32 (1965)
6.25   H.H. Love, D.A. Savitt: In *Associative Information Techniques*, ed.
       by E.L. Jacks (American Elsevier, New York 1971) p. 147
6.26   H.H. Love, R.A. Rutman: Hughes Aircraft, FR-68-11-1179, Dec. 1968
6.27   H.H. Love: Hughes Aircraft, FR-69-11-487, Jun. 1969
6.28   R.A. Rutman: Hughes Aircraft, FR-69-11-208, Feb. 1969
6.29   J.H. Holland: 1959 EJCC, p. 108
6.30   G.J. Lipovski: Proc. AFIPS 1970 SJCC, p. 385 (1970)
6.31   G.H. Schmitz: Final Rep. Contr. No. DAH 60-72-C0050 (1972)
6.32   Proc. 1972 Sagamore Comp. Conf. (IEEE, New York 1972)
6.33   W.S. Litzler: 1973 Swieeeco Record of Technical Papers, p. 482
6.34   E.C. Stanke II: RADC-TR-77-366 (1978)
6.35   J.A. Githens: "An Associative, Highly-Parallel Computer for Radar
       Data Processing", in *Parallel Processor Systems, Technologies, and
       Applications*, ed. by L.C. Hobbs, D.J. Theis, J. Trimble, H. Titus,
       I. Highberg (Spartan Books, New York 1970)
6.36   R.O. Berg, M.D. Johnson: Proc. IEEE 1970 Int. Comp. Group. Conf.,
       Washington, p. 336
6.37   J.A. Githens: Proc. NAECON 1970, p. 290
6.38   J.A. Githens: Proc. IEEE 1972 Int. Comp. Soc. Conf., p. 57
6.39   R.O. Berg, H.G. Schmitz, S.J. Nuspl: Proc. NAECON 1972, p. 312
6.40   J.A. Cornell: Proc. WESCON 1972, p. 1/3-1
6.41   J.A. Cornell: Proc. COMPCON 1972, p. 69
6.42   K.E. Batcher: WESCON Tech. Papers *16*, 1 (1972)
6.43   J.A. Rudolph: Proc. AFIPS 1972 FJCC, p. 229 (1972)

6.44   K.E. Batcher: Proc. 1974 Nat. Comp. Conf., p. 405

6.45   E.W. Davis: Proc. 1974 Nat. Comp. Conf., p. 17

6.46   Goodyear Aerospace Corp.: Doc. 8284 C (1974)

6.47   Goodyear Aerospace Corp.: Doc. GER-15636B (1974)

6.48   Goodyear Aerospace Corp.: Doc. GER-15637B (1974)

6.49   Goodyear Aerospace Corp.: Doc. GER-15643A (1974)

6.50   Goodyear Aerospace Corp.: Doc. GER-15644A (1974)

6.51   Goodyear Aerospace Corp.: Doc. GER-16139 (1974)

6.52   Goodyear Aerospace Corp.: Doc. AP-112286 (1965)

6.53   D. Brotherton, S. Domchick: Goodyear Aerospace Corp., Doc. GER-12318 (1966)

6.54   L.C. Fulmer, W.C. Meilander: Proc. 1970 IEEE Int. Comp. Group. Conf., p. 325

6.55   W.C. Meilander, J. Garrett, M. Bialer: "A Mission oriented associative processor using plated wire", in *Parallel Processor Systems, Technologies, and Applications*, ed. by L.C. Hobbs, D.J. Theis, J. Trimble, H. Titus, I. Highberg (Spartan Books, New York 1970) p. 153

6.56   W. Shooman: Proc. 1960 EJCC, p. 111

6.57   W. Shooman: U.S. Patent 3,277,449 (1966)

6.58   W. Shooman: "Orthogonal Processing", in *Parallel Processor Systems, Technologies, and Applications*, ed. by L.C. Hobbs, D.J. Theis, J. Trimble, H. Titus, I. Highberg (Spartan Books, New York 1970)

6.59   L.C. Highbie: COMPCON '72, p. 287

6.60   J.C. Murtha, R.L. Beadles: ONR Rep. No. 4755 (1964)

6.61   J.C. Murtha: "Highly Parallel Information Processing Systems", in *Advances in Computers*, Vol. 7 (Academic Press, New York 1966) p. 1

6.62   M.J. Flynn: Proc. IEEE *54*, 1901 (1966)

6.63   G.L. Hollander: Proc. AFIPS 1967 SJCC, p. 463 (1967)

6.64   L.C. Hobbs, D.J. Theis, J. Trimble, H. Titus, I. Highberg (eds.): *Parallel Processor Systems, Technologies, and Applications* (Spartan Books, New York 1970)

6.65   L.C. Higbie: Computer *6*, 48 (1973)

6.66   J.E. Shore: 1972 IEEE Int. Conv. Digest, p. 358

6.67   J.E. Shore: Comput. Electron. Eng. *1*, 95 (1973)

6.68   C.C. Foster: *Content Addressable Parallel Processors* (Van Nostrand, New York 1976)

6.69   K.J. Thurber: *Large Scale Computer Architecture* (Hayden, Rochelle Park, N.J. 1976)

6.70   L.C. Roberts: IEEE Spectrum *11*, 46 (1974)

6.71   B.H. McCormick: IEEE Trans. C-*12*, 791 (1963)

6.72   D.L. Slotnick: Proc. AFIPS 1967 SJCC, p. 477 (1967)

6.73   R.M. Barnes, R.M. Brown, M. Kato, D.J. Kuck, D.L. Slotnick, R.A. Stokes: IEEE Trans. C-*17*, 746 (1968)

6.74   P. Alsberg, J. Gaffney, C. Grossman, T. Mason: Illinois Univ. Illiac-IV-212, March 1969

6.75   R.L. Davis: IEEE Trans. C-*18*, 800 (1969)

6.76   B.H. McCormick, J.L. Divilbiss: Rept. No. 4031, Digital Comp. Lab., Univ. of Illinois, 1969

6.77   Burroughs Corp.: *ILLIAC IV Systems Characteristics and Programming Manual*, Contract Report AF 30 (602) 4144 (1973)

6.78   E.J. Jacks (ed.): *Associative Information Techniques* (American Elsevier, New York 1971)

6.79   Proc. 1973 Sagamore Comput. Conf. (IEEE, New York 1973)

6.80   Proc. 1974 Sagamore Comput. Conf. (IEEE, New York 1974)

6.81   Proc. 1975 Sagamore Comput. Conf. (IEEE, New York 1975)

6.82   M. Feilmeyer (Ed.): *Parallel Computers - Parallel Mathematics*. Proc. IMACS (AICA)-61 Symposium March 14-16, 1977, TU Munich (North-Holland, Amsterdam 1977)

6.83   *Proc. 1976 Int. Conf. on Parallel Processing*, Aug. 24-27, 1976
       (IEEE, New York 1976)
6.84   *Proc. 1977 Int. Conf. on Parallel Processing*, Aug. 23-26, 1977
       (IEEE, New York 1977)
6.85   Control Eng. *9*, 22 (1962)
6.86   R.H. Fuller: General Precision-Librascope Inc., Interim Rept.,
       AD-608 427, October 1964
6.87   R.H. Fuller: Comput. Des. *6*, 43 (1967)
6.88   R.H. Fuller: Proc. AFIPS 1967 SJCC, p. 471
6.89   R.H. Fuller: General Precision, ONR/RADC Seminar on Assoc. Proc. 1967
6.90   R.H. Fuller, R.M. Bird, J.N. Medick: "Associative Processor Study",
       Librascope Div. General Precision, Oct. 1964
6.91   R.H. Fuller, R.M. Bird, R.M. Worthy: RADC-TR-65 210, AD-621 516,
       August 1965
6.92   Westinghouse Defense and Space Center, Final Rept. June 1964,
       AD-602 693
6.93   J.A. Feldman: M.I.T. Lincoln Lab. Tech. Note 1965-13, April 1965,
       AD-614 634
6.94   General Precision Inc.: "Associative Processing Techniques"
       (Librascope Group, 1965)
6.95   D.L. Reich: "Associative Memories and Information Retrieval",
       in *Some Problems in Information Science*, ed. by M. Kochen
       (Scarecrow Press, New York 1965)
6.96   J.A. Dugan, R.S. Green, J. Minker, W.E. Shindle: Proc. ACM 21st
       Nat. Conf. 1966, p. 347
6.97   K.E. Knight: Datamation *12*, 40 (1966)
6.98   M.A. Knapp: "RADC Programs in Associative Processing", ONR/RADC
       Seminar on Assoc. Proc., May 1967
6.99   H.I. Jauvits: Interim Rept., Lab. For Electronics Inc., FFB, 1968,
       NASA-CR-86076
6.100 J.A. Rudolph: Proc. IEEE Region 6 Conf., Apr. 1969, p. 179
6.101 M.H. Cannell, A.J. Nickelson, M.F. Owens, K.W. Wadman, M.L. Urban:
       Mitre Corp., Repts. Nos. MTR-1735-Rev-1, MTR-863, AD-879 281
       Dec. 1970
6.102 L.C. Hobbs, D.J. Theis: "Survey of Parallel Processor Approaches and
       Techniques", Symp. on Parallel Proc. Systems Technologies and
       Applications, Monterey 1969 (Papers ed. by L.C. Hobbs et al. 1970)
6.103 J.C. Murtha: NAECON '70 Records, May 1970, p. 298
6.104 W.C. Meilander, R.G. Gall: "Evaluation of the Goodyear associative
       processor in an operational ATC environment", IEEE Comp. Soc. Conf.,
       Boston, Mass., Sep. 1971
6.105 M. Minsky, S. Papert: "On Some Associative, Parallel and Analog
       Computations", in *Associative Information Techniques*, ed. by E.J.
       Jacks (American Elsevier, New York 1971)
6.106 K.J. Thurber, R.O. Berg: Comput. Des. *10*, 103 (1971)
6.107 B. Parhami: "Design Techniques for Associative Memories and Processors",
       UCLA, Comput. Sci., Rept. No. PB-220 714 (1973)
6.108 K.J. Thurber, P.C. Patton: IEEE Trans. C-*22*, 1140 (1973)
6.109 R.M. Lea: Computer *8*, 25 (1975)
6.110 L.C. Higbie: Comput. Electr. Eng. *2*, 397 (1975)
6.111 L.C. Higbie: Comput. Des. *15*, 75 (1976)
6.112 B.W. Prentice, R. Katz, R. Komadja, H. Lee: Boeing Comp. Services Inc.,
       Seattle, Wash. Jan. 1975, RADC-TR-74-326, AD-A005 308
6.113 K.J. Thurber, L.D. Wald: Comput. Surv. *7*, 215 (1975)
6.114 D. Lewin: "Introduction to Associative Processors", Proc.Conf. Comp.
       Archit., St. Raphael, France, 12-24 Sept. 1976, ed by G.G. Boylaye,
       D.W. Lewin

6.115 M.W. Summers: Rome Air Devel. Cent. Rept. RADC-TR-75-318, Jan. 1976, AD-A021 232

6.116 *Proc. IEEE 1977 Int. Conf. on Parallel Processing*, Aug. 23-26, 1977 (IEEE, New York 1977)

6.117 Infotech. Int.: *Future Systems, State of the Art Rept.* (Maidenhead, England, 1977)

6.118 S.S. Yau, H.S. Fung: Comput. Surv. *9*, 3 (1977)

6.119 N.J. Zimmerman, H.J. Sips: Informatie (Netherlands) *20*, 3 (1978)

6.120 D.L. Slotnick, W.C. Borck, R.C. McReynolds: Proc. AFIPS 1962 FJCC *22*, 97 (1962)

6.121 Westinghouse Defence and Space Center: "Parallel Network Computer (SOLOMON) Applications Analyses", August 1964, AD-606 578

6.122 F.W. Weingarten, P.T. Rux, J.A. Boles: "On an Associative Memory for Nebula Computer", Dept. of Math., Oregon State Univ., In-House Doc., 1964

6.123 J.A. Boles: "The Logical Design of the Nebula Computer"; Ph. D. Thesis, Oregon State Univ. (1968) AD-673 990

6.124 IBM: "Project Lightning", AD-250 678 (1960)

6.125 IBM: "Project Lightning", U.S. Gov. Res. Repts. *36*, 124(A) (1961)

6.126 S.H. Unger: Proc. IRE *46*, 1744 (1958)

6.127 J.H. Holland: Proc. WJCC, 259 (1960)

6.128 W.T. Comfort: IBM Report No. 62-825-496 (1962)

6.129 E.A. Feigenbaum, H.A. Simon: Proc. IFIP Congr. 1962, p. 177

6.130 P.M. Davies: Proc. 1963 IEEE Pacific Comp. Conf. (1963) p. 109

6.131 P. Davies: "Associative Processors", IEEE Symp. on Search Memory, May 1964

6.132 P.M. Davies: U.S. Patent No. 3,320,594, May 16, 1967

6.133 E.V. Evreinov, Y.G. Kosarev: Kibernetika *4*, 3 (1963)

6.134 R.G. Ewing, P.M. Davies: Proc. FJCC *25*, 147 (1964)

6.135 B. Hasbrouck, N.S. Prywes, D. Lefkovitz, N. Kornfield: Comp. Command and Control Co., April 1965, AD-466 313

6.136 R.G. Gall: "Hybrid Associative Computer Study", Vol. I, AD-489 929 (Goodyear Aerospace Corp., 1966)

6.137 R.G. Gall: "Hybrid Associative Computer Study", Vol. II, AD-489 930 (Goodyear Aerospace Corp., 1966)

6.138 R.G. Gall, D.E. Brotherton: "Associative List Selector", AD-802 993 (Goodyear Aerospace Corp., 1966)

6.139 D.L. Rohrbacher: "Advanced Computer Organization Study", AD-631 870 and AD-631 387 (April 1966)

6.140 J.L. Cass: "Organization and Applications of Associative File Processors", ONR/RADC Seminar on Associative Processing, May 1967

6.141 T. Feng: "An Associative Processor"; Ph. D. Dissertation, Univ. of Michigan (1967)

6.142 T. Feng: "An Associative Processor", Tech. Rept., Systems Engineering Lab., Univ. of Michigan, Dec. 1967

6.143 T. Feng: "An Associative Processor", Michigan Univ. Rept. No. 06920-17-T, AD-682 353 (Jan. 1969)

6.144 T. Feng: Proc. Nat. Electron. Conf. XXIV, 257 (1968)

6.145 Auerbach Publ. Inc.: TECH Note 1374-TR-500-1 (AD-679 227) (1968)

6.146 W.A. Lea: NASA-TM-X1544, March 1968

6.147 R.M. Lea: Radio and Electron. Eng. *46*, 487 (1976)

6.148 R.M. Lea: Comput. J. *21*, 45 (1978)

6.149 MIT Lincoln Lab.: Rept. No. ESD-TR-6890 (1968)

6.150 H.H. Love: Hughes Aircraft Co., Rept. No. FR-69-11-487, AD-855 770 (1969)

6.151 H.H. Love: Proc. Sagamore Comput. Conf. Parallel Process., Aug. 22-24, 1973 (IEEE, New York 1973) p. 103

6.152 P.M. Melliar-Smith: Proc. FJCC 1969, p. 201

6.153   J.E. Shore, F.A. Polkinghorn: NRL Rept. NRL-6961, Nov. 1969, AD-702 394

6.154   W.S. Tuma: Goodyear Aerospace Corp., Rept. No. GER-14566, AD-862 134 (1969)

6.155   R.R. Kressler: Air Force Report No. AFAL-TR-70-142, Aug. 1970

6.156   R.O. Berg, K.J. Thurber: NAECON '71 Record, p. 206 (1971)

6.157   J.E. Shore, T.L. Collins: Rept. of NRL Progress, p. 15, March 1972

6.158   R.A. Urban: Nat. Electron. Conf. 1972, p. 318

6.159   R.D. Arnold: Colorado Univ. Rept. CU CS 051 74, NSF GH 660, August 1974

6.160   D.L. Baldauf: Mitre Corp., Bedford, Mass., MTR-2879, ESD-TR-74-199 (AD-A003 414), Nov. 1974

6.161   L.A. Gambino: Army Engineer. Topographic Labs., AD-A056 438, Jun. 1978

6.162   G.J. Lipovski: Proc. 5th Annual Symp. Comp. Archit. (IEEE, New York 1978) p. 31

6.163   S.Ya. Berkovich, Yu.Ya. Kochin, G.M. Lapir: Autom. Remote Control 35, 1342 (1974)

6.164   H.K. Resnick: California Univ., Livermore Lawrence Rad. Lab., Computer Inf. Center, Vol. 3, Publication No. 6 (1975)

6.165   L. Kerschberg, E.A. Ozkaharan, J.E.S. Pacheo: Proc. 2nd Int. Conf. Software Engineering, San Fransisco, Cal., 13-15 Oct., 1976 (IEEE, New York 1976) p. 505

6.166   C.Y. Hicks: ACM Comp. Sci. Conf., 31 Jan.-2 Feb., 1977, Atlanta, Georgia

6.167   R.R. Seeber, A.B. Lindquist: Proc. AFIPS 1963 FJCC 24, 489 (1963)

6.168   J.S. Squire, S.M. Paleis: Proc. AFIPS 1963 SJCC, 395 (1963)

6.169   R.S. Entner: "The Advanced Avionic Digital Computer", Symp. Parallel Processor Systems, Tech. & Appl., Monterey, June 1969

6.170   L.J. Koczela, G. Wang.: IEEE Electron. Comp., p. 520, June 1969

6.171   G.J. Lipovski: Report R-424, Coordinated Sci. Lab., Univ. of Illinois, July 1969 (AD-692 195)

6.172   C.C. Foster: Goodyear Aerospace Corp. Doc. GER-11772 (1964)

6.173   M.J. Kroeger: Goodyear Aerospace Corp. Doc. GER-16378, RADC-TR-76-352 (1976)

6.174   Z.H. Glanz: Int. Electr. Electron. Conf. and Expos., 29 Sep.-1 Oct., 1975, Toronto, Canada

6.175   B. Parhami, A. Avizienis: Symp. on Comput. Archit., Univ. of Florida, Gainesville, p. 141 (1973)

6.176   K.J. Thurber, P.C. Patton: COMPCON '72, p. 275 (1972)

6.177   G.J. Nutt: Acta Infor. 6, 211 (1976)

6.178   A.P. Kisylia: Illinois Univ. Rept. No. R-390, Aug. 1968 AD-675 310

6.179   R.R. Linde, R. Gaten, T.F. Peng: Proc. AFIPS Nat. Comp. Conf. 42, 187 (1973)

6.180   C.R. DeFiore: Datamation 16, 47 (1970)

6.181   C.R. DeFiore, N.J. Stillman, P.B. Berra: Proc. ACM Nat. Conf., Aug. 3-5, 1971, p. 28

6.182   V.L. Arlazarov, S.Ya. Berkovich, A.A. Leman, M.Z. Rosenfeld: Avtom. Telemekh. 12, 184 (1971)

6.183   G. Salton: Commun. ACM 15, 658 (1972)

6.184   C.R. DeFiore, P.B. Berra: Proc. AFIPS Conf. Nat. Comp. Composition and Exposition 42, 181 (1973)

6.185   C.R. DeFiore, P.B.Berra: IEEE Trans. C-23, 121 (1974)

6.186   R. Moulder: Proc. Sagamore Comput. Conf. Parallel Process., Sagamore Lake, N.Y. 1973 (IEEE, New York 1973) p. 161

6.187   E.A. Ozkarahan, S.A. Schuster, K.C. Smith: Proc. AFIPS Nat. Comput. Conf. Expo. 44, 379 (1975)

6.188   E.A. Ozkarahan, S.A. Schuster, K.C. Sevcik: ACM Trans. Database Syst. 2, 175 (1977)

6.189   I.S. Charnaya: Program. and Comput. Software *3*, 61 (1977)
6.190   R.E. Asratyan, V.T. Lysikov: Autom. Remote Control (USSR) *39*, 755 (1978)
6.191   R. Beaufils, J.P. Sansonnet: Euromicro J. (Netherlands) *4*, 275 (1978)
6.192   R.E. Birney, M.I. Davis, R.A. Hood: IBM Tech. Discl. Bull. *20*, 2972 (1978)
6.193   T. Ishikawa: AFIPS 1978 Nat. Comp. Conf., 5-8 June, 1978, Anaheim, Cal.
6.194   G.G. Langdon, Jr.: ACM Trans. Database Syst. *3*, 148 (1978)
6.195   S.A. Schuster, H.B. Nguyen, E.A. Ozkarahan, K.O. Smith: Proc. of the 5th Ann. Symp. on Comp. Archit., Palo Alto, 3-5 April, 1978 (IEEE, New York 1978)
6.196   J.G. Dyke, R.M. Lea: Digital Process. *1*, 89 (1975)
6.197   V.A. Pronina, A.A. Chudin: Avtom. Telemekh. *8*, 106 (1975)
6.198   H.H. Love, J. Baer: Proc. 1977 Int. Conf. Parallel Processing (IEEE, New York 1977) p. 153
6.199   R.M. Lea: Comput. J. *21*, 45 (1978)
6.200   R.H. Fuller, R.M. Bird: Proc. AFIPS 1965 FJCC *28*, 105 (1965)
6.201   C. Yang: Northwestern Univ. Tech. Rept. TR-66-103 (1966)
6.202   C. Yang: "Pattern Recognition by an Associative Memory", Northwestern Univ. 1966 (unpublished paper)
6.203   S.S. Yau, C.C. Yang: IEEE Trans. EC-*15*, 944 (1966)
6.204   S.S. Yau, C.C. Yang: IEEE Trans. EC-*15*, 938 (1966)
6.205   N.J. Stillman, C.R. DeFiore, P.R. Berra: Proc. AFIPS 1971 SJCC, 557 (1971)
6.206   B. Kruse: IEEE Trans. C-*22*, 1075 (1973)
6.207   E.C. Joseph, A. Kaplan: Proc. 6th Nat. MILECON, 1962, p. 255
6.208   Librascope Rept. Libi 6081, July 1966
6.209   E.E. Eddey: Proc. NAECON '67, 39 (1967)
6.210   E.E. Eddey: Proc. NAECON '70, 302 (1970)
6.211   W.C. Meilander: Proc. NAECON '68, 57 (1968)
6.212   L.E. Cannon: Ph. D. Thesis, Montana State Univ. (1969)
6.213   A. Costanzo, J. Garrett: Proc. NAECON '69, 107 (1969)
6.214   R.M. Bird: In *Parallel Processor Systems, Technologies, and Applications*, ed. by L.C. Hobbs (Spartan Books, Washington, D.C. 1970) p. 107
6.215   K.J. Thurber: Proc. AFIPS 1971 SJCC, 49 (1971)
6.216   L.D. Wald: Proc. 1972 Sagamore Comput. Conf., Aug. 1972, (IEEE, New York) p. 135
6.217   L.D. Wald: Proc. AFIPS 1974 Nat. Comput. Conf., May 1974, p. 133
6.218   E.E. Eddey, W.C. Meilander, T. Feng (ed.): *Parallel Processing* (Springer, Berlin, Heidelberg, New York 1975) p. 417
6.219   C.L. Morefield: Proc. Annu. Allerton Conf. Circuit Syst. Theory, Monticello, Ill. Sep. 29- Oct. 1, 1976, p. 1074 (1976)
6.220   H.G. Schmitz: Proc. ACM/ICST 15th Annu. Tech. Symp., June 17, 1976, *4*, 2338 (1976)
6.221   A.K. Singhania: Proc. IMACS(AICA)-GI-Symp. Parallel Comput. March 14-16, 1977, *5*, 1071 (North Holland, Amsterdam 1977)
6.222   S.M. Lamb, R. Vanderslice: J. Acoust. Soc. *64*, *S1*, 573 (1978)
6.223   G. Estrin, C.R. Viswanathan: J. ACM, Jan 1962, p. 41
6.224   J.H. Katz: In *Parallel Processor Systems, Technologies, and Applications*, ed. by L.C. Hobbs (Spartan, Washington, D.C., 1970) p. 131
6.225   P. Gilmore: Goodyear Aerospace Corp. Doc. GER-15260, June 1971
6.226   P.B. Berra, E. Oliver: AD-A049 617/4SL, Syracuse Univ., Dept. of Indust. Eng. and Operations Res., Dec. 1977
6.227   M.A. Wesley: IEEE Trans. AU-*17*, 162 (1969)
6.228   Yu.G. Naimark, G.M. Popova, I.V. Prangishvili: Avtom. Telemekh. No. 4, April 1972, p. 136
6.229   A.J. Krygiel: Proc. 1976 Int. Conf. Parallel Process (IEEE, New York 1976)

6.230 P.A. Gilmore: Proc. AFIPS 1971 FJCC, *39*, 411 (1971)
6.231 W.F. Beausoleil, R.M. Chittenden, G.H. Ottaway: IBM Tech. Discl.
       Bull. *20*, 2770 (1977)
6.232 W.C. Liles, J.C. Demmel, I.S. Reed, J.D. Mallett, L.E. Brennan:
       Rept. No. TSC-PD-8525-1-Vol.-1, Apr. 1978, AD-A054 357
6.233 W.C. Liles, J.C. Demmel, I.S. Reed, J.D. Mallett, L.E. Brennan:
       Rept. No. TSC-PD-8525-1-Vol.-2, Apr. 1978, AD-A054 358
6.234 M.E. Sherry: Amer. Document. Instit. 27th Ann. Meeting, 1964
6.235 M.A. Wesley, S.K. Chang, J.H. Mommens: Proc. AFIPS 1972 FJCC, 461
       (1972)
6.236 D.C. Gunderson: WESCON Tech. Papers (Session 9, 1966)
6.237 J.P. Hayes: Univ. of Illinois, Comput. Lab. Rept. 227, June 1967
6.238 J. Previte, E. Tippie: EMI-TM-67-1, Feb. 1967
6.239 V.A. Orlando, P.B. Berra: Proc. AFIPS 1972 FJCC, 859 (1972)
6.240 G.M. Popova, I.V. Prangishvili: Avtom. Telemekh. *1*, 171 (1972)
6.241 L.D. Wald, T.R. Armstrong, C.C. Huang, T.L. Saxton: RADC-TR-73-19
       Final Tech. Rept., Feb. 1973
6.242 W.T. Cheng, T.Y. Feng: Proc. 1974 Sagamore Comput. Conf. Parallel
       Processing, Aug. 20-23, 1974 (IEEE, New York) p. 53
6.243 W. Cheng, T. Feng: AD-A009 873, Syracuse Univ., Dept. of Electr.
       and Comput. Eng., March 1975 (RADC-TR-75-65)
6.244 H.O. Welch: Proc. 1977 Int. Conf. Parallel Processing, ed. by J. Baer
       p. 186 (IEEE, New York 1977)
6.245 D.D. Marshall: Proc. 1977 Int. Conf. Parallel Processing, ed. by J.
       Baer p. 199 (IEEE, New York 1977)
6.246 R. Napoli: Elettrotecnic. *65*, 641 (1978)
6.247 N.V. Findler: Cybernetica (Namur) *10*, 229 (1967)
6.248 N.V. Findler, W.R. McKinzie: Proc. Int. Joint Conf. Artificial
       Intelligence, May 1969, p. 259
6.249 C.C. Foster: Univ. of Mass., Comput. Sci. Dept., TNCS-00023,
       (Dec. 1970)
6.250 J.E. Shore: Rept. of NRL Prog., April 1972, p. 12
6.251 B.F. Meyers: 8th Hawaii Int. Conf. Syst. Sci., 1975, p. 113
6.252 W. Ash, E. Sibley: Univ. of Michigan, Tech. Rept. 5, June 1967
       AD-672 206
6.253 W.L. Ash, E.H. Sibley: Proc. ACM 23rd Nat. Conf., 1968, p. 143
6.254 W.L. Ash: Univ. of Michigan Rept. TR-17, May 1969 (AD-689 861)
6.255 E.H. Sibley, R.W. Taylor, D.G. Gordon: Proc. AFIPS 1968 FJCC *33*, 545
       (1968)
6.256 P.D. Rovner, J.A. Feldman: MIT, Lincoln Lab. (AD-655 810), April 1967
6.257 P.D. Rovner, J.A. Feldman: In *Information Processing 68* (North-Holland,
       Amsterdam 1969) p. 579
6.258 J.A. Feldman, P.D. Rovner: Stanford Univ. Rept. No. AI-Memo-66,
       Aug. 1968 (AD-675 037)
6.259 P.D. Rovner, D.A. Henderson, Jr.: Proc. Int. Joint Conf. Artificial
       Intelligence, May 1969, p. 9
6.260 J.A. Feldman, J.R. Low, D.C. Swinehart, R.H. Taylor: Proc. AFIPS
       1972 FJCC *41*, 1193 (1972)
6.261 J.A. Feldman: Abst. of Tech. Repts. Comput. Sci. Dept. of Univ.
       Rochester, TR9, Nov. 1976
7.1   R.M. Cowan, M.L. Griss: In *Symbolic and Algebraic Computation*, ed.
       by E.W. Ng (Springer, Berlin, Heidelberg 1979) p. 266
7.2   R. Devillers, G. Louchard: BIT *19*, 302 (1979)
7.3   J. Hemenway, E. Teja: EDN Mag. *24*, 108 (1979)
7.4   T. Gunji, E. Goto: J. Inf. Process. *3*, 1 (1980)
7.5   E. Goto, M. Sassa, Y. Kanada: J. Inf. Process. *3*, 13 (1980)
7.6   E. Goto, M. Terashima: J. Inf. Process. *3*, 23 (1980)

7.7   J.A. Kapecki: Kilobaud Microcomput., No. 11, 198 (1980)
7.8   G. Lyon: In *Proc. Conf. on Information Sciences and Systems*, Princeton, New Jersey, March 26-28, 1980, p. 443
7.9   S. Nishihara: Inf. Process. Soc. Jpn. *21*, 980 (1980)
7.10   R.T. Vizzone: Kilobaud. Microcomput., No. 5, 78 (1980)
7.11   R. Gall, M. Nagl: Elektron. Rechenanlagen *23*, 61 (1981)
7.12   T. Ida: Inf. Process. Soc. Jpn. *24*, 391 (1983)
7.13   M. Tamminen: BIT *25*, 135 (1985)
7.14   M. Ancona, S. Antoy: Calcolo 1978, *15*, No. 3, p. 225 (1979)
7.15   B.G. Blank: Program. and Comput. Software *5*, 340 (1979)
7.16   D. Ince: Comput. Age, No. 11, 46 (1980)
7.17   H. Mendelson, U. Yechiali: J. ACM *26*, 654 (1979)
7.18   Yu.Ya. Kochin: Autom. and Remote Control *41*, 723 (1980)
7.19   T.A. Standish: *Data Structure Techniques* (Addison-Wesley, Reading, MA 1980)
7.20   Yu.V. Trifonov: Program. and Comput. Software *6*, 111 (1980)
7.21   T.G. Lewis: *Software Engineering, Analysis and Verification* (Reston Publ. Company, Reston, VA 1982)
7.22   E. Mitchell: Creative Computing *8*, 160 (1982)
7.23   D. Jenkins, M.K. Crowe: In *Software Engineering for Microprocessor Systems*, ed. by P. Depledge (Peter Peregrinus Ltd., London 1984) p. 66
7.24   M. Regnier: BIT *25*, 335 (1985)
7.25   W. Litwin: In *Database Performance, State of the Art Report*, ed. by D.A. Bell (Pergamon Infotech, Maidenhead, Berks. 1984) p. 93
7.26   C.S. Ellis: IEEE Trans. C-*34*, 1178 (1985)
7.27   Anon.: Tech. Rundsc. *72*, 21 (1980)
7.28   Hewlett-Packard: Data Base Management —on a Desktop, in Computer Advances, No. 3 (1980)
7.29   F. Labek: Angew. Inf. *23*, 354 (1981)
7.30   A. Billionnet: Eur. J. Oper. Res. *11*, 167 (1982)
7.31   T. Osborn: Micro —The 6502/6809 Journal, No. 52, 47 (1982)
7.32   D.B. Lomet: SIGMOD Record *13*, 120 (1983)
7.33   S.M. Deen, D.A. Bell: In *Database Performance, State of the Art Report* (Pergamon Infotech, Maidenhead, Berks., England 1984) p. 39
7.34   J.B. Cannell: AD-D007 480/7, Rept. No. PAT-APPL-6-146-580, Dept. of the Army Washington, DC, May 5, 1980
7.35   J.L. Carter, M.N. Wegman: J. Comput. and Syst. Sci. *18*, 143 (1979)
7.36   D. Kinzer: Byte *4*, 200 (1979)
7.37   P.G. Sorenson, J.P. Tremblay: In *Auerbach Annual 1979 Best Computer Papers*, ed. by I.L. Auerbach (North-Holland, New York 1979) p. 111
7.38   M.N. Wegman, J.L. Carter: In *Annual Symposium on Foundations of Computer Science* (IEEE, New York 1979) p. 175
7.39   M.N. Wegman, J.L. Carter: J. Comp. and Syst. Sci. *22*, 265 (1981)
7.40   V.A. Litvinov, V.I. Ivanenko: Program. Comput. Software *5*, 247 (1980)
7.41   K. Maly, L. Kampa: In *Information Processing 80, Proc. IFIP Congress 80*, ed. by S. Lavington (North-Holland, Amsterdam 1980) p. 445
7.42   Anon.: IBM Tech. Discl. Bull. *28*, 4869 (1986)
7.43   C.H. Papadimitriou, P.A. Bernstein: ACM Trans., Program. Lang. and Syst. *2*, 77 (1980)
7.44   P.-Å. Larson: J. ACM *30*, 805 (1983)
7.45   A.C. Yao: J. ACM *32*, 687 (1985)
7.46   D. Comer, M.J. O'Donnell: Siam J. Comput. *11*, 217 (1982)
7.47   M.R. Anderson, M.G. Anderson: Commun. ACM *22*, 104 (1979)
7.48   G.V. Cormack, R.N.S. Horspool, M. Kaiserswerth: Comput. J. *28*, 54 (1985)
7.49   R.J. Cichelli: Commun. ACM *23*, 17 (1980)

7.50   G. Jaeschke: Commun. ACM *24*, 829 (1981)
7.51   C.R. Cook, R.R. Oldehoeft: Sigplan Not. *17*, 18 (1982)
7.52   C.C. Chang, R.C.T. Lee, M.W. Du: IEEE Trans. SE-*8*, 235 (1982)
7.53   C.C. Chang: Inf. Sci. *32*, 165 (1984)
7.54   C.C. Chang: Commun. ACM *27*, 384 (1984)
7.55   C.C. Chang, R.C.T. Lee: Comput. J. *29*, 277 (1986)
7.56   C.C. Chang, J.-C. Shieh: J. Chin. Inst. Eng. *8*, 285 (1985)
7.57   R.C. Bell, B. Floyd: Commun. ACM *26*, 924 (1983)
7.58   M.W. Du, T.M. Hsieh, D.W. Shieh: IEEE Trans. SE-*9*, 305 (1983)
7.59   F. Berman, M.E. Bock, E. Dittert, M.J. O'Donnell, D. Plank: Siam J. Comput. *15*, 604 (1986)
7.60   T. Yuba: Res. Electrotech. Lab., No. 823, p. 1 (1982)
7.61   R.E. Krichevsky: Inf. Control *62*, 64 (1984)
7.62   J. Schmidt, E. Shamir: Lecture Notes Computer Sci. *85*, 569 (1980)
7.63   P.-Å. Larson: BIT *19*, 223 (1979)
7.64   G.H. Gonnet: J. Comput. and Syst. Sci. *21*, 354 (1980)
7.65   H. Mendelson, U. Yechiali: J. ACM *27*, 474 (1980)
7.66   H. Murao: J. Inf. Process. *3*, 87 (1980)
7.67   G.H. Gonnet: J. ACM *28*, 289 (1981)
7.68   K.-C. Tai, A.L. Tharp: In *AFIPS Conf. Proc., 1980 National Computer Conference* (AFIPS Press, Arlington, VA 1980) p. 275
7.69   K.C. Tai, A.L. Tharp: Inf. Syst. *6*, 111 (1981)
7.70   J.S. Vitter: Commun. ACM *25*, 911 (1982)
7.71   J.S. Vitter: In *21st Annual Symp. on Foundations of Computer Science* (IEEE, New York 1980) p. 238
7.72   J.S. Vitter, W.C. Chen: Siam J. Comput. *14*, 490 (1985)
7.73   J.S. Vitter: In *22nd Annual Symp. on Foundations of Computer Science* (IEEE, New York 1981) p. 127
7.74   J.S. Vitter: J. Algorithms *3*, 261 (1982)
7.75   J.S. Vitter: Inf. Proc. Lett. *13*, 77 (1981)
7.76   J.S. Vitter: J. ACM *30*, 231 (1983)
7.77   W.-C. Chen, J.S. Vitter: Siam J. Computer *12*, 667 (1983)
7.78   W.-C. Chen, J.S. Vitter: ACM Trans. Database Syst. *9*, 616 (1984)
7.79   P.-Å. Larson: J. Algorithms *5*, 36 (1984)
7.80   A.A. Törn: BIT *24*, 317 (1984)
7.81   G. Lyon: Comput. J. *28*, 313 (1985)
7.82   G.E. Lyon: *Alternation-Tree Insertions for Open-Addressed Hash Buckets*, PB82-101312, National Bureau of Standards, Washington, DC, 1981, 4 p., in Proc. Information Sciences and Systems Ann. Conf. (15th), Baltimore, MD, 1981, p. 398
7.83   P. Quittner, S. Csoka, S. Halasz, D. Kotsis, K. Varnai: Commun. ACM *24*, 579 (1981)
7.84   A.B. Cremers, T.N. Hibbard: Math. Syst. Theory *12*, 151 (1978)
7.85   P. Scheuermann: Inf. Syst. *4*, 183 (1979)
7.86   P. Quittner: Software —Practice and Experience *13*, 471 (1983)
7.87   G.H. Gonnet, J.I. Munro: Siam J. Comput. *8*, 463 (1979)
7.88   S. Nishihara, K. Ikeda: Comm. ACM *26*, 1082 (1983)
7.89   E. Lodi, F. Luccio: Inf. Proc. Lett. *20*, 131 (1985)
7.90   W.A. Burkhard: In *Conference on Information Sciences and Systems* (John Hopkins Univ., Baltimore, MD 1978) p. 378
7.91   J.A.T. Maddison: Comput. J. *23*, 188 (1980)
7.92   P.-Å. Larson: BIT *20*, 25 (1980)
7.93   M. Scholl: Rapport de Recherche No. 354, I.R.I.A., Lab. Rech. Informat. Automat. (1979)
7.94   A.D. Astakhov: Programmirovanie, No. 3, 28 (1980)
7.95   G. Lyon: In *Conference on Information Sciences and Systems* (John Hopkins Univ., Baltimore, MD 1979) p. 262

7.96   J.R. Muehlbacher: Computing *26*, 9 (1981)
7.97   W. Litwin: In *Proc. of the Int. Conf. on Data Bases*, ed. by S.M.
        Deen, P. Hammersley (Heyden and Son, London 1980) p. 260
7.98   W. Litwin: In *Proc. of the 6th Int. Conf. on Very Large Data Bases*
        (IEEE, New York 1980) p. 212
7.99   J.K. Mullin: BIT *21*, 390 (1981)
7.100   P.-Å. Larson: In *Proc. of the 8th Int. Conf. on Very Large Data
        Bases* (VLDB Endownment, Saratoga, CA 1982) p. 300
7.101   P.-Å. Larson: ACM Trans. Database Syst. *7*, 566 (1982)
7.102   P.-Å. Larson: ACM Trans. Database Syst. *10*, 75 (1985)
7.103   P.-Å. Larson: Comput. J. *28*, 319 (1985)
7.104   K. Ramamohanarao, R. Sacks-Davis: ACM Trans. Database Syst. *9*, 369
        (1984)
7.105   M. Scholl: Rapport de Recherche No. 347, I.R.I.A., Lab. Rech.
        Informat. Automat. (1979)
7.106   M. Scholl: Tanulmanyok Magy. Tud. Akad. Szamitastech. and Autom.
        Kut. Intez., No. 100, p. 169 (1979)
7.107   M. Scholl: ACM Trans. Database Syst. *6*, 194 (1981)
7.108   M. Regnier: Inf. Proc. Lett. *13*, 64 (1981)
7.109   R.A. Frost: In *ICS 81*, Systems Architecture, Proc. of the 6th ACM
        European Regional Conf., London, 30 March - 1 April 1981 (IPC Sci.
        and Technol. Press Ltd., Guilford, Surrey, England 1981) p. 234
7.110   R.A. Frost, M.M. Peterson: Software —Practice and Experience *12*,
        163 (1982)
7.111   K. Ramamohanarao, J.W. Lloyd: Comput. J. *25*, 478 (1982)
7.112   J.K. Mullin: Comput. J. *28*, 330 (1985)
7.113   E. Veklerov: ACM Trans. Database Syst. *10*, 90 (1985)
7.114   K. Kawagoe: SIGMOD Record *14*, 201 (1985)
7.115   R. Fagin, J. Nievergelt, N. Pippenger, H.R. Strong: ACM Trans.
        Database Syst. *4*, 315 (1979)
7.116   M. Tamminen: Helsinki Univ. of Technology, Report-HTKK-TKO-A19 (1980)
7.117   M. Tamminen: Helsinki Univ. of Technology, Report-HTKK-TKO-A20 (1980)
7.118   M. Tamminen: Helsinki Univ. of Technology, Report-HTKK-TKO-B27 (1981)
7.119   M. Tamminen: Helsinki Univ. of Technology, Report-HTKK-TKO-B28 (1981)
7.120   M. Tamminen: Helsinki Univ. of Technology, Report-HTKK-TKO-B29 (1981)
7.121   M. Tamminen: Helsinki Univ. of Technology, Report-HTKK-TKO-B35 (1981)
7.122   M. Tamminen: BIT *21*, 419 (1981)
7.123   M. Tamminen: Inf. Proc. Lett. *15*, 227 (1982)
7.124   A.C. Yao: Inf. Process. *11*, 84 (1980)
7.125   J.W. Lloyd: BIT *22*, 150 (1982)
7.126   H. Mendelson: IEEE Trans. SE-*8*, 611 (1982)
7.127   P. Flajolet: Acta Informatica *20*, 345 (1983)
7.128   U. Bechtold, K. Küspert: Inf. Proc. Lett. *19*, 21 (1984)
7.129   R.M. Bryant: IBM Tech. Discl. Bull. *26*, 6046 (1984)
7.130   S.-H.S. Huang: Int. J. Comput. and Inf. Sci. *14*, 73 (1985)
7.131   R.J. Lipton, A.L. Rosenberg, A.C. Yao: J. ACM *27*, 81 (1980)
7.132   G.H. Gonnet, P.-Å. Larson: In *Symp. on Principles of Database Systems*
        (ACM, New York 1982) p. 256
7.133   D.A. Bell, S.M. Deen: Comput. J. *25*, 486 (1982)
7.134   P.-Å. Larson, M.V. Ramakrishna: SIGMOD Record *14*, 190 (1985)
7.135   A. Bolour: J. ACM *26*, 196 (1979)
7.136   A. Bolour: Siam J. Comput. *10*, 721 (1981)
7.137   J.C. Dorng, S.K. Chang: In *Proc. of COMPSAC 81*, IEEE Computer
        Society's Fifth Int. Computer Software and Applications Conf. (IEEE,
        New York 1981) p. 245
7.138   C.C. Chang: Inf. Sciences *34*, 199 (1984)
7.139   E.J. Otoo: SIGMOD Record *14*, 214 (1985)
7.140   P. Valduriez, Y. Viemont: SIGMOD Record *14*, 107 (1984)

7.141 W.A. Burkhard: ACM Trans. Database Syst. *4*, 228 (1979)
7.142 C.S. Roberts: Proc. IEEE *67*, 1624 (1979)
7.143 K. Ramamohanarao, J.W. Lloyd, J.A. Thom: ACM Trans. Database Syst. *8*, 552 (1983)
7.144 K. Ramamohanarao, R. Sacks-Davis: BIT *25*, 477 (1985)
7.145 R.M. Colomb: Aust. Comput. J. *17*, 181 (1985)
7.146 T. Kohonen, H. Riittinen, E. Reuhkala, S. Haltsonen: Inf. Sci. *33*, 3 (1984)
7.147 S.S. Thakkar, A.E. Knowles: Computer *19*, 8 (1986)
7.148 F. Ducoin: Comput. Sci. *13*, 225 (1979)
7.149 R.E. Tarjan, A. Chi-Chih Yao: Commun. ACM *22*, 606 (1979)
7.150 M.L. Fredman, E. Szemeredi, J. Komlos: J. ACM *31*, 538 (1984)
7.151 H.M. Blanken: In *Proc. Int. Conf. on Data Bases*, ed. by S.M. Deen, P. Hammersley (Heyden and Son, London 1980) p. 99
7.152 P. Willet: J. Doc. *35*, 296 (1979)
7.153 D. Cooper, M.F. Lynch, A.H.W. McLure: Program *15*, 66 (1981)
7.154 D. Cooper, M.F. Lynch, A.H.W. McLure: Program *15*, 226 (1981)
7.155 W.A. Burkhard: BIT *23*, 274 (1983)
7.156 D.J. Dodds: Commun. ACM *25*, 368 (1982)
7.157 D. Comer, V.Y. Shen: Software —Practice and Experience *12*, 669 (1982)
7.158 J. Radue: SIGPC Not. *6*, 197 (1983)
7.159 W. Barth, H. Nirschl: Angew. Inf. Appl. Inf. *27*, 152 (1985)
7.160 A. Zamora: J. Am. Soc. Inf. Sci. *31*, 51 (1980)
7.161 M. Mor, A.S. Fraenkel: Commun. ACM *25*, 935 (1982)
7.162 O. Ventä: In *Proc. of the First Conf. on Artificial Intelligence Applications* (IEEE Comput. Soc. Press, Silver Spring, MD 1984) p. 446
7.163 O. Ventä: In *IEEE Seventh Int. Conf. on Pattern Recognition Proc.* (IEEE Comput. Soc. Press, Silver Spring, MD 1984) p. 1240
7.164 O. Ventä, T. Kohonen: In *Eighth Int. Conf. on Pattern Recognition Proc.* (IEEE Comput. Soc. Press, Washington, DC 1986) p. 1214
7.165 M. Ajtai, M. Fredman, J. Komlos: In *24th Annual Symposium on Foundations of Computer Science* (IEEE Comput. Soc. Press, Silver Spring, MD 1983) p. 299
7.166 W. Doster: Wissensch. Berichte AEG-Telefunken *51*, 104 (1978)
7.167 L.O. Hill, D.A. Zein: IEEE Circuits & Devices Mag. *2*, 18 (1986)
7.168 A. Tharp, K.-C. Tai: Software —Practice and Experience *12*, 35 (1982)
7.169 S.N. Porter: Comput. and Secur. *1*, 54 (1982)
7.170 U. Suhl: Angew. Inf. *23*, 58 (1981)
7.171 J. Cocke, W.S. Jr. Worley: IBM Techn. Discl. Bull. *24*, 2724 (1981)
7.172 G. Frieder, G. Herman, C. Meyer: IEEE Software *2*, 37 (1985)
7.173 Z. Hejun, Z. Yuefang: Chi Hshieh Kung Ch'eng Hsueh Pao *19*, 22 (1983)
7.174 J.W. Braidt, J.M. Taylor: IBM Tech. Discl. Bull. *24*, 3531 (1981)
7.175 G. Wolf: In *Papers of the 5th Workshop on Computer Architecture for Non-Numeric Processing* (ACM, New York 1980) p. 70
7.176 K. Hiraki, K. Nishida, T. Shimada: IEEE Trans. C-*33*, 851 (1984)
7.177 H.D. Robinson, G.E. Tayler: IBM Tech. Discl. Bull. *24*, 5354 (1982)
7.178 Y. Dohi, H. Arisawa, Y. Izumida: Bull. Fac. Eng. Yokohama Natl. University *28*, 89 (1979)
7.179 M.T. Benhase: IBM Tech. Discl. Bull. *25*, 3760 (1982)
7.180 J.G.D. da Silva, I. Watson: IEEE Proc.-E *130*, 19 (1983)
7.181 T. Ida: In *Int. Workshop on High-Level Language Computer Architecture* (Univ. of Maryland, College Park, MD 1980) p. 99
7.182 F.J. Burkowski: IEEE Trans. C-*31*, 825 (1982)
7.183 K. Ramamohanarao, R. Sacks-Davis: Inf. Proc. Lett. *13*, 23 (1981)
7.184 S.S. Thakkar, A.E. Knowles: In *Proc. of the First Int. Conf. on Supercomputing Systems: SCS 85* (IEEE Comput. Soc. Press, Washington, DC 1985) p. 697

7.185 M. Kishi, H. Yasuhara, Y. Kawamura: In *10th Annual Int. Conf. on Computer Architecture Conf. Proc.* (IEEE, New York 1983) p. 236
7.186 J.N. Barry, R.G. George: IEEE Proc. E *133*, 1 (1986)
7.187 M.M. Cirovic: *Handbook of Semiconductor Memories* (Reston Publishing Company, Reston, VA 1981)
7.188 D.D. Howard: IBM Tech. Discl. Bull. *24*, 2554 (1981)
7.189 B.A. Denis: IBM Tech. Discl. Bull. *24*, 2197 (1981)
7.190 A.G. Aipperspach, P.T. Wu: IBM Tech. Discl. Bull. *26*, 4208 (1984)
7.191 M. Yoshida, A. Kawaji, M. Yoneyama, T. Uchiumi: Trans. Inst. Electron. & Commun. Eng. Jpn., *J65C*, 997 (1982)
7.192 M. Yoshida, A. Kawaji, M. Yoneyama, T. Uchiumi: Trans. Inst. Electron. & Commun. Eng. Jpn., *J66C*, 343 (1983)
7.193 J.C. Hou: IBM Tech. Discl. Bull. , 4331 (1984)
7.194 T.P. Haraszti: In *ESSCIRC' 85; 11th European Solid State Circuits Conf.* (Univ. Paul Sabatier, Toulouse 1985) p. 237
7.195 T. Nikaido, T. Ogura, S. Hamaguchi, S. Muramato: Jap. J. Appl. Phys. *22*, 51 (1983)
7.196 Y.J. Vernay, J.L. Lardy, R. Gerber: In *Int. Conf. on New Trends in Integrated Circuits* (Comite du Colloque Int. Nouvelles Orientations des Circuits Integres, Paris 1981) p. 43
7.197 T. Ogura, S.-I. Yamada, T. Nikaido: IEEE J. SC-*20*, 1277 (1985)
7.198 H. Kadota, J. Miyake, Y. Nishimichi, H. Kudoh, K. Kagawa: IEEE J. Solid-State Circuits SC-*20*, 951 (1985)
7.199 S.M. Lamb: Mini-Micro Syst. *16*, 237 (1983)
7.200 D.A. Pucknell, M.L.J. Raymond: In *Proc. of Microelectronics '82, A National Conf. on Microelectronics* (Instn. Eng. Australia, Barton, Act., Australia 1982) p. 107
7.201 D.A. Pucknell, M.L.J. Raymond: J. Electr. Electron. Eng. Aust. *4*, 155 (1984)
7.202 J.C. Perez: IBM Tech. Discl. Bull. *27*, 3365 (1984)
7.203 G. Giles, C. Hunter: In *International Test Conference 1985 Proceedings, The Future of Test* (IEEE Comput. Soc. Press, Washington, DC 1985) p. 471
7.204 S.E. Schuster: IBM Tech. Discl. Bull. *26*, 5364 (1984)
7.205 S.Y. Lee, H. Chang: IEEE Trans. C-*28*, 627 (1979)
7.206 N.R. Strader: IEEE Trans. C-*31*, 265 (1982)
7.207 D. Tavangarian: Elektron. Rechenanlagen *27*, 264 (1985)
7.208 H. Shirakawa, T. Kumagai: Mem. Res. Inst. Sci. & Eng. Ritsumeikan Univ. (Japan), No. 41, p. 27 (1982)
7.209 T.A. Nodes, J.L. Smith, R. Hecht-Nielsen: In *Proc. of ICASSP 85* (IEEE, New York 1985) p. 1511
7.210 Y. Hamazaki: Denshi Gijutsu Sogo Kenkyusho Iho *48*, 730 (1984)
7.211 B. Cohen, R. McGarity: IEEE Micro *6*, 13 (1986)
7.212 L. Schubert: Siemens Forsch.- & Entwicklungsber. *15*, 92 (1986)
7.213 H. Shin, M. Malek, D. Degroot: In *Proc. of the 1984 Int. Conf. on Parallel Processing* (IEEE Comput. Soc. Press, Washington, DC 1985) p. 369
7.214 Anon.: IBM Tech. Discl. Bull. *28*, 2562 (1985)
7.215 T.P. Haraszti: IEEE J. Solid-State Circuits SC-*17*, 539 (1982)
7.216 B.I. Gekht, D.P. Frolov: Sist. Avtom. Nauchn. Issled., No. 6, 95 (1984)
7.217 V.A.J. Maller: In *Fifth Generation Computer Project, State of the Art Report*, ed. by G.G. Scarrot (Pergamon Infotech, Maidenhead, Berks., England 1983) p. 89
7.218 M. Malms, R. Kubera, H. Röhl: Elektron. Rechenanlagen *26*, 179 (1984)
7.219 M.M. Mirsalehi, T.K. Gaylord: In *Proc. of the 16th Int. Symp. on Multiple-Valued Logic* (IEEE Comput. Soc. Press, Washington, DC 1986) p. 174

7.220   C.C. Guest, T.K. Gaylord: Appl. Opt. *19*, 1201 (1980)
7.221   C.C. Guest, M.M. Mirsalehi, T.K. Gaylord: IEEE Trans. C-*33*, 927 (1984)
7.222   M.M. Mirsalehi, T.K. Gaylord: Appl. Opt. *25*, 2277 (1986)
7.223   M.M. Mirsalehi, T.K. Gaylord: IEEE Trans. C-*35*, 829 (1986)
7.224   T.K. Gaylord, M.M. Mirsalehi, C.C. Guest: Opt. Eng. *24*, 48 (1985)
7.225   C.A. Papachristou: In *Proc. of 11th Int. Symp. on Multiple-Valued Logic* (IEEE, New York 1981) p. 62
7.226   J.T. Butler: In *Proc. of the 13th Int. Symposium on Multiple-Valued Logic* (IEEE, New York 1983) p. 94
7.227   G.A. Nikitin, B.V. Vinnikov, I.L. Kaftannikov: Autom. Control & Comput. Sci. *18*, 23 (1984)
7.228   C.A. Papachristou, Kai Hwang: In *Proc. of 7th Symposium on Computer Arithmetic* (IEEE Comput. Soc. Press, Silver Spring, MD 1985) p. 182
7.229   M. Malms: Regelungstechnische Praxis *25*, 270 (1983)
7.230   V.C. Bhavsar, T.Y.T. Chan, L. Goldfarb: In *1985 IEEE Computer Society Workshop on Computer Architecture for Pattern Analysis and Image Database Management* (IEEE Comput. Soc. Press, Washington, DC 1985) p. 126
7.231   F. Badi'i, F. Majd: In *1985 IEEE Computer Society Workshop on Computer Architecture for Pattern Analysis and Image Database Management* (IEEE Comput. Soc. Press, Washington, DC 1985) p. 183
7.232   W.E. Snyder, C.D. Savage: IEEE Trans. C-*31*, 963 (1982)
7.233   W. Snyder, A. Cowart: IEEE Trans. PAMI-*5*, 349 (1983)
7.234   B. Sinha, P.K. Srimani: Inf. Sci. *24*, 201 (1981)
7.235   F. Badi'i, J. Jayawardena: In *Proc. of 7th Int. Conf. on Pattern Recognition* (IEEE Comput Soc. Press, Silver Spring, MD 1984) p. 659
7.236   K. Nakamura: J. of Logic Progr. 4, Vol. 1, 285 (1984)
7.237   J.V. Oldfield: IEEE Proc. I *133*, 123 (1986)
7.238   M. Demange: IBM Tech. Discl. Bull. *26*, 267 (1983)
7.239   S. Bozinovski, C. Anderson: In *Proc. of MELECON '83, Mediterranean Electrotechnical Conference*, ed. by E.N. Protonotarios, G.I. Stassinopoulos, P.P. Civalleri (IEEE, New York 1983) p. 13
7.240   Anon.: IBM Tech. Discl. Bull. *27*, 7069 (1985)
7.241   H. Yamamoto, T. Furukawa: Trans. Inst. Electron. & Commun. Eng. Jpn., *J68A*, 524 (1985)
7.242   S. Kaczamarek, P. Gofta: In *6th European Conference on Electrotechnics —EUROCON 84, Computers in Communication and Control* (Peter Peregrinus, London 1984) p. 84
7.243   M.F. Deering: Byte *10*, 193 (1985)
7.244   E.J. Schuegraf: In *Communicating Information, Proc. of the 4rd ASIS Annual Meeting* (Knowledge Ind. Publications, White Plains, NY 1981) p. 329
7.245   T. Ichikawa, N. Kamibayashi: J. Inst. Electron. & Commun. Eng. Jpn. *64*, 609 (1981)
7.246   J. Koller: Elektronik *32*, 45 (1983)
7.247   T. Durham: Computing *8* (1983)
7.248   B. Svensson: Dissertation, Dept. of Computer Eng., University of Lund, Sweden (1983)
7.249   C. Fernström: Dissertation, Dept. of Computer Eng., University of Lund, Sweden (1983)
7.250   I. Kruzela: Dissertation, Dept. of Computer Eng., University of Lund, Sweden (1983)
7.251   C.C. Foster: Massachusetts Univ. Rep. AD-A123 028/3 (1982)
7.252   C. Weems, S. Levitan, C. Foster: In *Proc. IEEE Int. Conf. on Circuits and Computers ICCC '82* (IEEE, New York 1982) p. 236

371

7.253 S. Berkovich, J.M. Pullen: In *Proc. of the IEEE Int. Conf. on Computer Design: VLSI in Computers ICCD '84* (IEEE Comput. Soc. Press, Silver Spring, MD 1984) p. 382
7.254 K.E. Batcher: IEEE Trans. C-*31*, 377 (1982)
7.255 Anon.: Multiple Instruction Associative Processor (MIAP), PB80-980220, PB80-925110, PC E02 NTIS
7.256 S.A. Gerasimova, V.M. Zakharchenko: Sov. J. Opt. Technol. *48*, 404 (1981)
7.257 S. Kumar, S.N. Maheshwari, P.C.P. Bhatt: In *Proc. of the First Int. Conf. on Supercomputing Systems: SCS 85* (IEEE Comput. Soc. Press, Washington, DC 1985) p. 641
7.258 D. Parkinson, H.M. Liddell: IEEE Trans. C-*32*, 32 (1983)
7.259 O. Wing: In *Proc. IEEE Int. Conf. on Computer Design: VLSI in Computers ICCD '83* (IEEE Computer Soc. Press, Silver Spring, MD 1983) p. 247
7.260 L. Wallis: Electronic Design *32*, 217 (1984)
7.261 Y. Shimazu, T. Tamati: Trans. Inf. Process. Soc. Jpn. *26*, 53 (1985)
7.262 Computer, Vol. 12, No. 3 (1979)
7.263 L.A. Hollaar, J.J. Kuehn: In *Proc. of the Sixth Annual International ACM SIGIR Conference on Research and Development in Information Retrieval*, No. 24, p. 3 (1983)
7.264 A. Hurson: In *Proc. of IEEE Computer Society Workshop on Computer Architecture for Pattern Analysis and Image Database Management* (IEEE, New York 1981) p. 225
7.265 P. Hawthorn, D.J. De Witt: AD-A104 927/9, Report No. CSTR-383, Wisconsin Univ.-Madison, Dept. of Computer Sciences (1980)
7.266 D.J. De Witt: IEEE Trans. C-*28*, 59 (1979)
7.267 S.P. Kartashev, S.I. Kartashev: IEEE Trans. C-*33*, 28 (1984)
7.268 K. Goser, C. Foelster, U. Rueckert: Inf. Sci. *34*, 61 (1984)
7.269 D. Lawton, S. Levitan, C. Weems, E. Riseman, A. Hanson, M. Callahan: Proc. SPIE Int. Soc. Opt. Eng. *504*, 92 (1984)
7.270 C. Weems, D. Lawton, S. Levitan, E. Riseman, A. Hanson, M. Callahan: In *Proceedings CVPR '85: IEEE Computer Society Conference on Computer Vision and Pattern Recognition* (IEEE Comput. Soc. Press, Silver Spring, MD 1985) p. 598
7.271 A.M. Veronis: In *IEEE SOUTHEASTCON '83 Conference Proc.* (IEEE, New York 1983) p. 119
7.272 C. Weems, D.T. Lawton: Proc. SPIE Int. Soc. Opt. Eng. *435*, 121 (1983)
7.273 J.L. Potter: In *Proc. of the IEEE Int. Conf. on Computer Design: VLSI in Computers ICCD '84* (IEEE Comput. Soc. Press, Silver Spring, MD 1984) p. 520
7.274 R.M. Lea: IEE Proc. I *133*, 105 (1986)
7.275 M.E. Steenstrup, D.T. Lawton, C. Weems: In *Proc. of IEEE Comp. Soc. Conf. on Computer Vision and Pattern Recognition*, ed. by H.J. Siegel, L. Siegel (IEEE Comput. Soc. Press, Silver Spring, MD 1983) No. 2, p. 492
7.276 D.R. McGregor: In *IEEE Colloquium on VLSI Special Purpose Computer Architectures and Implementations* (IEE, London 1985) p. 6
7.277 W.R. Cyre: AD-A082 324/5, Control Data Corp., Minneapolis (1979)
7.278 T. Kohonen: *Self-Organization and Associative Memory*, Springer Ser. Inform. Sci., Vol. 8 (Springer, Berlin, Heidelberg 1984)
7.279 D. Psaltis, N. Farhat: Opt. Lett. *10*, 98 (1985)
7.280 IEEE Spectrum, Vol. 23, No. 8 (1986) (Special issue)
7.281 H. Mada: Appl. Opt. *24*, 2063 (1985)
7.282 D.A. Gregory, H.K. Liu: Appl. Opt. *23*, 4560 (1984)
7.283 H.J. Caulfield: Opt. Commun. *55*, 80 (1985)

7.284 A.D. Fisher, C.L. Giles: In *Proc. of the IEEE 1985 COMPCON Spring*
       (IEEE Computer Society Press, Silver Spring, MD 1985) p. 342
7.285 B.H. Soffer, G.J. Dunning, Y. Owechko, E. Marom: Opt. Lett. *11*, 118
       (1986)
7.286 P.J. Becker, H. Bolle, A. Keller, W. Kistner, W.D. Riecke: FB-DV-79-
       05, Bundesministerium für Forschung und Technologie, Bonn-Bad Godes-
       berg, FRG, March (1979)
7.287 S.A. Gerasimova, V.M. Zakharchenko: Sov. J. Opt. Technol. *48*, 404
       (1981)
7.288 A.A. Verbovetskii: Autom. Remote Control *45*, 1382 (1984)
7.289 C. Warde, J. Kottas: Appl. Opt. *25*, 940 (1986)
7.290 D.Z. Anderson: Opt. Lett. *11*, 56 (1986)
7.291 A. Yariv, S.-K. Kwong: Opt. Lett. *11*, 186 (1986)
7.292 *Digest of Technical Papers, Optical Society of America 1985 Annual
       Meeting*, Washington, DC, October 14-18, 1985
7.293 *Proc. of the SPIE Special Institute on Optical and Hybrid Computing*,
       Leesburg, VA, March 24-27, 1986 (in press)

Subject Index

Mask
register 142-143,265,304
vector 184
word 128
Masked search 128
Masking 6,13,135-136,143-144,149,
156-158,202-203,237
Master-slave flip-flop 296
Match
argument 283
flip-flop 283-284,288-290,302,
304,308
left (right), in DLM 291
operation, DLM 284,291
Matching
continuous-valued logic 24-25
dynamic 26
equality cf logical equivalence,
string
logical equivalence 9,21,24,
128-129
relation 13
score 26
string 19-20,26-27,165-171,282-288
Matrix 30-32
inversion 288
processing 298,325,329
Maximum-length sequence 49
probing 62-63
Maximum (minimum) search 4,159-160,
188-189
MBM: cf Magnetic-bubble memory
Mellin transform 27
Membership test 55-56
Memory
address-free 30-32
associative 1
biological 1
buffer organization 3

bubble 220-228,338
capacity 28
content-addressable 2
cryotronic 203-210
distributed 29,30-32
distributed-logic 281-292
dynamic 166-172,211,338
functional 172-183
head-per-track 166-172
hierarchy 241-244,262,268-271
holographic 231-239
human 1,18-19,28
infinite 28,29,31
linear-select 5,211-212
location-free 30-32
magnetic 210-215,220-228
magneto-optical 229-231
multiaccess 189
multilevel 256-257
neural 29
optoelectronic 242
rotating 166-172,268-270
semiconductor 191-203,215-219
selectivity 30
superimposition of traces 29
system-theoretical approach 30-37
thin-film 213-215,229-231,242
virtual 245-247
Memory function, of ASP
content-addressable search 303-304
context addressing 304-305
mass write 307
pulse 304
read 307
reset 304
search 303-304
write 304
Memory map 247-253,264-268
Memory matrix 184